MACHINE
DEVICES
and
COMPONENTS

ILLUSTRATED SOURCEBOOK

MACHINE DEVICES
and
COMPONENTS
ILLUSTRATED SOURCEBOOK

ROBERT O. PARMLEY, P.E.

Editor-in-Chief

McGraw-Hill

New York Chicago San Francisco Lisbon London Madrid
Mexico City Milan New Delhi San Juan Seoul
Singapore Sydney Toronto

The _McGraw·Hill_ Companies

Cataloging-in-Publication Data is on file with the Library of Congress.

1 2 3 4 5 6 7 8 9 0 QPD/QPD 0 1 0 9 8 7 6 5 4

ISBN 0-07-143687-1

The sponsoring editor for this book was Kenneth P. McCombs and the production supervisor was Pamela A. Pelton. The art director for the cover was Anthony Landi.

Text Design by Wayne C. Parmley.

Printed and bound by Quebecor World, Dubuque.

This book was printed on acid-free paper.

McGraw-Hill books are available at special quantity discounts to use as premiums and sales promotions, or for use in corporate training programs. For more information, please write to the Director of Special Sales, Professional Publishing, McGraw-Hill, Two Penn Plaza, New York, NY 10121-2298. Or contact your local bookstore.

DEDICATED TO:
Regin & Spencer

CONTENTS

PREFACE

Preservation of information, especially technical data, is essential for continued progress of any discipline of technology. Without the knowledge of basic elements, engineers, designers, craftsmen, and technicians are handicapped. In some cases, they would literally have to "reinvent the wheel"; thus wasting valuable time, resources and energy that could and should be spent on developing new designs.

We are told by respected archaeological experts that the pyramids of ancient Egypt and prehistoric South America were built without the use of pulleys and gears; both indispensable mechanical components since early Greek and Roman times. However, these magnificent structures were constructed, but apparently no recorded information exists describing their construction methods or erection tools they employed. Perhaps some unknown component, device, or mechanism was used by those ancient builders that remains unknown even to this day.

Basic or standard designs are invaluable and often stimulate the creative process, which can lead to new components and mechanisms. But if they are not properly recorded and available for future review, those ideas can easily become lost. Fortunately, modern engineering literature has faithfully published handbooks, manuals and codes describing most standard designs. However, innovative devices and unusual component applications often escape a permanent place in technical literature. A classic example of this is the two-page illustrated design files featured in *Product Engineering* magazine. This bi-weekly publication, over the decades, contained thousands of innovative mechanical designs and applications. Unfortunately, this magazine ceased publication in the early 1970s, but some of the original articles were reprinted in Greenwood's books in the late 1950s and 1960s. Chironis' *Mechanisms & Mechanical Devices Sourcebook* and the recently published book entitled, *Illustrated Sourcebook of Mechanical Components*, the latter of which I had the honor to serve as Editor-in-Chief, also contained many selections from *Product Engineering*. Other technical magazines periodically include novel mechanical designs, as do various technical reports from professional societies. Too many of these articles and their innovative designs fade into obscurity.

With the foregoing discussion in mind, it was proposed to produce a practical sourcebook of selected innovative material that machine designers could use as a reference. Therefore, this sourcebook is a modified and condensed version of the massive *Illustrated Sourcebook of Mechanical Components* with the emphasis on machine devices and unusual applications of components. Significant data was culled from that book and rearranged to fit into a new format. Additional material was obtained from other sources and blended into the manuscript to round out the presentation.

The reader will notice a wide range of drawing styles and techniques throughout the pages of this sourcebook. This material was prepared over many decades and the sources were very broad. It is the opinion of the Editor in Chief that the range of drafting modes adds authentically and character to the collection of devices and components.

As always, a sourcebook of this kind draws on the talents, skills and knowledge of many individuals, organizations, consulting firms, publications, and technical societies. This effort is no exception. The sources, where known, have been faithfully recorded on the appropriate pages throughout the sourcebook. We thank them all.

My son, Wayne, has again served as the graphic designer for this (our 6th) book. As always, his skills and professionalism have been top quality.

In summary, it is hoped that this illustrated sourcebook will continue the tradition of its predecessors. Preservation and dissemination of this type of material is a professional obligation and should not be taken lightly. We trust that we have been true to that mission.

ROBERT O. PARMLEY, P.E.
Ladysmith, Wisconsin
May 2004

INTRODUCTION

As previously stated in the Preface, the major portion of the material contained throughout this sourcebook has been culled from over five decades of technical publications. Thus, the reader will certainly notice the wide range of drafting techniques and printing styles. Because these differences do not affect the technical data, we have opted to let these variations stand, as originally printed, and believe they reveal a historical flavor to the overall presentation.

Before the reader or user of the sourcebook commences to explore the pages, it should be stated that both the cross-referenced Index and Table of Contents (located at the opening of each section) were included to assist in finding specific items. This format has been time-tested and insures user-friendliness.

The sections of this sourcebook have been arranged into three general categories. They are: assemblies, power transmission and individual components.

The first five sections (1 thru 5) are devoted to innovative mechanisms, creative assemblies, linkages, connections and related locking devices. The end product is the final assembly of various mechanical components into a mechanism, device, machine or system that performs a desired function.

The next five sections (6 thru 10) illustrate mechanical power transmission; i.e., gears, gearing, clutches, chains, sprockets, ratchets, belts, belting, shafts and couplings. These sections include some of the essential mechanical combinations used in transporting power from its source to other locations. Some materials are basic while other data illustrates some novel designs.

The third and concluding category is devoted to individual mechanical components. Sections 11 thru 20 depict some universal and innovative uses of standard mechanical components. These single components are the building blocks of mechanical mechanisms and assemblies.

In every machine or mechanism, each component must be properly selected and precisely arranged in a predetermined position to result in a successfully functioning unit or device. As each assembly is connected to larger and more complex machines, the individual components become less noticeable, until the system fails. Then the component that malfunctioned becomes the focus of attention. Therefore, the designer must always bear in mind that every element of a machine or mechanism is extremely important.

This sourcebook is not a textbook or standard handbook of machine design. Rather, it is a creative reference for designers of machine devices. The material contained herein is an illustrated collection of unique designs and novel applications extracted from hard-to-locate technical journals, out-of-print publications and private consultants whose specialized topics limited their dissemination to the general engineering community.

Good design and creative innovations rarely are spontaneous. They are usually developed over time and generated from previous developments. Therefore, it is the core purpose of this sourcebook to provide the reader with a broad based assortment of unique designs and unusual component applications. Hopefully, these illustrations will inspire the readers' creative thought process and ultimately produce solutions to their respective design problems.

It is the professional opinion of the Editor in Chief that to develop into a good designer of machine devices one must have access to a broad resource of mechanical data. This sourcebook aims to be a key element in the designer's library. From quick surveys to full-blown systematic searches of the material contained within these pages should inspire the user to develop innovative devices and cost-effective solutions to various design challenges.

The hundreds of illustrations displayed on the following pages were developed by a long list of engineers, designers, inventors, technicians, and artisans over many decades. Consult these drawings and let their practical ideas speak to you. Let the collective ideas rearrange themselves into new and innovative designs. This in itself will honor those individuals who took the time to faithfully record the original material and thereby preserve their ideas and concepts.

ROBERT O. PARMLEY, P.E.
Editor in Chief

SECTION 1

INGENIOUS MECHANISMS

Modified Geneva Drives and Special Mechanisms

These sketches were selected as practical examples of uncommon, but often useful mechanisms. Most of them serve to add a varying velocity component to the conventional Geneva motion.

Sigmund Rappaport

Fig. 1—(Below) In the conventional external Geneva drive, a constant-velocity input produces an output consisting of a varying velocity period plus a dwell. In this modified Geneva, the motion period has a constant-velocity interval which can be varied within limits. When spring-loaded driving roller *a* enters the fixed cam *b*, the output-shaft velocity is zero. As the roller travels along the cam path, the output velocity rises to some constant value, which is less than the maximum output of an unmodified Geneva with the same number of slots; the duration of constant-velocity output is arbitrary within limits. When the roller leaves the cam, the output velocity is zero; then the output shaft dwells until the roller re-enters the cam. The spring produces a variable radial distance of the driving roller from the input shaft which accounts for the described motions. The locus of the roller's path during the constant-velocity output is based on the velocity-ratio desired.

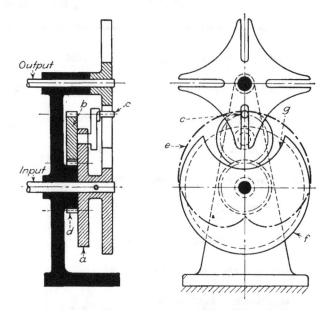

Fig. 2—(Above) This design incorporates a planet gear in the drive mechanism. The motion period of the output shaft is decreased and the maximum angular velocity is increased over that of an unmodified Geneva with the same number of slots. Crank wheel *a* drives the unit composed of plant gear *b* and driving roller *c*. The axis of the driving roller coincides with a point on the pitch circle of the planet gear; since the planet gear rolls around the fixed sun gear *d*, the axis of roller *c* describes a cardioid *e*. To prevent the roller from interfering with the locking disk *f*, the clearance arc *g* must be larger than required for unmodified Genevas.

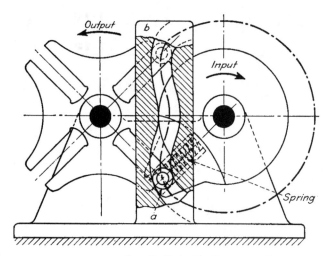

Saxonian Carton Machine Co., Dresden, Germany

Fig. 3—A motion curve similar to that of Fig. 2 can be derived by driving a Geneva wheel by means of a two-crank linkage. Input crank *a* drives crank *b* through link *c*. The variable angular velocity of driving roller *d*, mounted on *b*, depends on the center distance *L*, and on the radii *M* and *N* of the crank arms. This velocity is about equivalent to what would be produced if the input shaft were driven by elliptical gears.

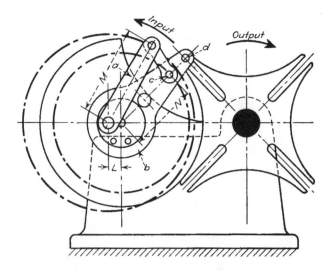

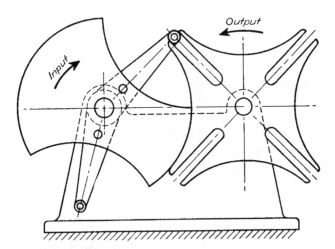

Fig. 4—(Left) The duration of the dwell periods is changed by arranging the driving rollers unsymmetrically around the input shaft. This does not affect the duration of the motion periods. If unequal motion periods are desired as well as unequal dwell periods, then the roller crank-arms must be unequal in length and the star must be suitably modified; such a mechanism is called an "irregular Geneva drive."

Fig. 5—(Below) In this intermittent drive, the two rollers drive the output shaft as well as lock it during dwell periods. For each revolution of the input shaft the output shaft has two motion periods. The output displacement ϕ is determined by the number of teeth; the driving angle, ψ, may be chosen within limits. Gear a is driven intermittently by two driving rollers mounted on input wheel b, which is bearing-mounted on frame c. During the dwell period the rollers circle around the top of a tooth. During the motion period, a roller's path d relative to the driven gear is a straight line inclined towards the output shaft. The tooth profile is a curve parallel to path d. The top land of a tooth becomes the arc of a circle of radius R, the arc approximating part of the path of a roller.

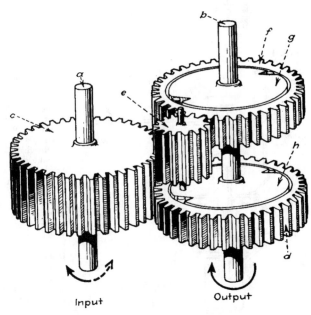

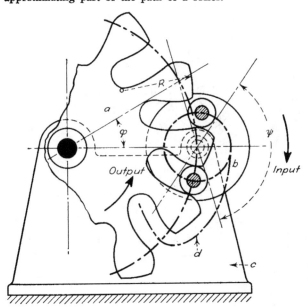

Fig. 6—This uni-directional drive was developed by the author and to his knowledge is novel. The output shaft rotates in the same direction at all times, without regard to the direction of the rotation of the input shaft; angular velocity of the output shaft is directly proportional to the angular velocity of the input shaft. Input shaft a carries spur gear c, which has approximately twice the face width of spur gears f and d mounted on output shaft b. Spur gear c meshes with idler e and with spur gear d. Idler e meshes with spur gears c and f. The output shaft b carries two free-wheel disks g and h, which are oriented uni-directionally.

When the input shaft rotates clockwise (bold arrow), spur gear d rotates counter-clockwise and idles around free-wheel disk h. At the same time idler e, which is also rotating counter-clockwise, causes spur gear f to turn clockwise and engage the rollers on free-wheel disk g; thus, shaft b is made to rotate clockwise. On the other hand, if the input shaft turns counter-clockwise (dotted arrow), then spur gear f will idle while spur gear d engages free-wheel disk h, again causing shaft b to rotate clockwise.

Overriding Spring Mechanisms for Low-Torque Drives

Henry L. Milo, Jr.

Extensive use is made of overriding spring mechanisms in the design of instruments and controls. Anyone of the arrangements illustrated allows an incoming motion to override the outgoing motion whose limit has been reached. In an instrument, for example, the spring device can be placed between

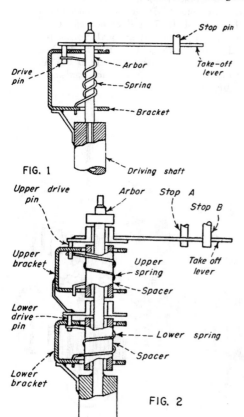

FIG. 1

FIG. 2

Fig. 1—Unidirectional Override. The take-off lever of this mechanism can rotate nearly 360 deg. It's movement is limited by only one stop pin. In one direction, motion of the driving shaft also is impeded by the stop pin. But in the reverse direction the driving shaft is capable of rotating approximately 270 deg past the stop pin. In operation, as the driving shaft is turned clockwise, motion is transmitted through the bracket to the take-off lever. The spring serves to hold the bracket against the drive pin. When the take-off lever has traveled the desired limit, it strikes the adjustable stop pin. However, the drive pin can continue its rotation by moving the bracket away from the drive pin and winding up the spring. An overriding mechanism is essential in instruments employing powerful driving elements, such as bimetallic elements, to prevent damage in the overrange regions.

Fig. 2—Two-directional Override. This mechanism is similar to that described under Fig. 1, except that two stop pins limit the travel of the take-off lever. Also, the incoming motion can override the outgoing motion in either direction. With this device, only a small part of the total rotation of the driving shaft need be transmitted to the take-off lever and this small part may be anywhere in the range. The motion of the driving shaft is transmitted through the lower bracket to the lower drive pin, which is held against the bracket by means of the spring. In turn, the lower drive pin transfers the motion through the upper bracket to the upper drive pin. A second spring holds this pin against the upper drive bracket. Since the upper drive pin is attached to the take-off lever, any rotation of the drive shaft is transmitted to the lever, provided it is not against either stop A or B. When the driving shaft turns in a counterclockwise direction, the take-off lever finally strikes against the adjustable stop A. The upper bracket then moves away from the upper drive pin and the upper spring starts to wind up. When the driving shaft is rotated in a clockwise direction, the take-off lever hits adjustable stop B and the lower bracket moves away from the lower drive pin, winding up the other spring. Although the principal uses for overriding spring arrangements are in the field of instrumentation, it is feasible to apply these devices in the drives of major machines by beefing up the springs and other members.

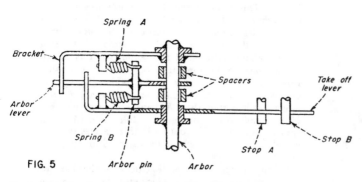

FIG. 5

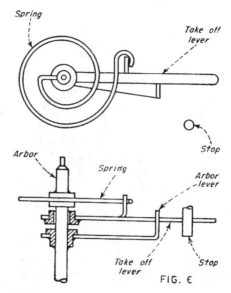

FIG. 6

Fig. 5—Two-directional, 90 Degree Override. This double overriding mechanism allows a maximum overtravel of 90 deg in either direction. As the arbor turns, the motion is carried from the bracket to the arbor lever, then to the take-off lever. Both the bracket and the take-off lever are held against the arbor lever by means of springs A and B. When the arbor is rotated counterclockwise, the take-off lever hits stop A. The arbor lever is held stationary in contact with the take-off lever. The bracket, which is soldered to the arbor, rotates away from the arbor lever, putting spring A in tension. When the arbor is rotated in a clockwise direction, the take-off lever comes against stop B and the bracket picks up the arbor lever, putting spring B in tension.

the sensing and indicating elements to provide over-range protection. The dial pointer is driven positively up to its limit, then stops; while the input shaft is free to continue its travel. Six of the mechanisms described here are for rotary motion of varying amounts. The last is for small linear movements.

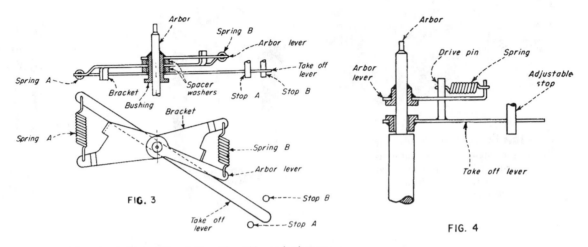

FIG. 3

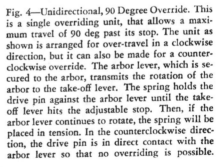

FIG. 4

Fig. 3—Two-directional, Limited-Travel Override. This mechanism performs the same function as that shown in Fig. 2, except that the maximum override in either direction is limited to about 40 deg, whereas the unit shown in Fig. 2 is capable of 270 deg movement. This device is suited for uses where most of the incoming motion is to be utilized and only a small amount of travel past the stops in either direction is required. As the arbor is rotated, the motion is transmitted through the arbor lever to the bracket. The arbor lever and the bracket are held in contact by means of spring B. The motion of the bracket is then transmitted to the take-off lever in a similar manner, with spring A holding the take-off lever and the bracket together. Thus the rotation of the arbor is imparted to the take-off lever until the lever engages either stops A or B. When the arbor is rotated in a counterclockwise direction, the take-off lever eventually comes up against the stop B. If the arbor lever continues to drive the bracket, spring A will be put in tension.

Fig. 4—Unidirectional, 90 Degree Override. This is a single overriding unit, that allows a maximum travel of 90 deg past its stop. The unit as shown is arranged for over-travel in a clockwise direction, but it can also be made for a counterclockwise override. The arbor lever, which is secured to the arbor, transmits the rotation of the arbor to the take-off lever. The spring holds the drive pin against the arbor lever until the take-off lever hits the adjustable stop. Then, if the arbor lever continues to rotate, the spring will be placed in tension. In the counterclockwise direction, the drive pin is in direct contact with the arbor lever so that no overriding is possible.

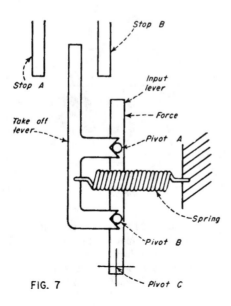

FIG. 7

Fig. 6—Unidirectional, 90 Degree Override. This mechanism operates exactly the same as that shown in Fig. 4. However, it is equipped with a flat spiral spring in place of the helical coil spring used in the previous version. The advantage of the flat spiral spring is that it allows for a greater override and minimizes the space required. The spring holds the take-off lever in contact with the arbor lever. When the take-off lever comes in contact with the stop, the arbor lever can continue to rotate and the arbor winds up the spring.

Fig. 7—Two-directional Override, Linear Motion. The previous mechanisms were overrides for rotary motion. The device in Fig. 7 is primarily a double override for small linear travel although it could be used on rotary motion. When a force is applied to the input lever, which pivots about point C, the motion is transmitted directly to the take-off lever through the two pivot posts A and B. The take-off lever is held against these posts by means of the spring. When the travel is such the take-off lever hits the adjustable stop A, the take-off lever revolves about pivot post A, pulling away from pivot post B and putting additional tension in the spring. When the force is diminished, the input lever moves in the opposite direction, until the take-off lever contacts the stop B. This causes the take-off lever to rotate about pivot post B, and pivot post A is moved away from the take-off lever.

10 Ways to Amplify Mechanical Movements

How levers, membranes, cams, and gears are arranged to measure, weigh, gage, adjust, and govern.

Federico Strasser

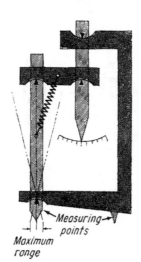

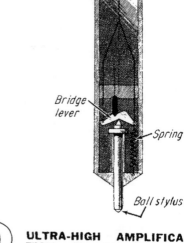

1 **HIGH AMPLIFICATION** for simple measuring instruments is provided by double lever action. Accuracy can be as high as 0.0001 in.

2 **PIVOTED LEVERS** allow extremely sensitive action in comparator-type measuring device shown here. The range, however, is small.

3 **ULTRA-HIGH AMPLIFICATION,** with only one lever, is provided in the Hirth-Minimeter shown here. Again, the range is small.

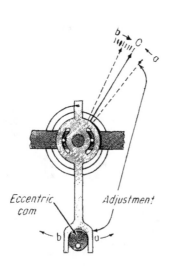

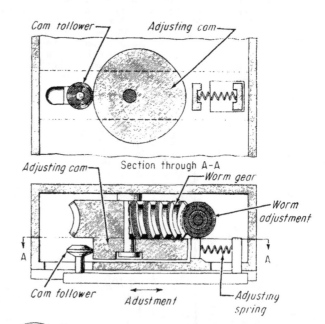

7 **FOR CLOSE ADJUSTMENT,** electrical measuring instruments employ eccentric cams. Here movement is reduced, not amplified.

8 **MICROSCOPIC ADJUSTMENT** is achieved here by employing a large eccentric-cam coupled to a worm-gear drive. Smooth, fine adjustment result.

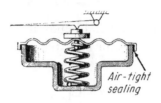

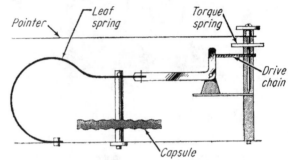

5 **CAPSULE UNIT** for gas-pressure indicators should be provided with a compression spring to preload the membrane for more positive action.

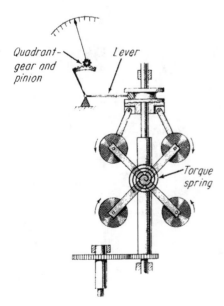

4 **LEVER - ACTUATED** weigh-scale needs no springs to maintain balance. The lever system, mounted on knife edges, is extremely sensitive.

6 **AMPLIFIED MEMBRANE MOVEMENT** can be gained by the arrangement shown here. A small chain-driven gear links the lever system.

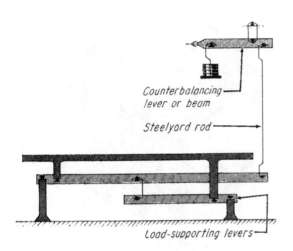

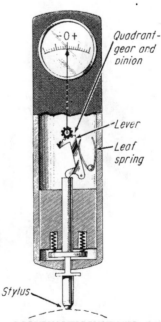

9 **QUADRANT-GEAR AND PINION** coupled to an L-lever provide ample movement of indicator needle for small changes in governor speed.

10 **COMBINATION LEVER AND GEARED** quadrant are used here to give the comparator maximum sensitivity combined with ruggedness.

10 Ways to Amplify Mechanical Action

Levers, wires, hair, and metal bands are arranged to give high velocity ratios for adjusting and measuring.

Federico Strasser

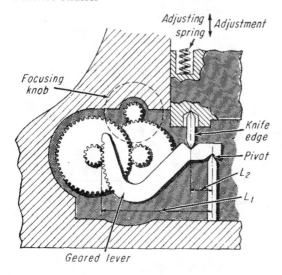

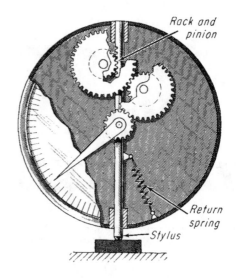

1 **LEVER AND GEAR** train amplify the microscope control-knob movement. Knife edges provide frictionless pivots for lever.

2 **DIAL INDICATOR** starts with rack and pinion amplified by gear train. The return-spring takes out backlash.

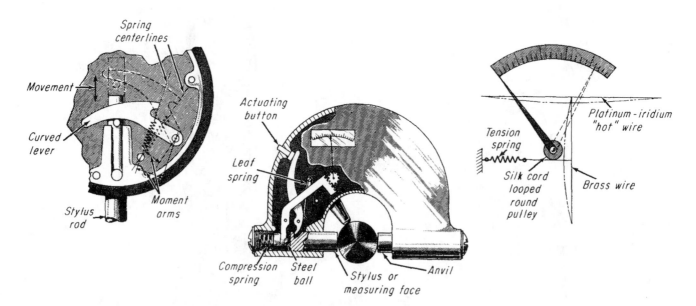

3 **CURVED LEVER** is so shaped and pivoted that the force exerted on the stylus rod, and thus stylus pressure, remains constant.

4 **ZEISS COMPARATOR** is provided with a special lever to move the stylus clear of the work. A steel ball greatly reduces friction.

5 **"HOT-WIRE" AMMETER** relies on the thermal expansion of a current-carrying wire. A relatively large needle movement occurs.

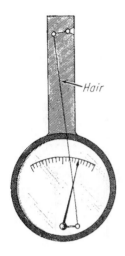

Hair

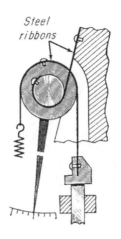

Steel ribbons

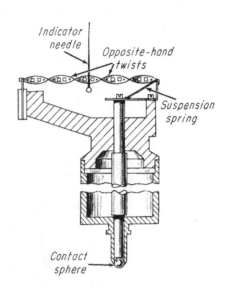

Indicator needle

Opposite-hand twists

Suspension spring

Contact sphere

HYGROMETER is actuated by a hair. When humidity causes expansion of the hair, its movement is amplified by a lever.

STEEL RIBBONS transmit movement without the slightest backlash. The movement is amplified by differences in diameter.

METAL BAND is twisted and supported at each end. Small movement of contact sphere produces large needle movement.

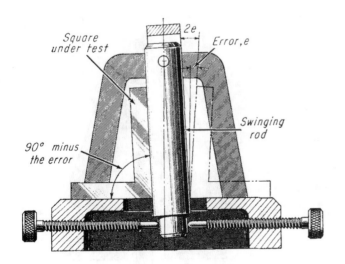

Square under test

2e

Error, e

Swinging rod

90° minus the error

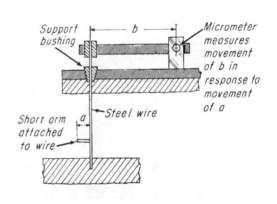

Support bushing

b

Micrometer measures movement of b in response to movement of a

Short arm attached to wire

a

Steel wire

ACCURACY of 90° squares can be checked with a device shown here. The rod makes the error much more apparent.

TORSIONAL deflection of the short arm is transmitted with low friction to the longer arm for micrometer measurement.

How to Damp Axial and Rotational Motion

Fluid-friction devices include two hydraulic and two pneumatic actions; swinging-vane arrangements dissipate energy and govern speed.

Federico Strasser

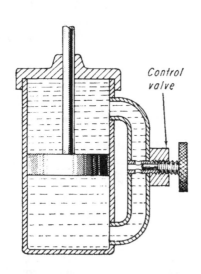

Control valve

1 **ADJUSTABLE BYPASS** between the two sides of the piston controls speed at which fluid can flow when piston is moved.

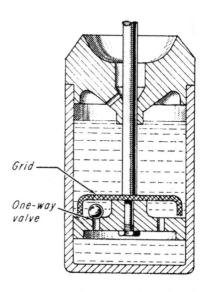

Grid

One-way valve

2 **CHECK VALVE** in piston lets speed be controlled so that the piston moves faster in one direction than in the other.

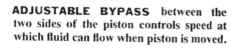

Spring-grip attachment

Locating notch in shaft

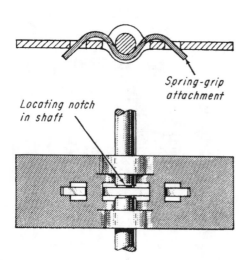

5 **ROTATING VANES** are resisted by the air as they revolve. Make allowance for sudden stops by providing a spring.

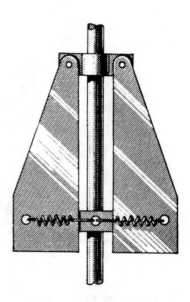

6 **SWINGING VANES** create increased wind drag as centrifugal force opens them to a larger radius.

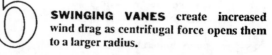

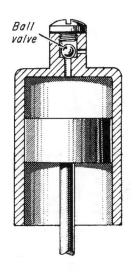

Ball valve

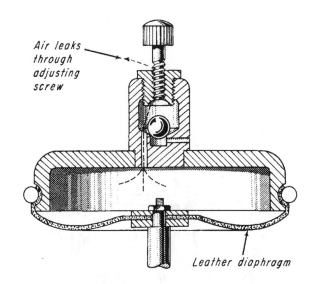

Air leaks through adjusting screw

Leather diaphragm

3 PNEUMATIC CHECK VALVE acts in manner similar to that of previous device. Vertical position, of course, is necessary.

4 FLEXIBLE DIAPHRAGM controls short movements. Speed is fast in one direction, but greatly slowed in return direction.

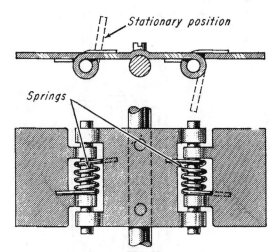

Stationary position

Springs

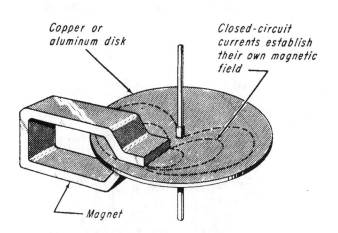

Copper or aluminum disk

Closed-circuit currents establish their own magnetic field

Magnet

7 VANE AREA INCREASES when the spring-loaded vanes swing out. Forces differ for motion into or against the wind.

8 EDDY CURRENTS are induced in disk when it is moved through a magnetic field. Braking is directly proportional to speed.

Make Diaphragms Work for You

Diaphragms have more uses than you think. Here's a display of applications that simple fabric-elastomer diaphragms can handle economically and with a minimum of design problems.

John F. Taplin

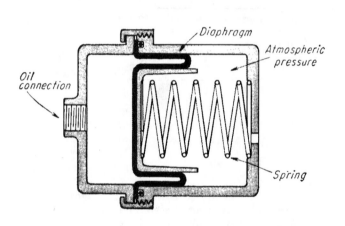

2 Expansion compensator for liquid-filled systems handles thermal expansion of the liquid as well as any system losses.

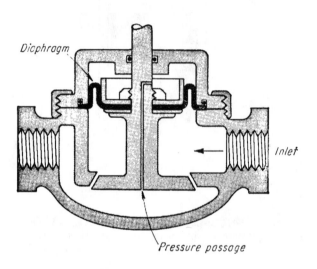

3 A balanced valve uses a fabric-elastomer diaphragm to hydrostatically balance the valve poppet as well as the valve head.

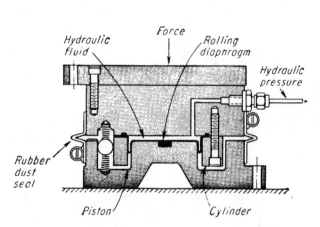

6 A force-balance load cell converts the weight or force of any object into an accurate reading at a remote point.

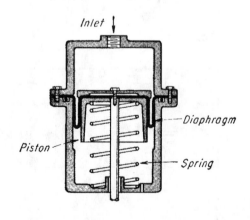

7 Linear actuator converts gas or fluid pressure into a linear stroke without leakage or break-out friction effects.

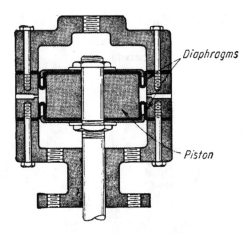

1 Double-acting actuator provides for thrust in either direction by placing two diaphragm assemblies back to back.

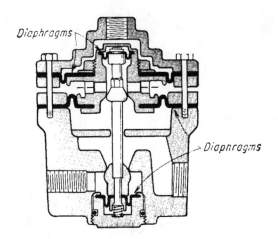

4 Regulating valve controls the value of air pressure by means of a diaphragm-balanced valve and two control diaphragms.

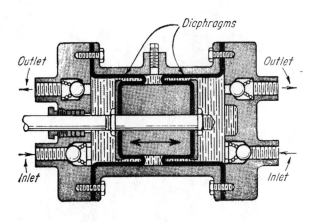

5 Double-acting pump has two diaphragms to give smooth and continuous flow of fluid to equipment at a safe working pressure.

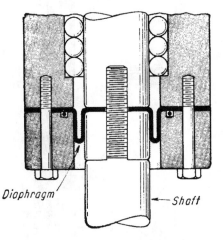

8 Shaft seal uses lubricant pressure to force the sidewall of the diaphragm to roll against the shaft and housing.

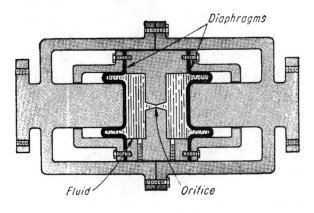

9 Damping mechanism prevents abrupt or sudden motion in a machine. Damping amount is controlled by orifice size.

4 Ways to Eliminate Backlash

Wedges take up freedom in threads and gears, hold shaft snug against bearing.

L. Kasper

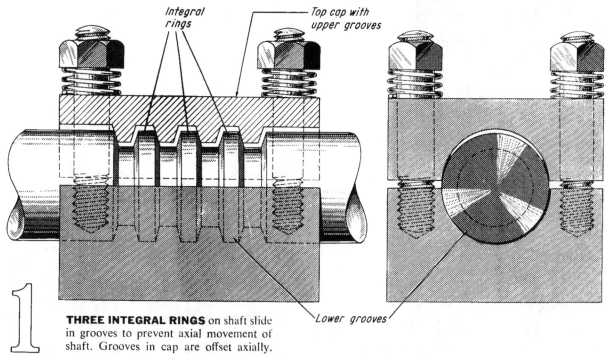

THREE INTEGRAL RINGS on shaft slide in grooves to prevent axial movement of shaft. Grooves in cap are offset axially.

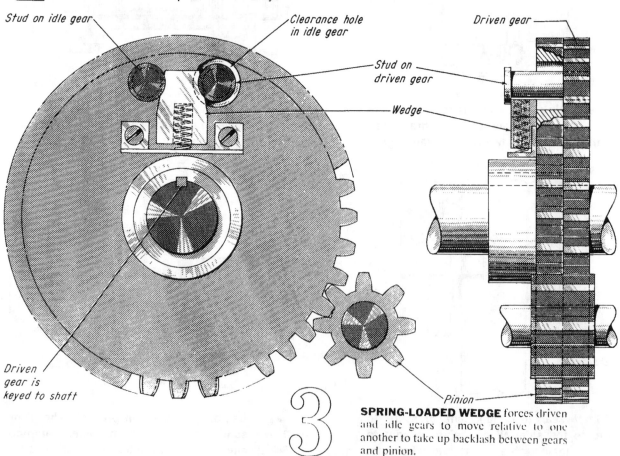

SPRING-LOADED WEDGE forces driven and idle gears to move relative to one another to take up backlash between gears and pinion.

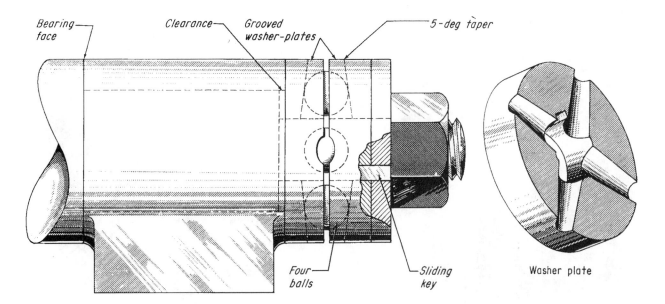

Bearing face — Clearance — Grooved washer-plates — 5-deg taper

Four balls — Sliding key

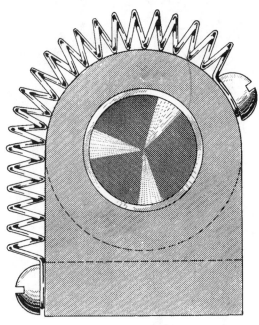

Washer plate

CENTRIFUGAL FORCE causes balls to exert force on grooved washer-plates when shaft rotates, pulling it against bearing face.

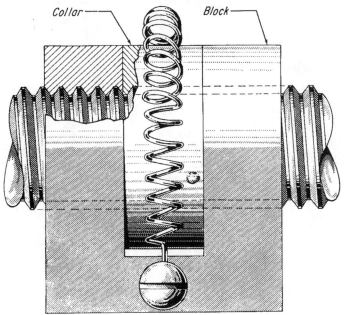

Collar — Block

COLLAR AND BLOCK have continuous V-thread. When wear takes place in lead screw, the collar always maintains pressure on threads.

4 More Ways to Prevent Backlash

Springs combine with wedging action to ensure that threads, gears and toggles respond smoothly.

L. Kasper

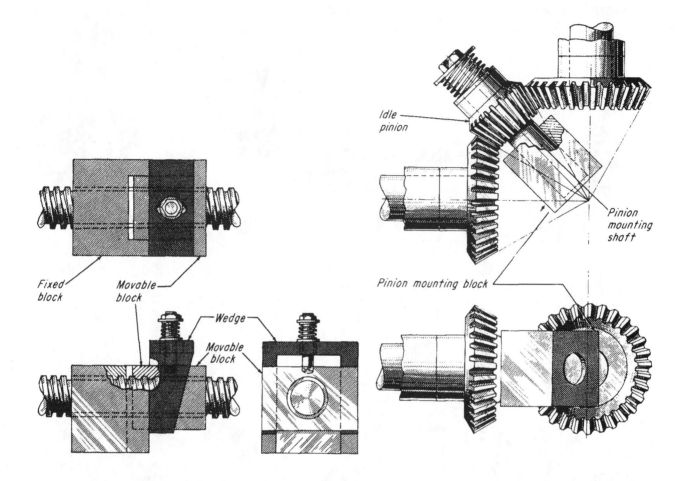

Fixed block

Movable block

Wedge

Movable block

Idle pinion

Pinion mounting shaft

Pinion mounting block

1 MOVABLE BLOCK is forced away from fixed block by spring-loaded wedge. Pressure is applied to both sides of lead screw, thus ensuring snug fit.

2 SPRING-LOADED PINION is mounted on a shaft located so that the spring forces pinion teeth into gear teeth to take up lost motion or backlash.

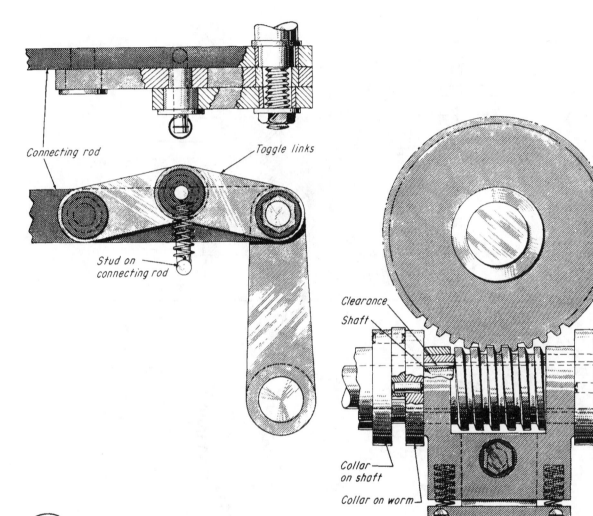

Connecting rod

Toggle links

Stud on
connecting rod

Clearance

Shaft

Collar
on shaft

Collar on worm

Link

TOGGLE LINKS are spring-loaded and approach alignment to take up lost motion as wear in the joint takes place. Smooth response is thus gained.

HOLLOW WORM has clearance for shaft, which drives worm through pinned collars and links. As wear occurs, springs move worm into teeth.

Limit-Switch Backlash

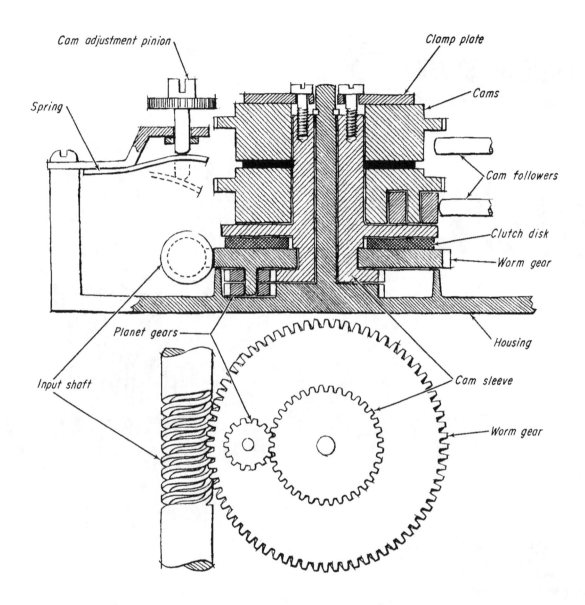

Cam adjustment pinion

Clamp plate

Cams

Spring

Cam followers

Clutch disk

Worm gear

Housing

Planet gears

Cam sleeve

Input shaft

Worm gear

SWITCH-ACTUATING CAMS are driven by double-reduction gearing. The first pass is the input worm and its worm gear. The second reduction consists of a planetary system with two keyed planets pivoted on the worm gear. The two planets do not have the same number of teeth. When the worm gear is rotated the planet gears move around a sun gear cast into the base of the housing. The upper planet meshes with gear teeth on the sleeve. The cams clamped to the sleeve actuate the switches. The ratio of the planetary reduction can be altered by changing the planets.

A friction clutch between the sleeve flange and the worm gear makes the switch exceedingly sensitive to reversals at the input worm. When a switch is actuated to reverse input direction, the cams are driven directly by the input worm and gear through the friction clutch until the backlash has been taken up. At this point the clutch begins to slip. The immediate reversal of the cams resets switches in $\frac{1}{3}$ to 1 revolution depending on the worm-gear ratio.

In some of the reduction ratios available a deliberate mismatch is employed in planetary gear sizes. This intentional mismatch creates no problems at the pitch velocities produced, since the 3500 rpm maximum at the input shaft is reduced by the step-down of the worm and worm gear. The low torque requirements of the switch-operating cams eliminate any overstressing due to mismatch. The increased backlash obtained by the mismatch is desirable in the higher reduction ratios to allow the friction clutch to reset the switches before the backlash is taken up. This permits switch reset in less than 1 rpm despite the higher input gear reductions of as much as 1280:1.

SECTION 2

CREATIVE ASSEMBLIES

Rotary Piston Engine

Warren Ogren, Inventor
Robert Parmley, Draftsman

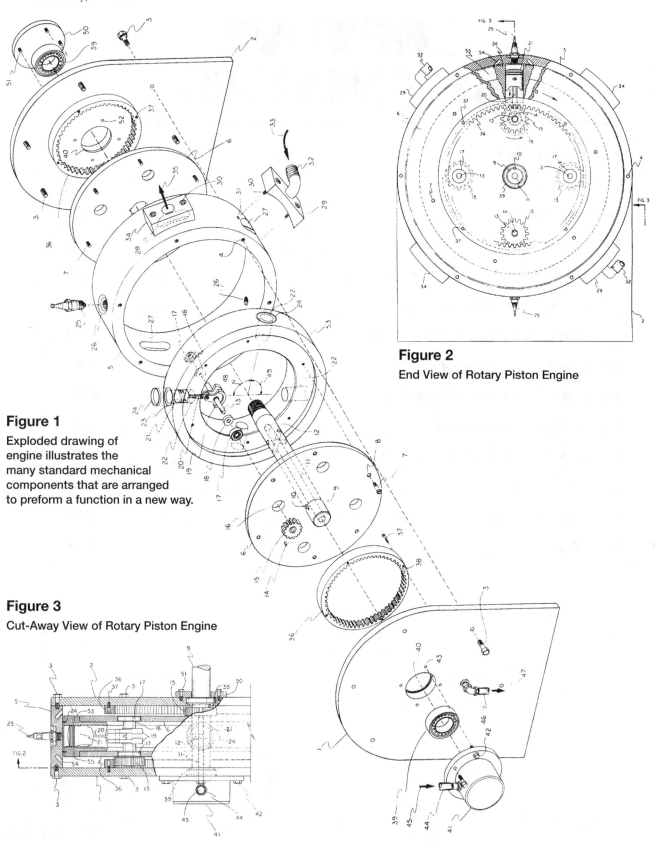

Figure 2

End View of Rotary Piston Engine

Figure 1

Exploded drawing of
engine illustrates the
many standard mechanical
components that are arranged
to preform a function in a new way.

Figure 3

Cut-Away View of Rotary Piston Engine

Courtesy: Warren Ogren & Morgan & Parmley, Ltd.

Milk Transfer System

Drawn by: Robert O. Parmley

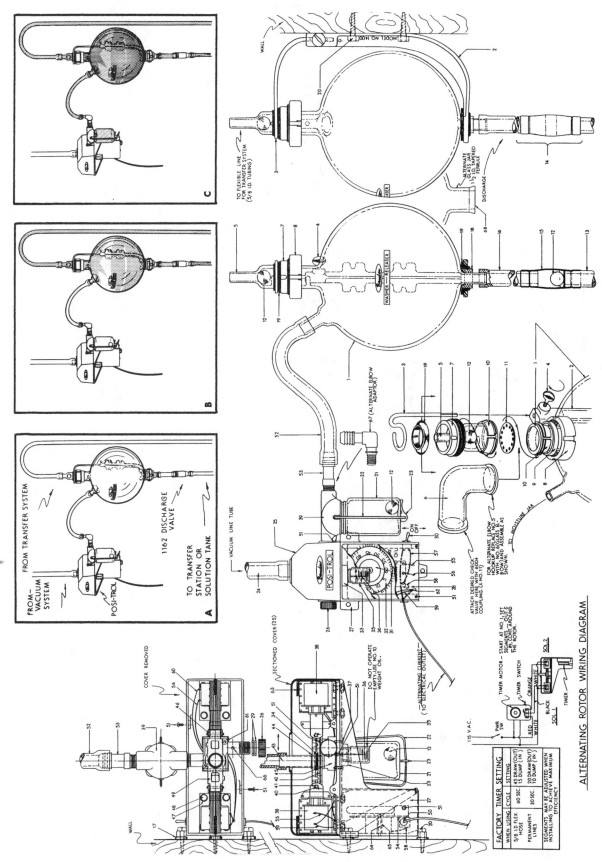

Courtesy: Bender Machine Works, Inc.

Hydraulic Motor

Drawn by: Robert O. Parmley

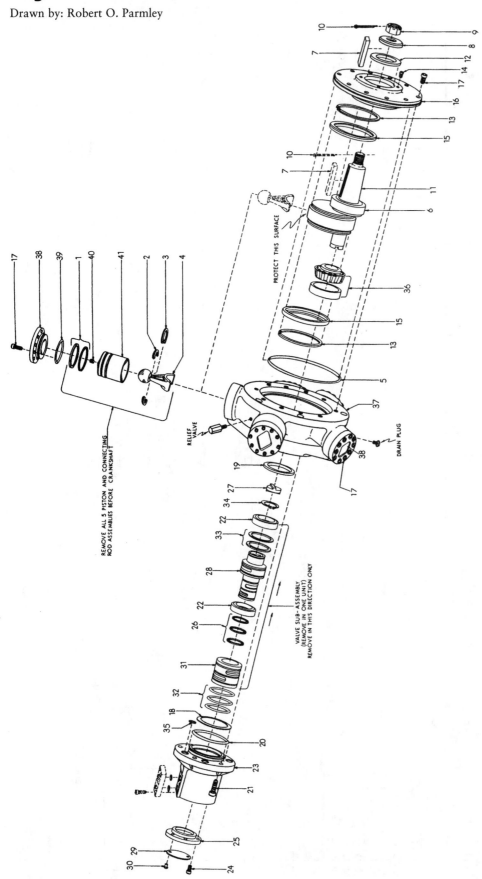

Slash Errors with Sensitive Balance

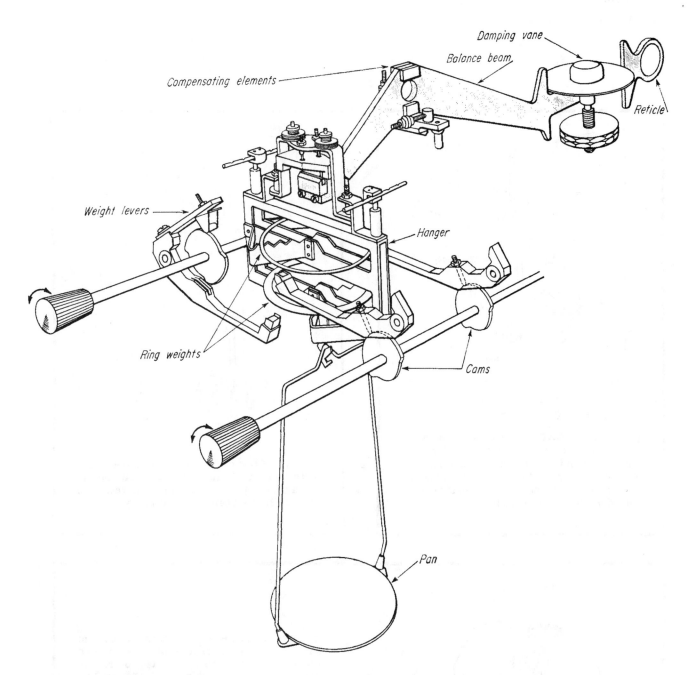

Compensating elements

Damping vane

Balance beam

Reticle

Weight levers

Hanger

Ring weights

Cams

Pan

SENSITIVITY OF BALANCE is independent of temperature fluctuations. To keep the center of gravity constant, two temperature-sensitive elements are riveted to aluminumalloy balance beam, bridging a slot which is directly over the balance point. Their coefficient of expansion compensates for beam deflection caused by variations in temperature.

Enclosed in a cylindrical canister at the rear of the balance beam is a vane that damps its movement, preventing oscillation. The hanger at the front of the scale carries sets of ring weights which are lifted by camoperated levers. The shafts on which the cams are mounted are connected to the mechanical readout.

The scale in effect weighs by sub-traction since it is balanced, when empty, by all the ring weights resting on the hanger. To weigh an unknown, the ring weights are lifted from the hanger. The sum of the raised weights is shown on the mechanical counter, which displays the first three digits. The complete total is displayed by the mechanical plus the optical system that projects through the reticle.

Control-Locked Thwart Vibration and Shock

Critical adjustments stay put-safe against accidental turning or deliberate fiddling with them.

Frank William Wood, Jr.

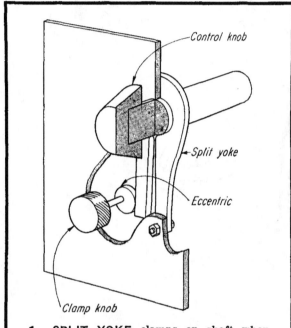

1..SPLIT YOKE clamps on shaft when eccentric squeezes ends of yoke together. Knurled knob is handy for constant use, and eliminates need for tool. Another advantage is high torque capacity. But this design needs considerable space on panel.

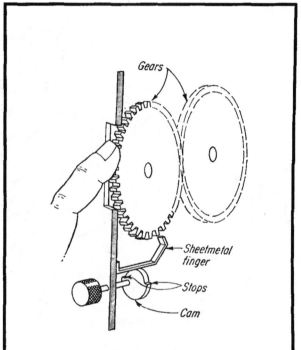

2..FINGER springs into place between gear teeth at turn of cam. Although gear lock is ideally suited for right-angle drives, size of teeth limits positioning accuracy.

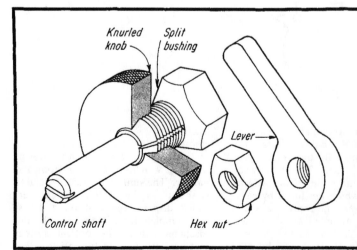

3..SPLIT BUSHING tightens on control shaft, because knurled knob has tapered thread. Bushing also mounts control to panel, so requires just one hole. Lever, like knob, does away with tools, but locks tighter and faster. For controls adjusted infrequently, hex nut turns a fault into an advantage. Although it takes a wrench to turn the nut, added difficulty guards against knob-twisters.

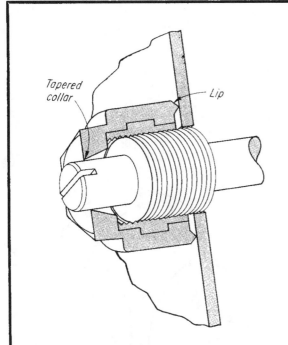

4..CONSTANT DRAG of tapered collar on shaft makes control stiff, so it doesn't need locking and unlocking. Compressed lip both seals out dust and keeps molded locking nut from rotating.

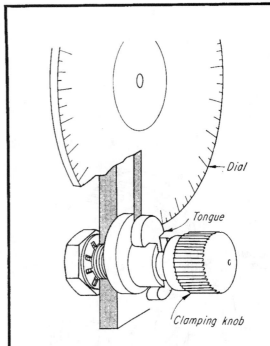

5..TONGUE slides in groove, clamps down on edge of dial. If clamp is not tight, it can scratch the face.

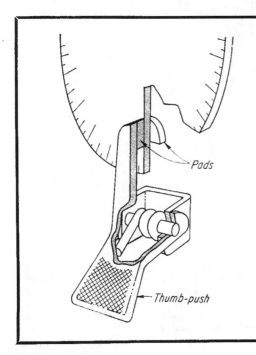

6..SPOT-BRAKE clamp is self-locking, which means it takes two hands to make an adjustment, one to hold the clamp open and one to turn the dial.

1-Way Output from Speed Reducers

When input reverses, these five slow-down mechanisms continue supplying a non-reversing rotation.

Louis Slegel

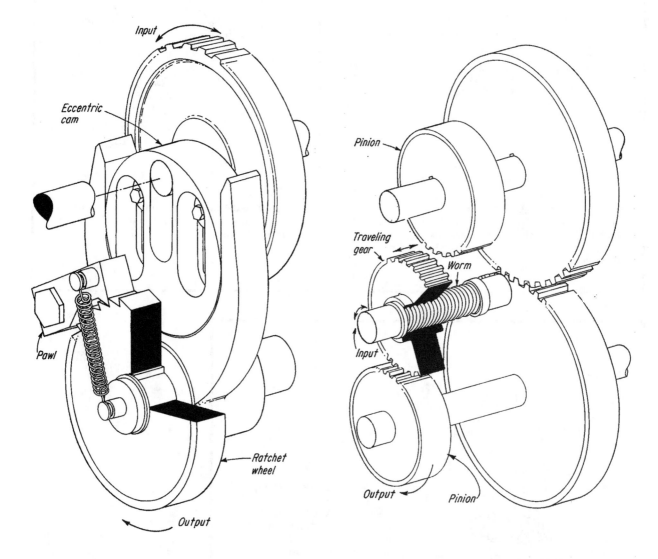

1 ECCENTRIC CAM adjusts over a range of high-reduction ratios, but unbalance limits it to low speeds. When direction of input changes, there is no lag in output rotation. Output shaft moves in steps because of ratchet drive through pawl which is attached to U-follower.

2 TRAVELING GEAR moves along worm and transfers drive to other pinion when input rotation changes direction. To ease engagement, gear teeth are tapered at ends. Output rotation is smooth, but there is a lag after direction changes as gear shifts. Gear cannot be wider than axial offset between pinions, or there will be interference.

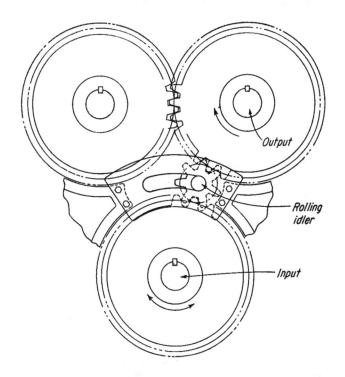

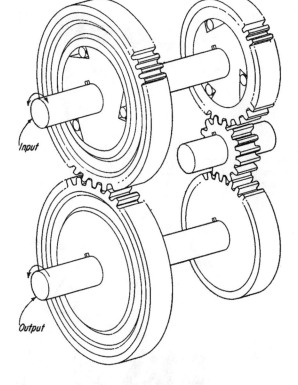

3 ROLLING IDLER also gives smooth output and slight lag after input direction changes. Small drag on idler is necessary, so that it will transfer into engagement with other gear and not sit spinning in between.

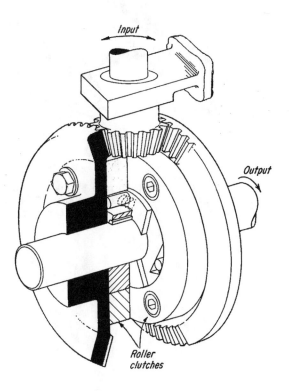

4 TWO BEVEL GEARS drive through roller clutches. One clutch catches in one direction; the other catches in the opposite direction. There is negligible interruption of smooth output rotation when input direction changes.

5 ROLLER CLUTCHES are on input gears in this drive, again giving smooth output speed and little output lag as input direction changes.

Torque-Limiters Protect Light-Duty Drives

In such drives the light parts break easily when overloaded. These eight devices disconnect them from dangerous torque surges.

L. Kasper

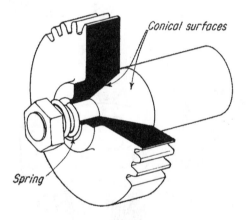

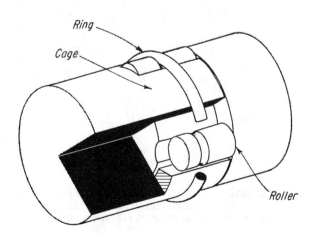

1
MAGNETS transmit torque according to their number and size. In-place control is limited to lowering torque capacity by removing magnets.

2
CONE CLUTCH is formed by mating taper on shaft to beveled hole through gear. Tightening down on nut increases torque capacity.

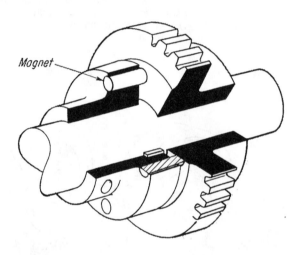

3
RING fights natural tendency of rollers to jump out of grooves cut in reduced end of one shaft. Slotted end of hollow shaft is like a cage.

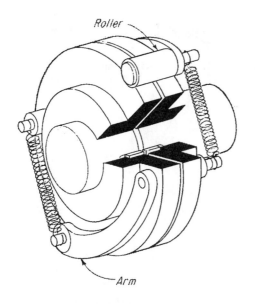

Roller

Arm

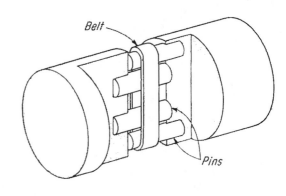

Belt

Pins

4
ARMS hold rollers in slots which are cut across disks mounted on ends of butting shafts. Springs keep rollers in slots; over-torque forces them out.

5
FLEXIBLE BELT wrapped around four pins transmits only lightest loads. Outer pins are smaller than inner pins to ensure contact.

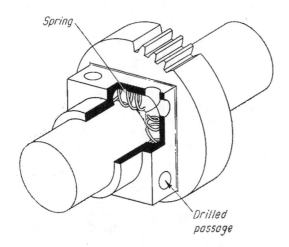

Spring

Drilled passage

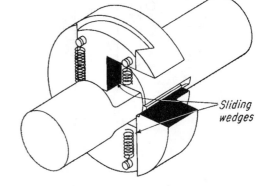

Sliding wedges

6
SPRINGS inside drilled block grip the shaft because they distort during mounting of gear.

7
SLIDING WEDGES clamp down on flattened end of shaft; spread apart when torque gets too high. Strength of springs which hold wedges together sets torque limit.

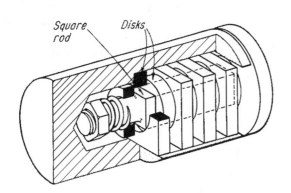

Square rod

Disks

8
FRICTION DISKS are compressed by adjustable spring. Square disks lock into square hole in left shaft; round ones lock onto square rod on right shaft.

6 Ways to Prevent Overloading

These "safety valves" give way if machinery jams, thus preventing serious damage.

Peter C. Noy

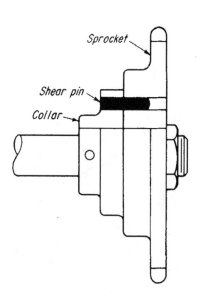

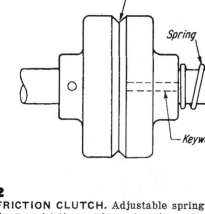

2
FRICTION CLUTCH. Adjustable spring tension that holds the two friction surfaces together sets overload limit. As soon as overload is removed the clutch reengages. One drawback is that a slipping clutch can destroy itself if unnoticed.

1
SHEAR PIN is simple to design and reliable in service. However, after an overload, replacing the pin takes a relatively long time; and new pins aren't always available.

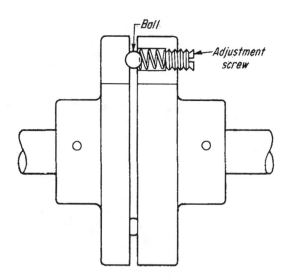

3
MECHANICAL KEYS. Spring holds ball in dimple in opposite face until overload forces the ball out. Once slip begins, wear is rapid, so device is poor when overload is common.

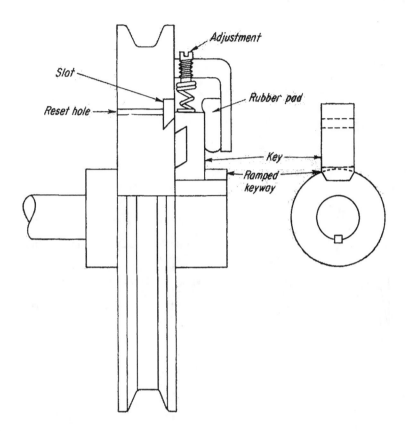

4
RETRACTING KEY. Ramped sides of keyway force key outward against adjustabe spring. As key moves outward, a rubber pad—or another spring—forces the key into a slot in the sheave. This holds the key out of engagement and prevents wear. To reset, push key out of slot by using hole in sheave.

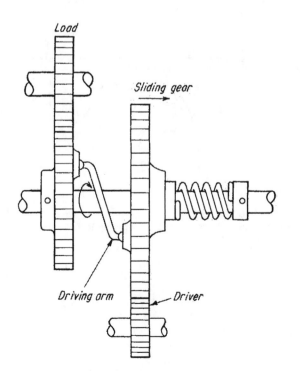

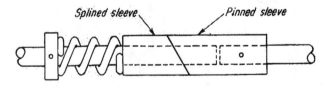

5
ANGLE-CUT CYLINDER. With just one tooth, this is a simplified version of the jaw clutch. Spring tension sets load limit.

6
DISENGAGING GEARS. Axial forces of spring and driving arm balance. Overload overcomes spring force to slide gears out of engagement. Gears can strip once overloading is removed, unless a stop holds gears out of engagement.

7 More Ways to Prevent Overloading

For the designer who must anticipate the unexpected, here are ways to guard machinery against carelessness or accident.

Peter C. Noy

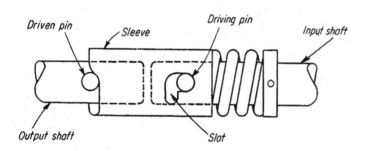

1
CAMMED SLEEVE connects input and output shafts. Driven pin pushes sleeve to right against spring. When overload occurs, driving pin drops into slot to keep shaft disengaged. Turning shaft backwards resets.

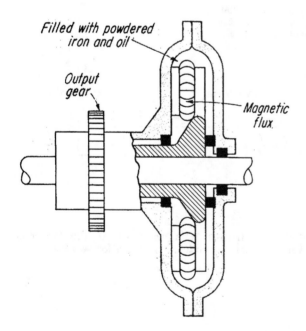

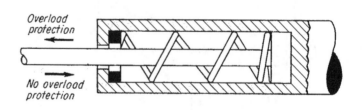

3
SPRING PLUNGER is for reciprocating motion with possible overload only when rod is moving left. Spring compresses under overload.

2
MAGNETIC FLUID COUPLING is filled with slurry made of iron or nickel powder in oil. Controlled magnetic flux that passes through fluid varies slurry viscosity, and thus maximum load over a wide range. Slip ring carries field current to vanes.

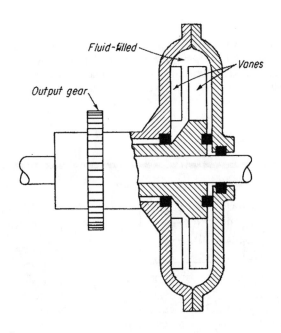

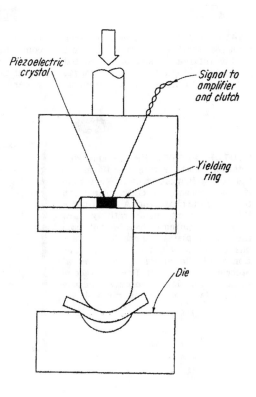

4
FLUID COUPLING. Maximum load can be closely controlled by varying viscosity and level of fluid. Other advantages are smooth transmission and low heat rise during slip.

5
TENSION RELEASE. When toggle-operated blade shears soft pin, jaws open to release eye. A spring that opposes the spreading jaws can replace the shear pin.

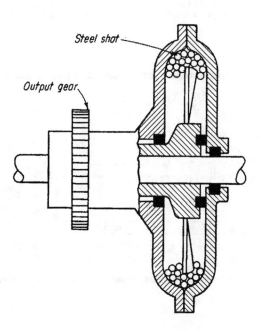

6
STEEL-SHOT COUPLING transmits more torque as speed increases. Centrifugal force compresses steel shot against case, increasing resistance to slip. Adding more steel shot also increases resistance to slip.

7
PIEZOELECTRIC CRYSTAL sends output signal that varies with pressure. Clutch at receiving end of signal disengages when pressure on the crystal reaches preset limit. Yielding ring controls compression of crystal.

7 Ways to Limit Shaft Rotation

Traveling nuts, clutch plates, gear fingers, and pinning members are the bases of these ingenious mechanisms.

I. M. Abeles

Mechanical stops are often required in automatic machinery and servomechanisms to limit shaft rotation to a given number of turns. Two problems to guard against, however, are: Excessive forces caused by abrupt stops; large torque requirements when rotation is reversed after being stopped.

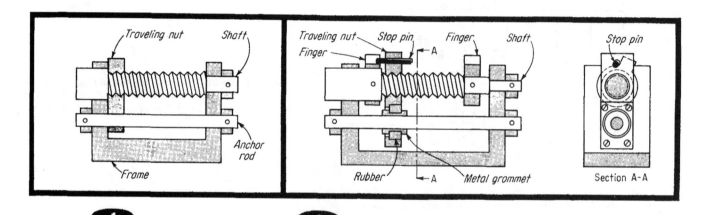

1

2

TRAVELING NUT moves (1) along threaded shaft until frame prevents further rotation. A simple device, but nut jams so tight that a large torque is required to move the shaft from its stopped position. This fault is overcome at the expense of increased length by providing a stop pin in the traveling nut (2). Engagement between pin and rotating finger must be shorter than the thread pitch so pin can clear finger on the first reverse-turn. The rubber ring and grommet lessen impact, provide a sliding surface. The grommet can be oil-impregnated metal.

3

CLUTCH PLATES tighten and stop rotation as the rotating shaft moves the nut against the washer. When rotation is reversed, the clutch plates can turn with the shaft from A to B. During this movement comparatively low torque is required to free the nut from the clutch plates. Thereafter, subsequent movement is free of clutch friction until the action is repeated at other end of the shaft. Device is recommended for large torques because clutch plates absorb energy well.

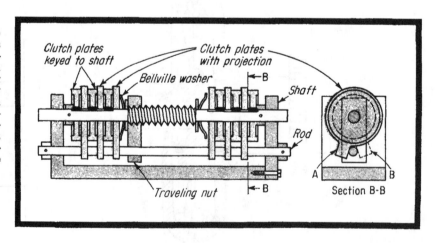

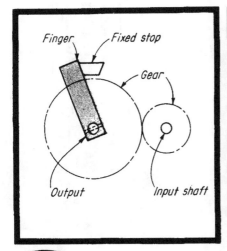

4

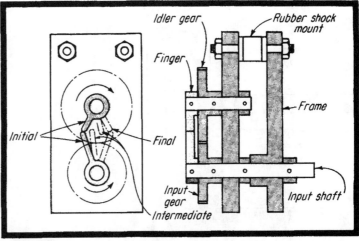

5

SHAFT FINGER on output shaft hits resilient stop after making less than one revolution. Force on stop depends upon gear ratio. Device is, therefore, limited to low ratios and few turns unless a worm-gear setup is used.

TWO FINGERS butt together at initial and final positions, prevent rotation beyond these limits. Rubber shock-mount absorbs impact load. Gear ratio of almost 1:1 ensures that fingers will be out of phase with one another until they meet on the final turn. Example: Gears with 30 to 32 teeth limit shaft rotation to 25 turns. Space is saved here but gears are costly.

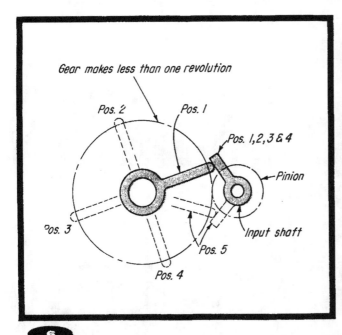

6

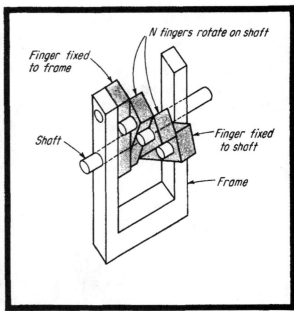

7

LARGE GEAR RATIO limits idler gear to less than one turn. Sometimes stop fingers can be added to already existing gears in a train, making this design simplest of all. Input gear, however, is limited to a maximum of about 5 turns.

PINNED FINGERS limit shaft turns to approximately $N+1$ revolutions in either direction. Resilient pin-bushings would help reduce impact force.

Devices for Indexing or Holding Mechanical Movements

Louis Dodger

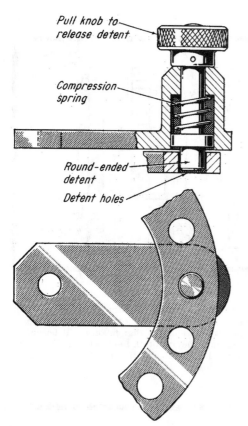

Pull knob to release detent

Compression spring

Round-ended detent

Detent holes

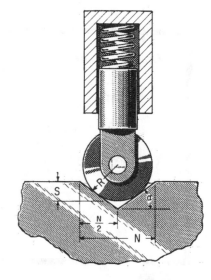

ROLLER DETENT POSITIONS IN A NOTCH:

RISE, $S = \dfrac{N \tan a}{2} - R \times \dfrac{1 - \cos a}{\cos a}$

ROLLER RADIUS, $R = \left(\dfrac{N \tan a}{2} - S\right)\left(\dfrac{\cos a}{1 - \cos a}\right)$

3

1 AXIAL POSITIONING (INDEXING) BY MEANS OF SPACED HOLES IN INDEX BASE

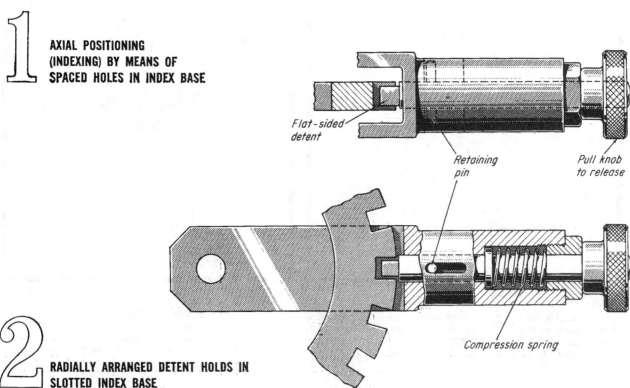

Flat-sided detent

Retaining pin

Pull knob to release

Compression spring

2 RADIALLY ARRANGED DETENT HOLDS IN SLOTTED INDEX BASE

Saw-Matic Mechanism

Leo Heikkinen, Inventor
Robert Parmley, Draftsman

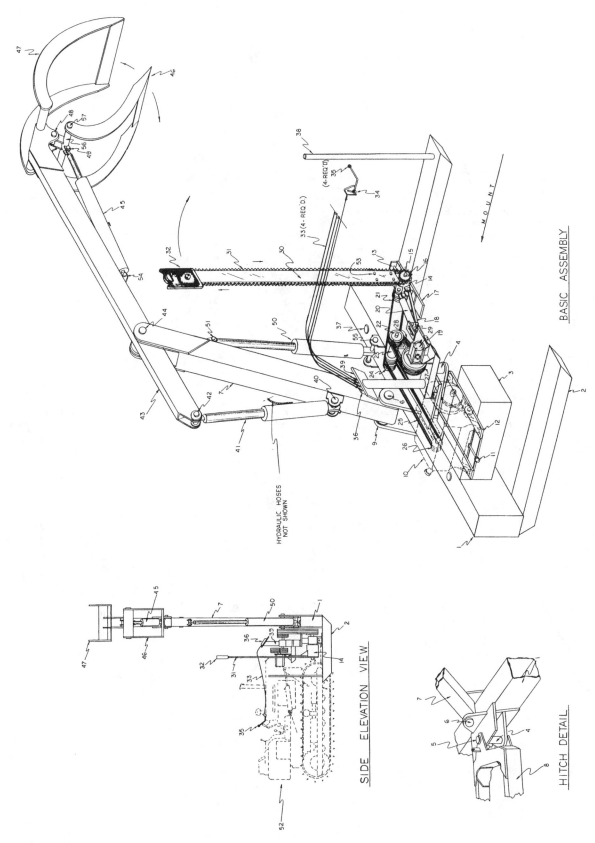

BASIC ASSEMBLY

SIDE ELEVATION VIEW

HITCH DETAIL

Courtesy: Dale Heikkinen

Piping Assembly for Sewage Lift Station Control Vault

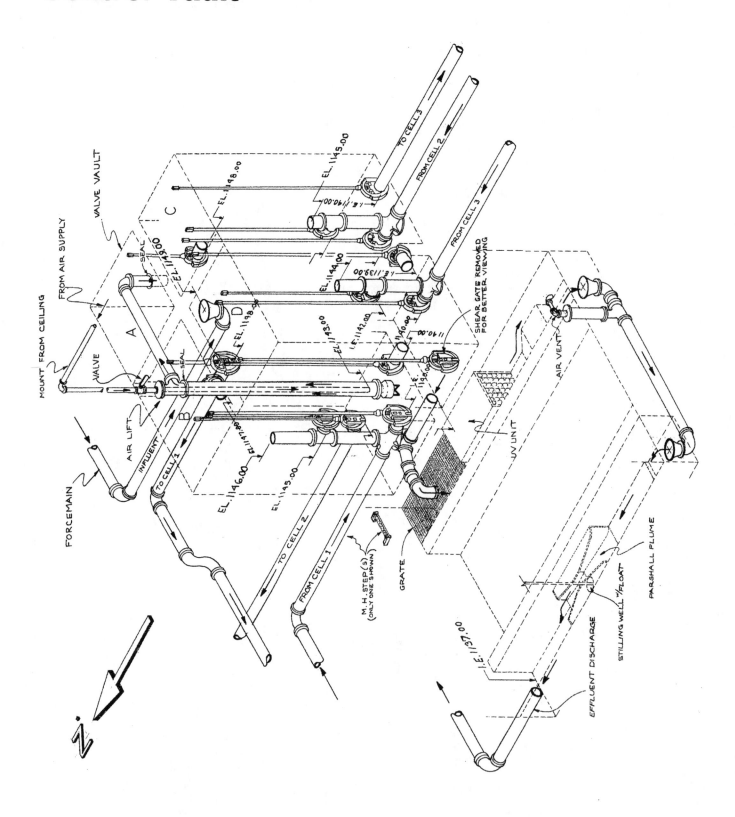

Courtesy: Morgan & Parmley, Ltd.

Water Bike

Drawn by: Robert O. Parmley

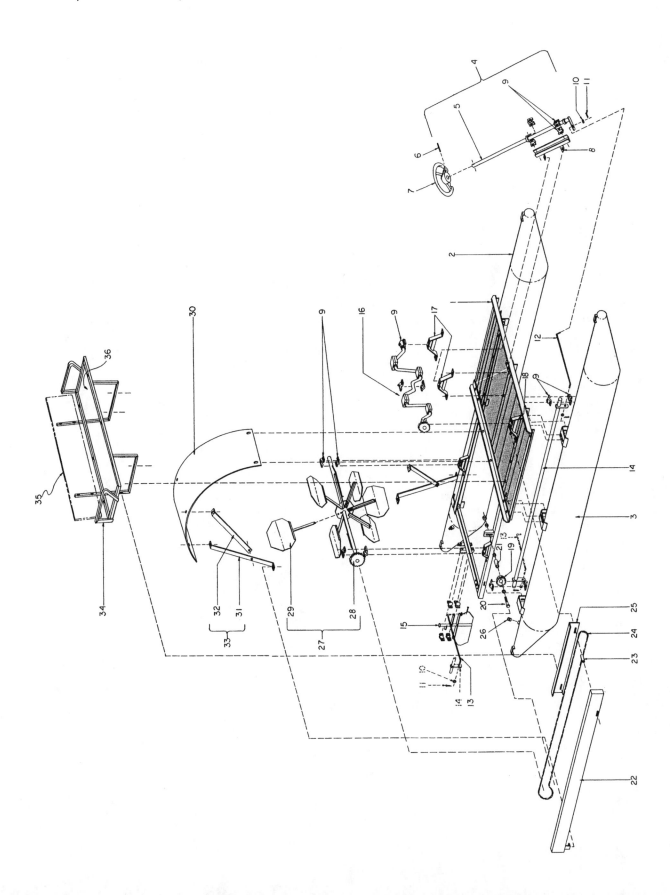

SECTION 3

LINKAGES

8 Basic Push-Pull Linkages

These arrangements are invariably the root of all linkage devices.

Frank William Wood, Jr.

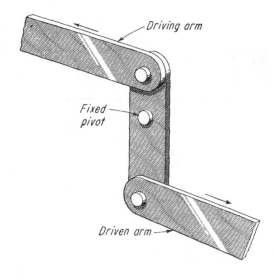

Driving arm

Fixed pivot

Driven arm

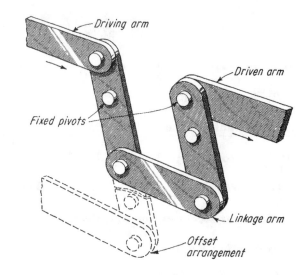

Driving arm

Driven arm

Fixed pivots

Linkage arm

Offset arrangement

1 **FIXED PIVOTS** on arm lengths are located to control ratio of input and output movements of this push-pull-actuated linkage. Mechanism can be either flat bars or round rods of adequate thickness to prevent bowing under compression.

2 **PUSH-PULL LINKAGE** for same direction of motion can be obtained by adding linkage arm to previous design. In both cases, if arms are bars it might be best to make them forked rather than merely flatted at their linkage ends.

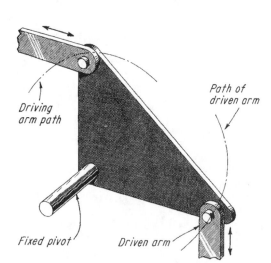

Driving arm path

Path of driven arm

Fixed pivot

Driven arm

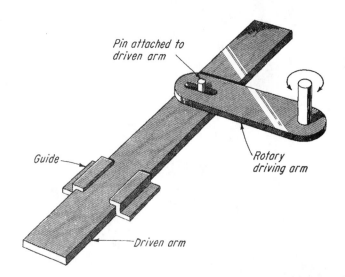

Pin attached to driven arm

Guide

Rotary driving arm

Driven arm

5 **VERTICAL OUTPUT MOVEMENT** from horizontal input is gained with this push-pull linkage. Although the triangular-shaped plate could be substituted by an L-shaped arm, the plate gives greater freedom of driving- and driven-arm location.

6 **FOR LIMITED STRAIGHT-LINE** 2-direction motion use this rotary-actuated linkage. Friction between the pin and sides of slot limit this design to small loads. A bearing on the pin will reduce friction and slot wear to negligible proportions.

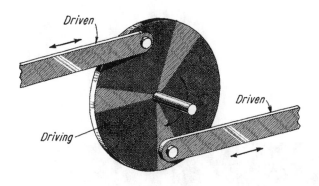

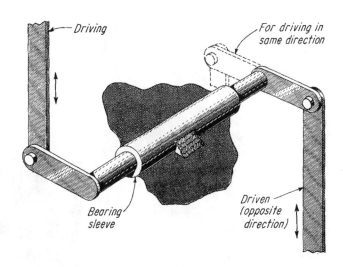

3 **ROTARY-ACTUATED LINKAGE** gives opposite direction of motion and can be obtained by using 3-bar linkage with pivot point of middle link located at midpoint of arm length. Disk should be adequately strengthened for heavy loads.

4 **SAME-DIRECTION MOTION** is given by this rotary-actuated linkage when end arms are located on the same sides; for opposite-direction motion, locate the arms on opposite sides. Use when a crossover is required between input and output.

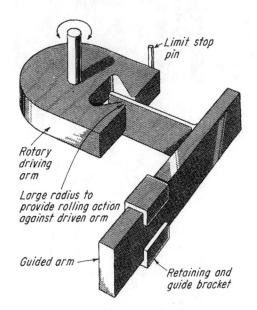

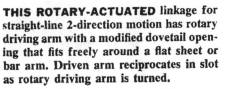

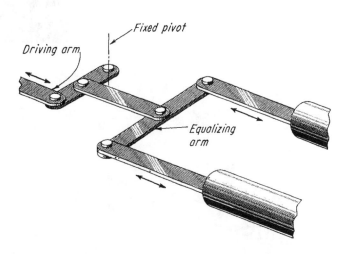

7 **THIS ROTARY-ACTUATED** linkage for straight-line 2-direction motion has rotary driving arm with a modified dovetail opening that fits freely around a flat sheet or bar arm. Driven arm reciprocates in slot as rotary driving arm is turned.

8 **EQUALIZING LINKAGE** here has an equalizing arm that balances the input force to two output arms. This arrangement is most suitable for air or hydraulic systems where equal force is to be exerted on the pistons of separate cylinders.

5 Linkages for Straight-Line Motion

These devices convert rotary to straight-line motion without the need for guides.

Sigmund Rappaport

1

Evans' linkage . . .

has oscillating drive-arm that should have a maximum operating angle of about 40°. For a relatively short guide-way, the reciprocating output stroke is large. Output motion is on a true straight line in true harmonic motion. If an exact straight-line motion is not required, however, a link can replace the slide. The longer this link, the closer does the output motion approach that of a true straight line—if link-length equals output stroke, deviation from straight-line motion is only 0.03% of output stroke.

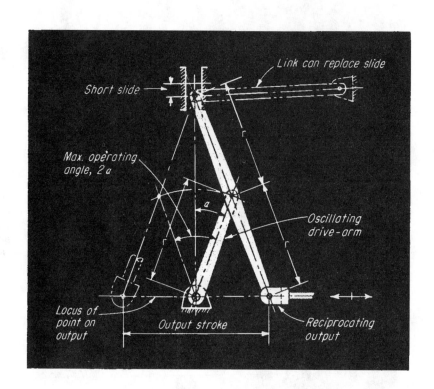

2

Simplified Watt's linkage . . .

generates an approximate straight-line motion. If the two arms are equally long, the tracing point describes a symmetrical figure 8 with an almost straight line throughout the stroke length. The straightest and longest stroke occurs when the connecting-link length is about 2/3 of the stroke, and arm length is 1.5S. Offset should equal half the connecting-link length. If the arms are unequal, one branch of the figure-8 curve is straighter than the other. It is straightest when a/b equals (arm 2)/(arm 1).

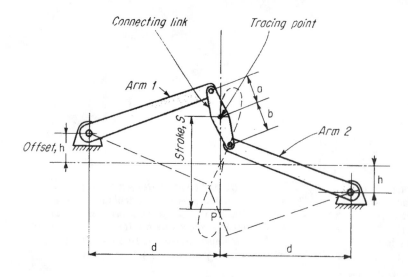

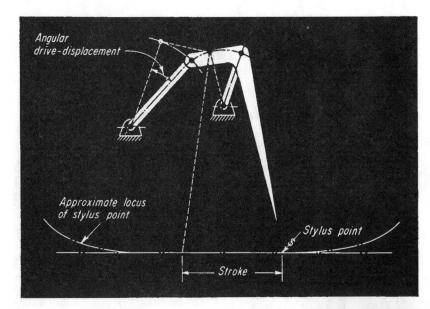

Four-bar linkage . . .

produces approximately straight-line motion. This arrangement provides motion for the stylus on self-registering measuring instruments. A comparatively small drive-displacement results in a long, almost-straight line.

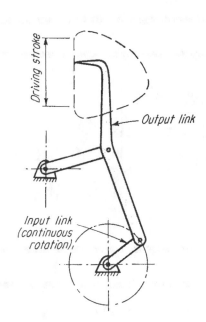

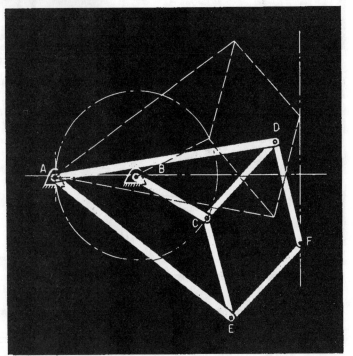

D-drive . . .

results when linkage arms are arranged as shown here. Output-link point describes a path resembling the letter D, thus it contains a straight portion as part of its cycle. Motion is ideal for quick engagement and disengagement before and after a straight driving-stroke. Example, the intermittent film-drive in movie-film projectors.

The "Peaucellier cell" . . .

was first solution to the classical problem of generating a straight line with a linkage. Its basis: within the physical limits of the motion, $AC \times AF$ remains constant. Curves described by C and F are, therefore, inverse; if C describes a circle that goes through A, then F will describe a circle of infinite radius—a straight line, perpendicular to AB. The only requirements are: $AB = BC$; $AD = AE$; and CD, DF, FE, EC are all equal. The linkage can be used to generate circular arcs of large radius by locating A outside the circular path of C.

10 Ways to Change Straight-Line Direction

Arrangements of linkages, slides, friction drives and gears that can be the basis of many ingenious devices.

Federico Strasser

Linkages

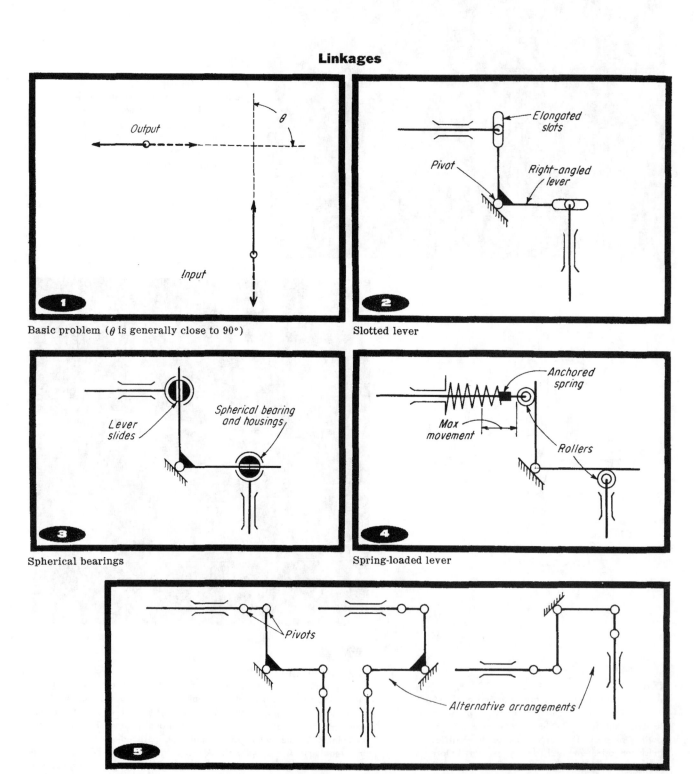

Basic problem (θ is generally close to 90°)

Slotted lever

Spherical bearings

Spring-loaded lever

Pivoted levers with alternative arrangements

Guides

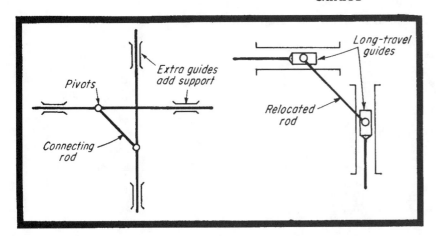

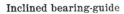

6

Single connecting rod (left) is relocated (right) to get around need for extra guides

Friction Drives

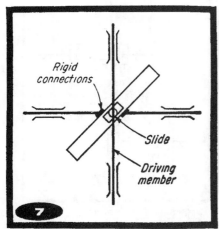

7

Inclined bearing-guide

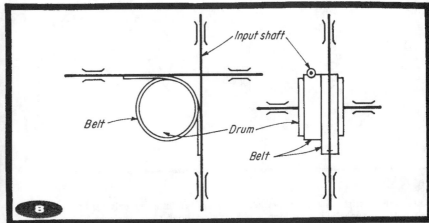

8

Belt, steel band, or rope around drum, fastened to driving and driven members; sprocket-wheels and chain can replace drum and belt

Gears

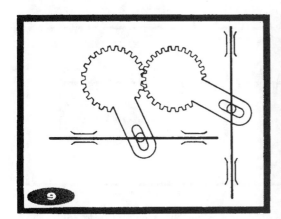

9

Matching gear-segments

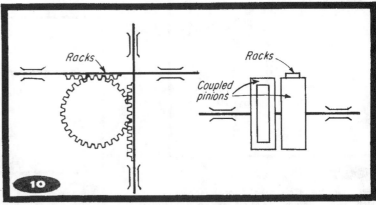

10

Racks and coupled pinions (can be substituted by friction surfaces for low-cost setup)

9 More Ways to Change Straight-Line Direction

These devices, using gears, cams, pistons, and solenoids, supplement similar arrangements employing linkages, slides, friction drives, and gears, shown.

Federico Strasser

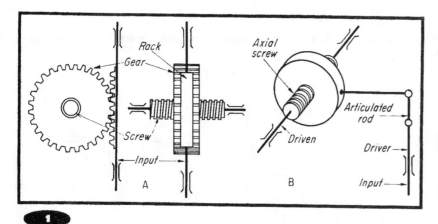

1

Axial screw with rack-actuated gear (A) and articulated driving rod (B) are both irreversible movements, i.e. driver must always drive.

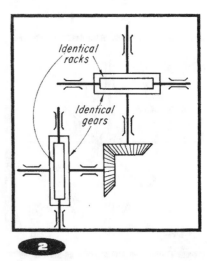

2

Rack-actuated gear with associated bevel gears is reversible.

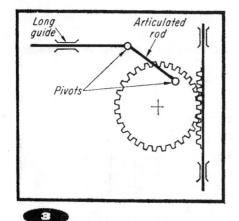

3

Articulated rod on crank-type gear with rack driver. Action is restricted to comparatively short movements.

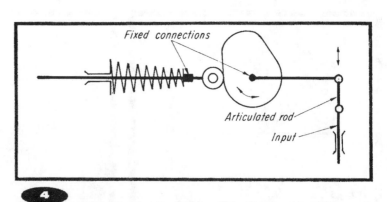

4

Cam and spring-loaded follower allow input/output ratio to be varied according to cam rise. Movement is usually irreversible.

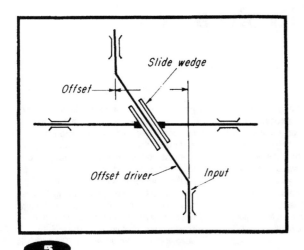

5

Offset driver actuates driven member by wedge action. Lubrication and low coefficient of friction help to allow max offset.

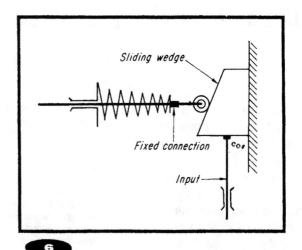

6

Sliding wedge is similar to previous example but requires spring-loaded follower; also, low friction is less essential with roller follower.

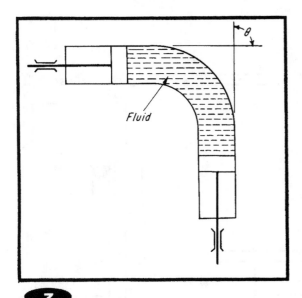

7

Fluid coupling is simple, allows motion to be transmitted through any angle. Leak problems and accurate piston-fitting can make method more expensive than it appears to be. Also, although action is reversible it must always be a compressive one for best results.

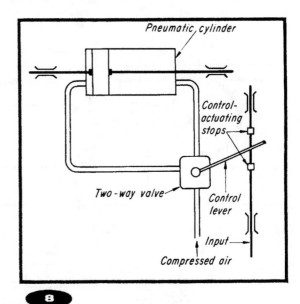

8

Pneumatic system with two-way valve is ideal when only two extreme positions are required. Action is irreversible. Speed of driven member can be adjusted by controlling input of air to cylinder.

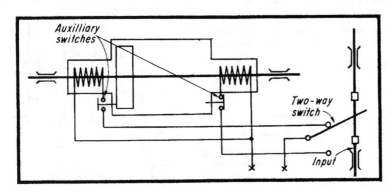

9

Solenoids and two-way switch are here arranged in analogous device to previous example. Contact to energized solenoid is broken at end of stroke. Again, action is irreversible.

Linkages for Accelerating and Decelerating Linear Strokes

When ordinary rotary cams cannot be conveniently applied, the mechanisms presented here, or adaptations of them, offer a variety of interesting possibilities for obtaining either acceleration or deceleration, or both.

Fig. 1   A slide block with a pinion and shaft and a pin for link B reciprocates at a constant rate. The pinion has a crankpin for mounting link D, and it also engages a stationary rack. The pinion can make one complete revolution at each forward stroke of the slide block and another as the slide block returns in the opposite direction. However, if the slide block is not moved through its normal travel range, the pinion turns only a fraction of a revolution. The mechanism can be made variable by making the connection link for F adjustable along the length of the element that connects links B and D. Alternatively, the crankpin for link D can be made adjustable along the radius of the pinion, or both the connection link and the crankpin can be made adjustable.

Fig. 2   A drive rod, reciprocating at a constant rate, rocks link BC about a pivot on a stationary block. A toggle between arm B and the stationary block contacts an abutment. Motion of the drive rod through the toggle causes deceleration of driven link B. As the drive rod moves toward the right, the toggle is actuated by encountering the abutment. The slotted link BC slides on its pivot while turning. This lengthens arm B and shortens arm C of link BC. The result is deceleration of the driven link. The toggle is returned by a spring (not shown) on the return stroke, and its effect is to accelerate the driven link on its return stroke.

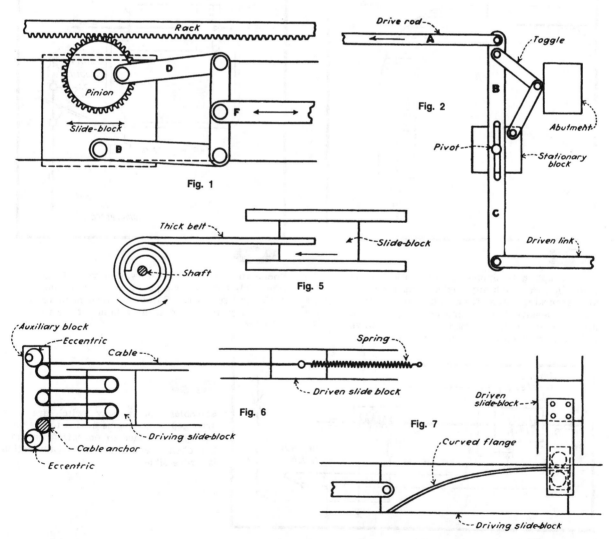

Fig. 1

Fig. 2

Fig. 5

Fig. 6

Fig. 7

Source: *Mechanisms and Mechanical Devices Sourcebook*, 3E, by Chironis & Sclater, © 2001 McGraw-Hill

Fig. 3 **The same direction** of travel for both the drive rod and the drive link is provided by the variation of the Fig. 2 mechanism. Here, acceleration is in the direction of the arrows, and deceleration occurs on the return stroke. The effect of acceleration decreases as the toggle flattens.

Fig. 4 **A bellcrank motion** is accelerated as the rollers are spread apart by a curved member on the end of the drive rod, thereby accelerating the motion of the slide block. The driven elements must be returned by spring to close the system.

Fig. 5 **A constant-speed shaft** winds up a thick belt or similar flexible connecting member, and its effective increase in radius causes the slide block to accelerate. It must be returned by a spring or weight on its reversal.

Fig. 6 **An auxiliary block** that carries sheaves for a cable which runs between the driving and driven slide block is mounted on two synchronized eccentrics. The motion of the driven block is equal to the length of the cable paid out over the sheaves, resulting from the additive motions of the driving and auxiliary blocks.

Fig. 7 A curved flange on the driving slide block is straddled by rollers that are pivotally mounted in a member connected to the driven slide block. The flange can be curved to give the desired acceleration or deceleration, and the mechanism returns by itself.

Fig. 8 The stepped acceleration of the driven block is accomplished as each of the three reciprocating sheaves progressively engages the cable. When the third acceleration step is reached, the driven slide block moves six times faster than the drive rod.

Fig. 9 A form-turned nut, slotted to travel on a rider, is propelled by reversing its screw shaft, thus moving the concave roller up and down to accelerate or decelerate the slide block.

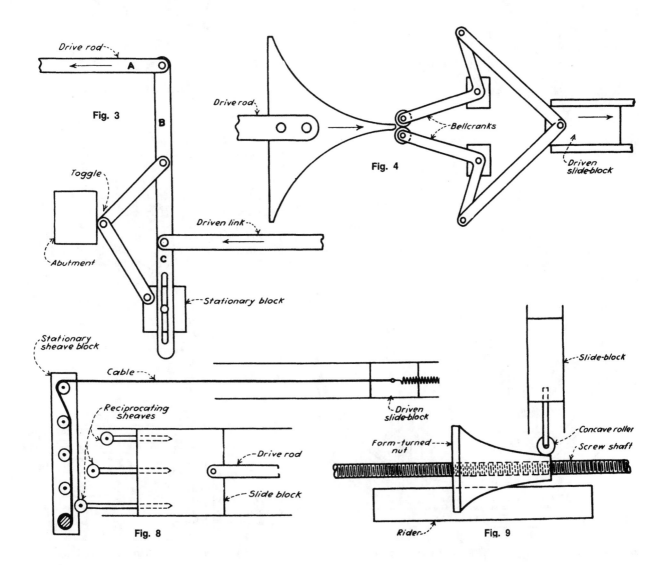

Linkages for Multiplying Short Motions

The accompanying sketches show typical linkages for multiplying short linear motions, usually converting the linear motion into rotation. Although the particular mechanisms shown are designed to multiply the movements of diaphragms or bellows, the same or similar constructions have possible applications wherever it is required to obtain greatly multiplied motions. These transmissions depend on cams, sector gears and pinions, levers and cranks, cord or chain, spiral or screw feed, magnetic attraction, or combinations of these mechanical elements.

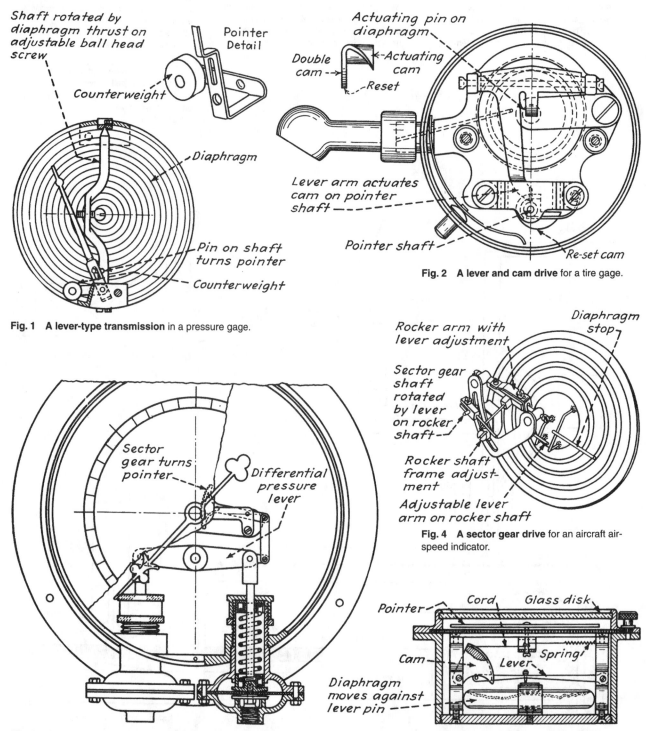

Fig. 1   A lever-type transmission in a pressure gage.

Fig. 2   A lever and cam drive for a tire gage.

Fig. 3   A lever and sector gear in a differential pressure gage.

Fig. 4   A sector gear drive for an aircraft air-speed indicator.

Fig. 5   A lever, cam, and cord transmission in a barometer.

Source: *Mechanisms and Mechanical Devices Sourcebook*, 3E, by Chironis & Sclater, © 2001 McGraw-Hill

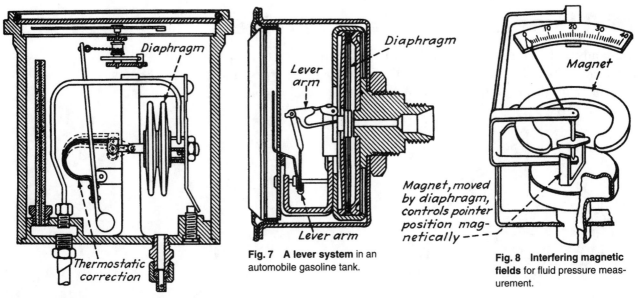

Fig. 6 **A link and chain transmission** for an aircraft rate of climb instrument.

Fig. 7 **A lever system** in an automobile gasoline tank.

Fig. 8 **Interfering magnetic fields** for fluid pressure measurement.

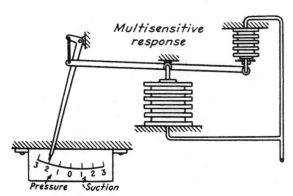

Fig. 9 **A lever system** for measuring atmospheric pressure variations.

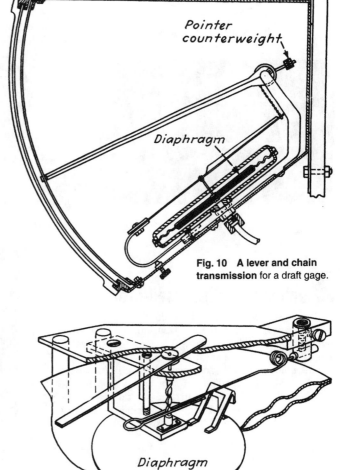

Fig. 10 **A lever and chain transmission** for a draft gage.

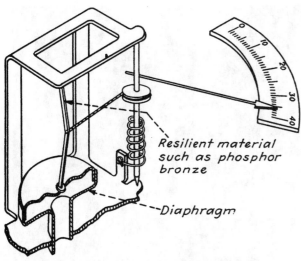

Fig. 11 **A toggle and cord drive** for a fluid pressure measuring instrument.

Fig. 12 **A spiral feed transmission** for a general purpose analog instrument.

Seven Popular Types of Three-Dimensional Drives

Main advantage of three-dimensional drives is their ability to transit motion between nonparallel shafts. They can also generate other types of helpful motion. With this roundup are descriptions of industrial applications.

Dr. W. Meyer Zur Capellen

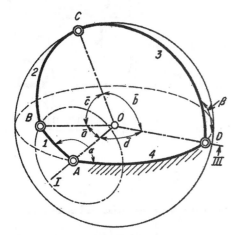

The Spherical Crank

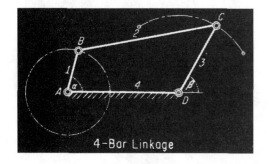

4-Bar Linkage

1 spherical crank drive

This type of drive is the basis for most 3-D linkages, much as the common 4-bar linkage is the basis for the two-dimensional field. Both mechanisms operate on similar principles. (In the accompanying sketches, α is the input angle, and β the output angle. This notation has been used throughout the article.)

In the 4-bar linkage, the rotary motion of driving crank 1 is transformed into an oscillating motion of output link 3. If the fixed link is made the shortest of all, then you have a double-crank mechanism, in which both the driving and driven members make full rotations.

In the spherical crank drive, link 1 is the input, link 3 the output. The axes of rotation intersect at point O; the lines connecting AB, BC, CD and DA can be thought of as part of great circles of a sphere. The length of the link is best represented by angles a, b, c and d.

2 spherical-slide oscillator

The two-dimensional slider crank is obtained from a 4-bar linkage by making the oscillating arm infinitely long. By making an analogous change in the spherical crank, you can obtain the spherical slider crank shown at right.

The uniform rotation of input shaft I is transferred into a nonuniform oscillating or rotating motion of output shaft III. These shafts intersect at an angle δ corresponding to the frame link 4 of the spherical crank. Angle γ corresponds to length of link 1. Axis II is at right angle to axis III.

The output oscillates when γ is smaller than δ; the output rotates when γ is larger than δ.

Relation between input angle α and output angle β is (as designated in skewed Hook's joint, below)

$$\tan \beta = \frac{(\tan \gamma)(\sin \alpha)}{\sin \delta + (\tan \gamma)(\cos \delta)(\cos \alpha)}$$

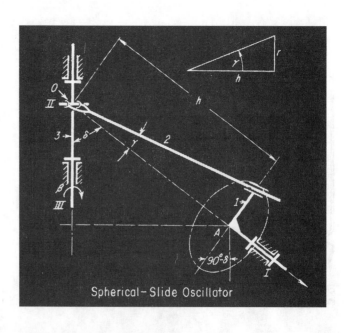

Spherical-Slide Oscillator

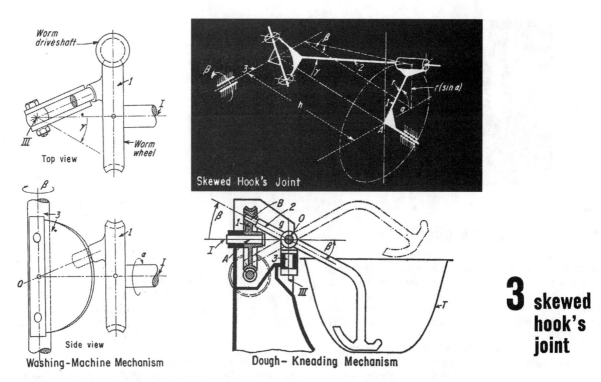

Top view

Worm driveshaft

Worm wheel

Skewed Hook's Joint

Side view

Washing-Machine Mechanism

Dough- Kneading Mechanism

3 skewed hook's joint

This variation of the spherical crank is often used where an almost linear relation is desired between input and output angles for a large part of the motion cycle.

The equation defining the output in terms of the input can be obtained from the above equation by making $\delta = 90°$. Thus $\sin \delta = 1$, $\cos \delta = 0$, and

$$\tan \beta = \tan \gamma \sin \alpha$$

The principle of the skewed Hook's joint has been recently applied to the drive of a washing machine (see sketch at left).

Here, the driveshaft drives the worm wheel 1 which has a crank fashioned at an angle γ. The crank rides between two plates and causes the output shaft III to oscillate in accordance with the equation above.

The dough-kneading mechanism at right is also based on the Hook's joint, but utilizes the path of link 2 to give a wobbling motion that kneads dough in the tank.

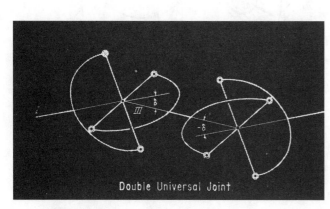

Double Universal Joint

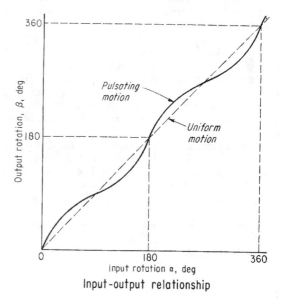

Input-output relationship

4 the universal joint

The universal joint is a variation of the spherical-slide oscillator, but with angle $\gamma = 90°$. This drive provides a totally rotating output and can be operated as a pair, as shown above.

Equation relating input with output for a single universal joint, where δ is angle between connecting link and shaft I:

$$\tan \beta = \tan \alpha \cos \delta$$

Output motion is pulsating (see curve) unless the joints are operated as pairs to provide a uniform motion.

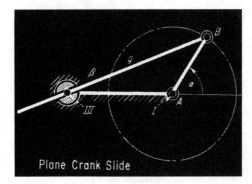

Plane Crank Slide

5 the 3-D crank slide

The three-dimensional crank slide is a variation of a plane crank slide (see sketch), with a ball point through which link g always slides, while a point B on link g describes a circle. A 3-D crank is obtained from this mechanism by making output shaft III not normal to the plane of the circle; another way is to make shafts I and III nonparallel.

A practical variation of the 3-D crank slide is the agitator mechanism (right). As input gear I rotates, link g swivels around (and also lifts) shaft III. Hence, vertical link has both an oscillating rotary motion and a sinusoidal

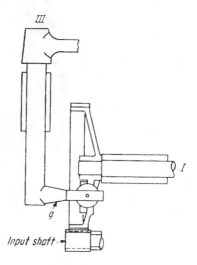

Agitator Mechanism

harmonic translation in the direction of its axis of rotation. The link performs what is essentially a screw motion in each cycle.

6 the elliptical slide

The output motion, β, of a spherical slide oscillator, p 23-12, can be duplicated by means of a two-dimensional "elliptical slide." The mechanism has a link g which slides through a pivot point D and is fastened to a point P moving along an elliptical path. The ellipse can be generated by a Cardan drive, which is a planetary gear system with the planet gear half the diameter of the internal gear. The center of the planet, point M, describes a circle; any point on its periphery describes a straight line, and any point in between, such as point P, describes an ellipse.

There are certain relationships between the dimensions of the 3-D spherical slide and the 2-D elliptical slide: $\tan \gamma / \sin \delta = a/d$ and $\tan \gamma / \cot \delta = b/d$, where a is the major half-axis, b the minor half-axis of the ellipse, and d is the length of the fixed link DN. The minor axis lies along this link.

If point D is moved within the ellipse, a completely rotating output is obtained, corresponding to the rotating spherical crank slide.

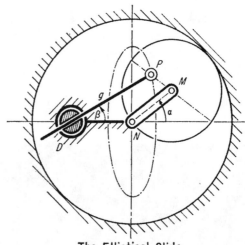

The Elliptical Slide

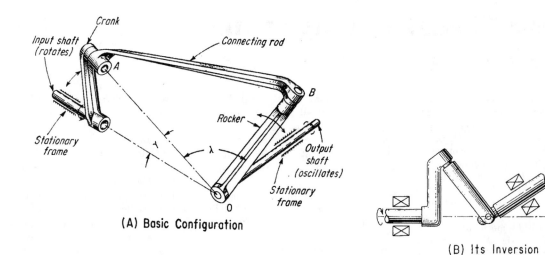

(A) Basic Configuration

(B) Its Inversion

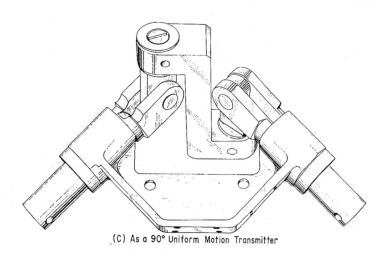

(C) As a 90° Uniform Motion Transmitter

7 the space crank

One of the most recent developments in 3-D linkages is the space crank shown in (A) see also PE—"Introducing the Space Crank—a New 3-D Mechanism," Mar 2 '59. It resembles the spherical crank discussed on page 76, but has different output characteristics. Relationship between input and output displacements is:

$$\cos \beta = (\tan \gamma)(\cos \alpha)(\sin \beta) - \frac{\cos \lambda}{\cos \gamma}$$

Velocity ratio is:

$$\frac{\omega_o}{w_i} = \frac{\tan \gamma \sin \alpha}{1 + \tan \gamma \cos \alpha \cot \beta}$$

where ω_n is the output velocity and ω_i is the constant input velocity.

An inversion of the space crank is shown in (B). It can couple intersecting shafts, and permits either shaft to be driven with full rotations. Motion is transmitted up to $37\frac{1}{2}°$ misalignment.

By combining two inversions, (C), a method for transmitting an exact motion pattern around a 90° bend is obtained. This unit can also act as a coupler or, if the center link is replaced by a gear, it can drive two output shafts; in addition, it can be used to transmit uniform motion around two bends.

Power Thrust Linkages and Their Applications

POWERED STRAIGHT LINE MOTION over short distances is applicable to many types of machines or devices for performing specialized services. These motions can be produced by a steam, pneumatic or hydraulic cylinder, or by a self-contained electric powered unit such as the General Electric Thrustors shown herewith. These Thrustors may be actuated manually by pushbuttons, or automatically by mechanical devices or the photo-electric relay as with a door opener. These illustrations will suggest many other arrangements.

Fig. 1—Transfer motion to distant point.

Fig. 2—Double throw by momentary applications.

Fig. 3—Trammel plate divides effort and changes directions of motion.

Fig. 4—Constant thrust toggle. Pressure is same at all points of throw.

Fig. 5—Multiplying motion, 6 to 1, might be used for screen shift.

Fig. 6—Bell crank and toggle may be applied in embossing press, extruder or die-caster.

Fig. 7—Horizontal pull used for clay pigeon traps, hopper trips, and sliding elements with spring or counterweighted return.

Fig. 8—Shipper rod for multiple and distant operation as series of valves.

Fig. 9—Door opener. Upthrust of helical racks rotates gear and arm.

Fig. 10—Accelerated motion by shape of cam such as on a forging hammer.

Fig. 11—Intermittent lift as applied to lifting pipe from well.

Fig. 12—Straight-line motion multiplied by pinion and racks.

Fig. 13—Rotary motion with cylindrical cam. Operates gate on conveyor belt.

Fig. 14—Thrust motions and "dwells" regulated by cam.

Fig. 15—Four positive positions with two Thrustors.

Fig. 16—Toggle increasing thrust at right angle.

Fig. 17—Horizontal straight-line motion as applied to a door opener.

Fig. 18—Thrusts in three directions with two Thrustors.

Fig. 19—Fast rotary motion using step screw and nut.

Fig. 20—Intermittent rotary motion. Operated by successive pushing of operating button, either manually or automatically.

Fig. 21—Powerful rotary motion with worm driven by rack and pinion.

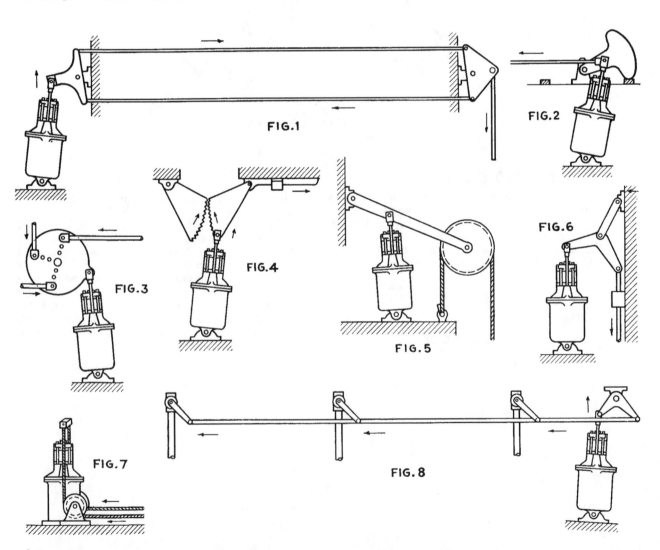

FIG.1

FIG.2

FIG.3

FIG.4

FIG.5

FIG.6

FIG.7

FIG.8

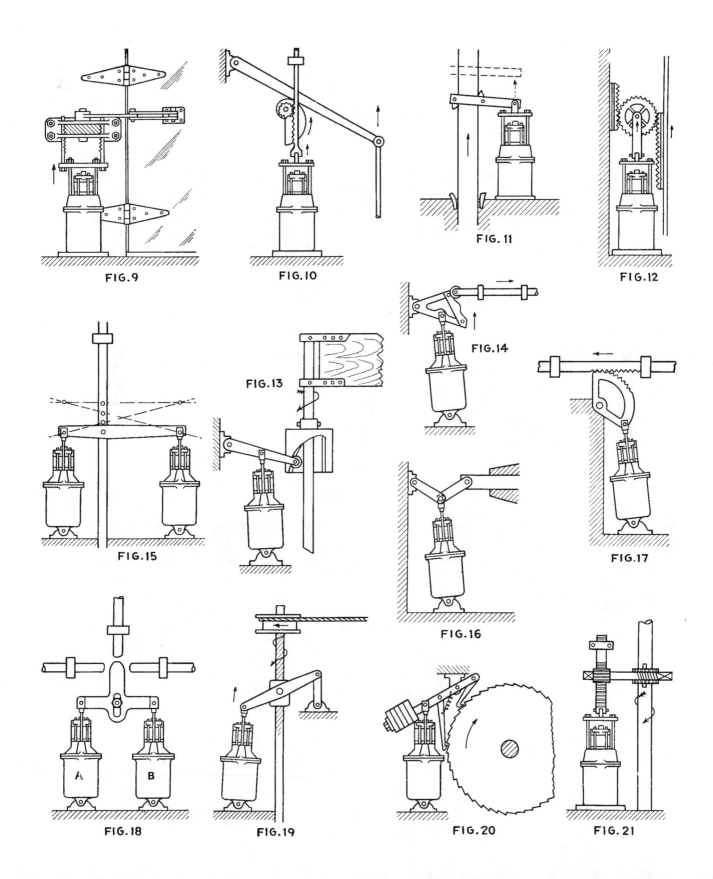

FIG. 9

FIG. 10

FIG. 11

FIG. 12

FIG. 13

FIG. 14

FIG. 15

FIG. 16

FIG. 17

FIG. 18

FIG. 19

FIG. 20

FIG. 21

Toggle Linkage Applications in Different Mechanisms

Thomas P. Goodman

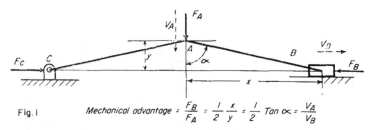

Fig. I

$$\text{Mechanical advantage} = \frac{F_B}{F_A} = \frac{1}{2}\frac{x}{y} = \frac{1}{2}\,Tan\,\alpha = \frac{V_A}{V_B}$$

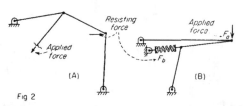

Fig 2

MANY MECHANICAL LINKAGES are based on the simple toggle which consists of two links that tend to line-up in a straight line at one point in their motion. The mechanical advantage is the velocity ratio of the input point A to the output point B: or V_A/V_B. As the angle α approaches 90 deg, the links come into toggle and the mechanical advantage and velocity ratio both approach infinity. However, frictional effects reduce the forces to much less than infinity although still quite high.

FORCES CAN BE APPLIED through other links, and need not be perpendicular to each other. (A) One toggle link can be attached to another link rather than to a fixed point or slider. (B) Two toggle links can come into toggle by lining up on top of each other rather than as an extension of each other. Resisting force can be a spring force.

HIGH MECHANICAL ADVANTAGE

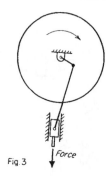

IN PUNCH PRESSES, large forces are needed at the lower end of the work-stroke, however little force is required during the remainder. Crank and connecting rod come into toggle at the lower end of the punch stroke, giving a high mechanical advantage at exactly the time it is most needed.

Fig. 3

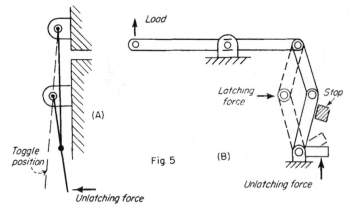

Fig. 5

LOCKING LATCHES produce a high mechanical advantage when in the toggle portion of the stroke. (A) Simple latch exerts a large force in the locked position. (B) For positive locking, closed position of latch is slightly beyond toggle position. Small unlatching force opens linkage.

COLD-HEADING RIVET MACHINE is designed to give each rivet two successive blows. Following the first blow (point 2) the hammer moves upward a short distance (to point 3), to provide clearance for moving the workpiece. Following the second blow (at point 4), the hammer then moves upward a longer distance (to point 1). Both strokes are produced by one revolution of the crank and at the lowest point of each stroke (points 2 and 4) the links are in toggle.

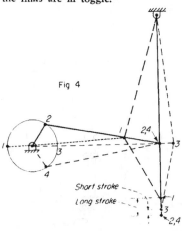

Fig 4

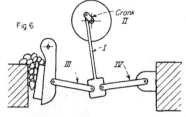

Fig 6

STONE CRUSHER uses two toggle linkages in series to obtain a high mechanical advantage. When the vertical link I reaches the top of its stroke, it comes into toggle with the driving crank II; at the same time, link III comes into toggle with link IV. This multiplication results in a very large crushing force.

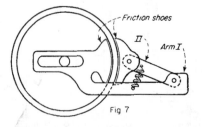

Fig 7

FRICTION RATCHET is mounted on a wheel; light spring keeps friction shoes in contact with the flange. This device permits clockwise motion of the arm I. However, reverse rotation causes friction to force link II into toggle with the shoes which greatly increases the locking pressure.

HIGH VELOCITY RATIO

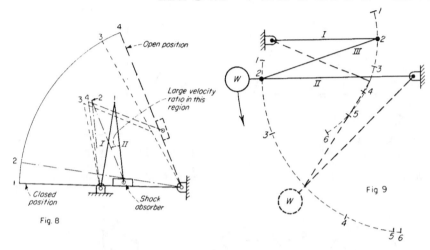

Fig. 8

Fig 9

DOOR CHECK LINKAGE gives a high velocity ratio at one point in the stroke. As the door swings closed, connecting link I comes into toggle with the shock absorber arm II, giving it a large angular velocity. Thus, the shock absorber is more effective retarding motion near the closed position.

IMPACT REDUCER used on some large circuit breakers. Crank I rotates at constant velocity while lower crank moves slowly at the beginning and end of the stroke. It moves rapidly at the mid stroke when arm II and link III are in toggle. Falling weight absorbs energy and returns it to the system when it slows down.

VARIABLE MECHANICAL ADVANTAGE

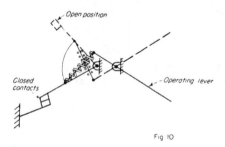

Fig 10

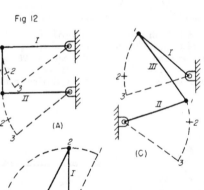

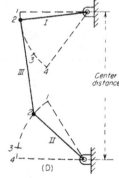

Fig 12

TOASTER SWITCH uses an increasing mechanical advantage to aid in compressing a spring. In the closed position, spring holds contacts closed and the operating lever in the down position. As the lever is moved upward, the spring is compressed and comes into toggle with both the contact arm and the lever. Little effort is required to move the links through the toggle position; beyond this point, the spring snaps the contacts closed.

FOUR-BAR LINKAGES can be altered to give variable velocity ratio (or mechanical advantage). (A) Since the cranks I and II both come into toggle with the connecting link III at the same time, there is no mechanical advantage. (B) Increasing the length of link III gives an increased mechanical advantage between positions 1 and 2, since crank I and connecting link III are near toggle. (C) Placing one pivot at the left produces similar effects as in (B). (D) Increasing the center distance puts crank II and link III near toggle at position 1; crank I and link III approach toggle position at 4.

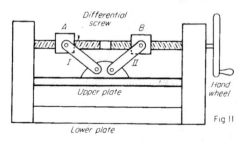

Fig 11

TOGGLE PRESS has an increasing mechanical advantage to counteract the resistance of the material being compressed. Rotating handwheel with differential screw moves nuts A and B together and links I and II are brought into toggle.

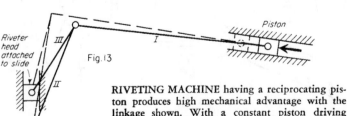

Fig. 13

RIVETING MACHINE having a reciprocating piston produces high mechanical advantage with the linkage shown. With a constant piston driving force, the force of the head increases to a maximum value when links II and III come into toggle.

Four-Bar Linkages and Typical Industrial Applications

All mechanisms can be broken down into equivalent four-bar linkages. They can be thought of as the basic mechanisms and are useful in many mechanical operations.

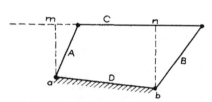

FOUR-BAR LINKAGE—Two cranks, a connecting rod and a line between the fixed centers of the cranks make up the basic four-bar linkage. Cranks can rotated if A is smaller than B or C or D. Link motion can be predicted.

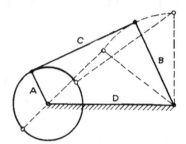

CRANK AND ROCKER— Following relations must hold for operation: $A+B+C>D$; $A+D+B>C$; $A+C-B<D$, and $C-A+B>D$.

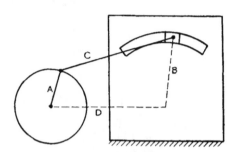

FOUR-BAR LINK WITH SLIDING MEMBER—One crank replaced by circular slot with effective crank distance of B.

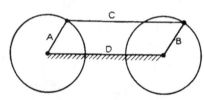

PARALLEL CRANK FOUR-BAR—Both cranks of the parallel crank four-bar linkage always turn at the same angular speed but they have two positions where the crank cannot be effective. They are used on locomotive drivers.

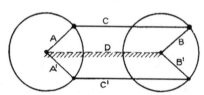

DOUBLE PARALLEL CRANK—This mechanism avoids dead center position by having two sets of cranks at 90 deg advancement. Connecting rods are always parallel. Sometimes used on driving wheels of locomotives.

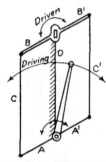

PARALLEL CRANKS—Steam control linkage assures equal valve openings.

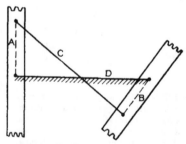

NON-PARALLEL EQUAL CRANK—The centrodes are formed as gears for passing dead center and can replace ellipticals.

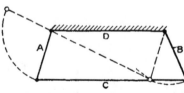

SLOW MOTION LINK—As crank A is rotated upward it imparts motion to crank B. When A reaches dead center position, the angular velocity of crank B decreases to zero. This mechanism is used on the Corliss valve.

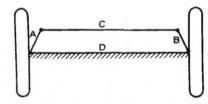

TRAPAZOIDAL LINKAGE—This linkage is not used for complete rotation but can be used for special control. Inside moves through larger angle than outside with normals intersecting on extension of rear axle in cars.

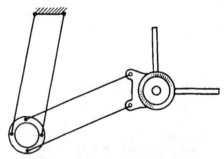

DOUBLE PARALLEL CRANK MECHANISM—This mechanism forms the basis for the universal drafting machine.

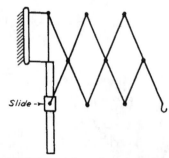

ISOSCELES DRAG LINKS—"Lazy-Tong" device made of several isosceles links; used for movable lamp support.

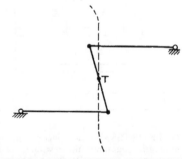

WATT'S STRAIGHT-LINE MECHANISM—Point T describes line perpendicular to parallel position of cranks.

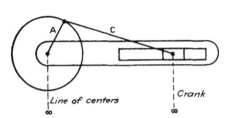

STRAIGHT SLIDING LINK—This is the form in which a slide is usually used to replace a link. The line of centers and the crank B are both of infinite length.

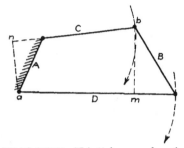

DRAG LINK—This linkage used as the drive for slotter machines. For complete rotation: $B>A+D-C$ and $B<D+C-A$.

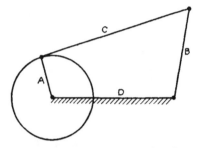

ROTATING CRANK MECHANISM—This linkage is frequently used to change a rotary motion to swinging movement.

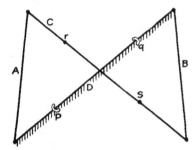

NON-PARALLEL EQUAL CRANK—If crank A has uniform angular speed, B will vary.

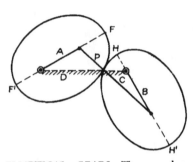

ELLIPTICAL GEARS—They produce same motion as non-parallel equal cranks.

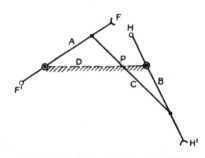

NON-PARALLEL EQUAL CRANK—Same as first but with crossover points on link ends.

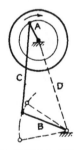

TREADLE DRIVE—This four-bar linkage is used in driving grindwheels and sewing machines.

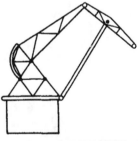

DOUBLE LEVER MECHANISM—Slewing crane can move load in horizontal direction by using D-shaped portion of top curve.

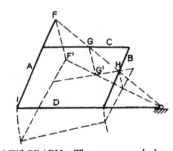

PANTOGRAPH—The pantograph is a parallelogram in which lines through F, G and H must always intersect at a common point.

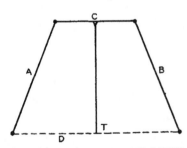

ROBERT'S STRAIGHT-LINE MECHANISM—The lengths of cranks A and B should not be less than 0.6 D; C is one half D.

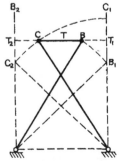

TCHEBICHEFF'S—Links made in proportion: $AB=CD=20$, $AD=16$, $BC=8$.

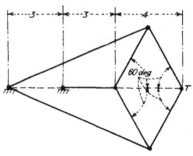

PEUCELLIER'S CELL—When proportioned as shown, the tracing point T forms a straight line perpendicular to the axis.

SECTION 4

CONNECTIONS

14 Ways to Fasten Hubs to Shafts

M. Levine

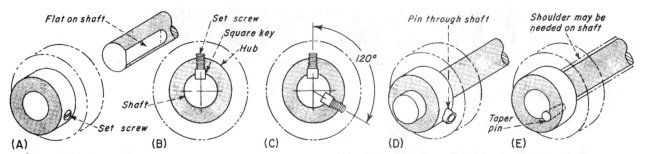

1 Cup-point setscrew...

in hub (A) bears against flat on shaft. Fastening suitable for fractional horsepower drives with low shock loads. Unsuitable when frequent removal and assembly necessary. Key with set-screw (B) prevents shaft marring from frequent removal and assembly. Not suitable where high concentricity is required.

Can withstand high shock loads. Two keys 120° apart (C) transmit extra heavy loads. Straight or tapered pin (D) prevents end play. For experimental setups expanding pin is positive yet easy to remove. Gear-pinning machines are available. Taper pin (E) parallel to shaft may require shoulder on shaft. Can be used when gear or pulley has no hub.

4 Splined shafts...

are frequently used when gear must slide. Square splines can be ground to close minor diameter fits but involute splines are self-centering and stronger. Non-sliding gears may be pinned to shaft if provided with hub.

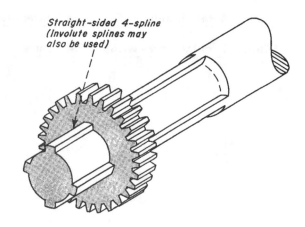

7 Interlocking...

tapered rings hold hub tightly to shaft when nut is tightened. Coarse tolerance machining of hub and shaft does not effect concentricity as in pinned and keyed assemblies. Shoulder is required (A) for end-of-shaft mounting; end plates and four bolts (B) allow hub to be mounted anywhere on shaft.

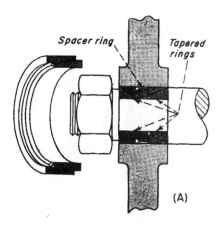

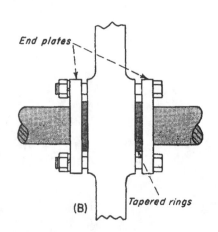

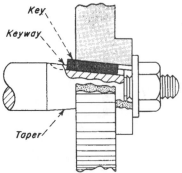

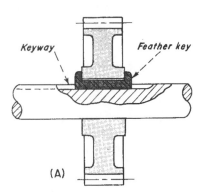

2 Tapered shaft...

with key and threaded end provides rigid, concentric assembly. Suitable for heavy-duty applications, yet can be easily dissasembled.

3 Feather key...

(A) allows axial movement. Keyway must be milled to end of shaft. For blind keyway (B) hub and key must be drilled and tapped, but design allows gear to be mounted anywhere on shaft with only a short keyway.

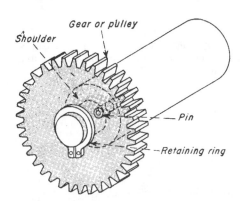

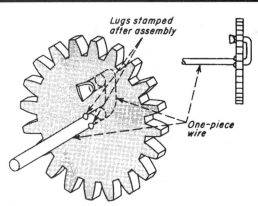

5 Retaining ring...

allows quick removal in light load applications. Shoulder on shaft necessary. Pin securing gear to shaft can be shear-pin if protection against excessive load required.

6 Stamped gear...

and formed wire shaft used mostly in toys. Lugs stamped on both legs of wire to prevent disassembly. Bend radii of shaft should be small enough to allow gear to seat.

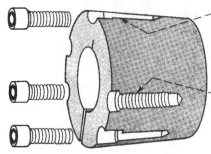

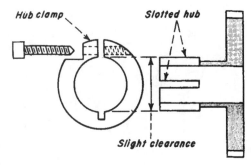

8 Split bushing...

has tapered outer diameter. Split holes in bushing align with split holes in hub. For tightening, hub half of hole is tapped, bushing half is un-tapped. Screw therefore pulls bushing into hub as screw is screwed into hub. Bushing is jacked from hub by a reverse procedure. Sizes of bushings avaliable for ½- to 10-in. dia shafts. Adapters are available for untapered hubs.

9 Split hub...

of stock precision gear is clamped onto shaft with separate hub clamp. Manufacturers list correctly dimensioned hubs and clamps so that efficient fastening can be made based on precision ground shaft. Ideal for experimental set-ups.

Attaching Hubless Gears to Shafts

Thin gears and cams save space—but how to fasten them to their shafts? These illustrated methods give simple, effective answers.

L. Kasper

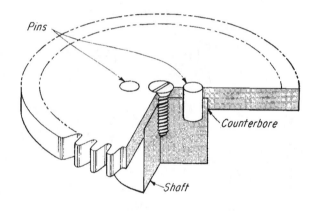

1 COUNTERBORE with close fit on shaft ensures concentric mounting. Torque is transmitted by pins; positive fastening is provided by flathead screw.

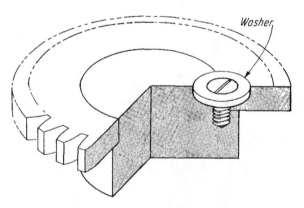

2 TIGHT-FITTING washer in counterbored hole carries the radial load; its shear area is large enough to ensure ample strength.

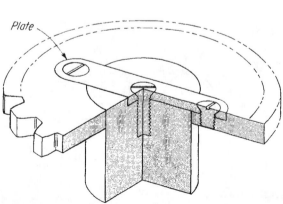

3 PLATE gives greater resistance to shear when radial loads are likely to be heavy. When the gear is mounted, the plate becomes the driver; the center screw merely acts as a retainer.

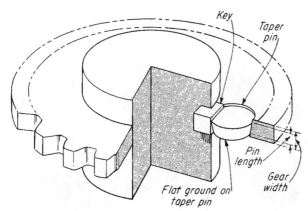

4 KEY AND FLATTED TAPER-PIN should not protrude above surface of gear; pin length should be slightly shorter than gear width. Note that this attachment is not positive—gear retention is by friction only.

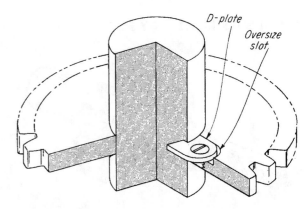

5 D-PLATE keys gear to shaft; optimum slot depth in shaft will depend upon torque forces and stop-and-start requirements—low, constant torque requires only minimum depth and groove length; heavy-duty operation requires enough depth to provide longer bearing surface.

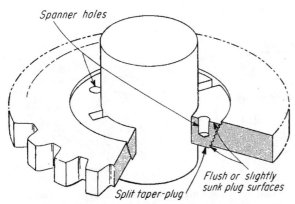

6 TAPERED PLUG is another friction holding device. This type mounting should be used so that the radial load will tend to tighten rather than loosen the thread. For added security, thread can be lefthand to reduce tampering risk.

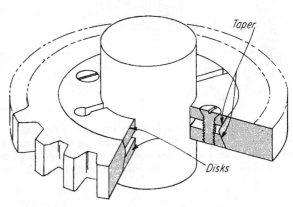

7 TWO FRICTION DISKS, tapered to about 5° included angle on their rims, are bored to fit the shaft. Flathead screws provide clamping force, which can be quickly eased to allow axial or radial adjustment of gear.

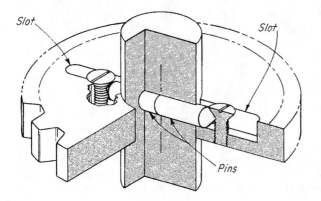

8 TWO PINS in radial hole of shaft provide positive drive that can be easily disassembled. Pins with conical end are forced tightly together by flathead screws. Slot length should be sufficient to allow pins to be withdrawn while gear is in place if backside of gear is "tight" against housing.

10 Different Types of Splined Connections

W. W. Heath

CYLINDRICAL TYPES

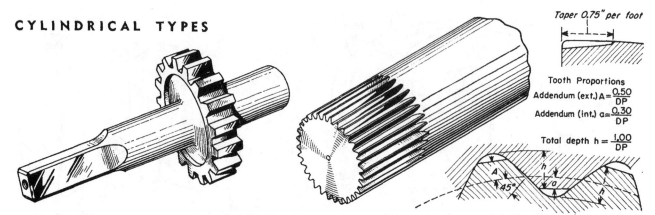

1 SQUARE SPLINES make a simple connection and are used mainly for applications of light loads, where accurate positioning is not important. This type is commonly used on machine tools; a cap screw is necessary to hold the enveloping member.

2 SERRATIONS of small size are used mostly for applications of light loads. Forcing this shaft into a hole of softer material makes an inexpensive connection. Originally straight-sided and limited to small pitches, 45 deg serrations have been standardized (SAE) with large pitches up to 10 in. dia. For tight fits, serrations are tapered.

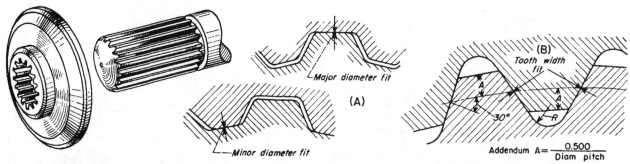

5 INVOLUTE-FORM splines are used where high loads are to be transmitted. Tooth proportions are based on a 30 deg stub tooth form. (A) Splined members may be positioned either by close fitting major or minor diameters. (B) Use of the tooth width or side positioning has the advantage of a full fillet radius at the roots. Splines may be parallel or helical. Contact stresses of 4,000 psi are used for accurate, hardened splines. Diametral pitch above is the ratio of teeth to the pitch diameter.

FACE TYPES

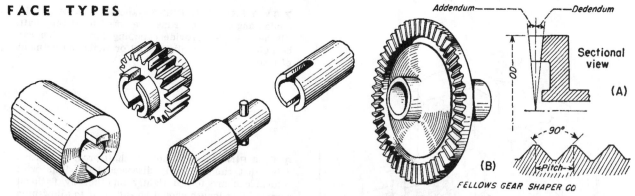

8 MILLED SLOTS in hubs or shafts make an inexpensive connection. This type is limited to moderate loads and requires a locking device to maintain positive engagement. Pin and sleeve method is used for light torques and where accurate positioning is not required.

9 RADIAL SERRATIONS by milling or shaping the teeth make a simple connection. (A) Tooth proportions decrease radially. (B) Teeth may be straight-sided (castellated) or inclined; a 90 deg angle is common.

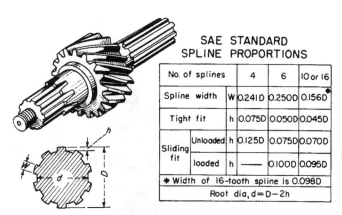

SAE STANDARD
SPLINE PROPORTIONS

No. of splines		4	6	10 or 16
Spline width	W	0.241D	0.250D	0.156D *
Tight fit	h	0.075D	0.050D	0.045D
Sliding fit	Unloaded h	0.125D	0.075D	0.070D
	loaded h	——	0.100D	0.095D
* Width of 16-tooth spline is 0.098D				
Root dia, d=D−2h				

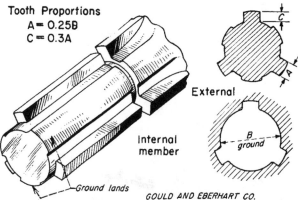

Tooth Proportions
A = 0.25B
C = 0.3A

External

Internal member

Ground lands

B ground

GOULD AND EBERHART CO.

3 STRAIGHT-SIDED splines have been widely used in the automotive field. Such splines are often used for sliding members. The sharp corner at the root limits the torque capacity to pressures of approximately 1,000 psi on the spline projected area. For different applications, tooth height is altered as shown in the table above.

4 MACHINE-TOOL spline has a wide gap between splines to permit accurate cylindrical grinding of the lands—for precise positioning. Internal parts can be ground readily so that they will fit closely with the lands of the external member.

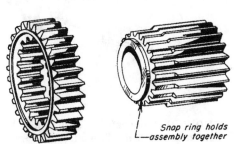

Snap ring holds assembly together

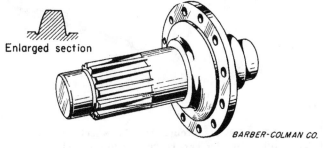

Enlarged section

BARBER-COLMAN CO.

6 SPECIAL INVOLUTE splines are made by using gear tooth proportions. With full depth teeth, greater contact area is possible. A compound pinion is shown made by cropping the smaller pinion teeth and internally splining the larger pinion.

7 TAPER-ROOT splines are for drives which require positive positioning. This method holds mating parts securely. With a 30 deg involute stub tooth, this type is stronger than parallel root splines and can be hobbed with a range of tapers.

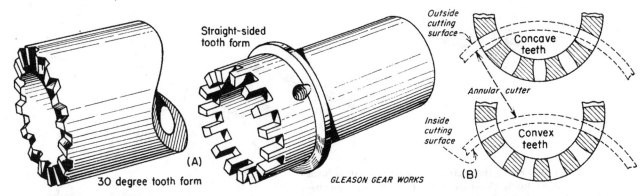

Straight-sided tooth form

30 degree tooth form

(A)

Outside cutting surface

Concave teeth

Annular cutter

Inside cutting surface

Convex teeth

(B)

GLEASON GEAR WORKS

10 CURVIC COUPLING teeth are machined by a face-mill type of cutter. When hardened parts are used which require accurate positioning, the teeth can be ground. (A) This process produces teeth with uniform depth and can be cut at any pressure angle, although 30 deg is most common. (B) Due to the cutting action, the shape of the teeth will be concave (hour-glass) on one member and convex on the other—the member with which it will be assembled.

Alternates for Doweled Fasteners

Some simple ways to fasten or locate round or flat parts without having to use dowels or other pins.

Federico Strasser

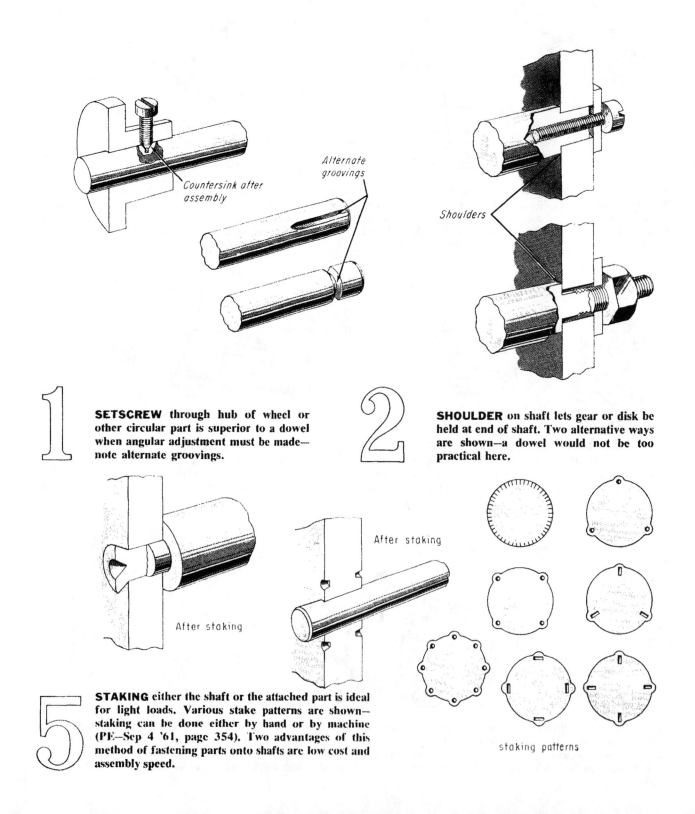

Countersink after assembly

Alternate groovings

Shoulders

After staking

After staking

staking patterns

1 SETSCREW through hub of wheel or other circular part is superior to a dowel when angular adjustment must be made—note alternate groovings.

2 SHOULDER on shaft lets gear or disk be held at end of shaft. Two alternative ways are shown—a dowel would not be too practical here.

5 STAKING either the shaft or the attached part is ideal for light loads. Various stake patterns are shown—staking can be done either by hand or by machine (PE—Sep 4 '61, page 354). Two advantages of this method of fastening parts onto shafts are low cost and assembly speed.

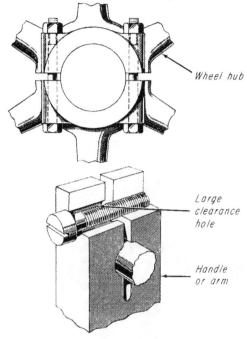

Wheel hub

Large
clearance
hole

Handle
or arm

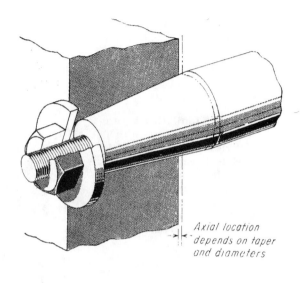

Axial location
depends on taper
and diameters

PRESSURE JOINTS are best when large composite wheels or similar parts are to be fastened to their shafts with only one or two screws.

TAPERED JOINTS are ideal when no clearance can be allowed between hub and shaft. Dowelling would be impracticable because of fit.

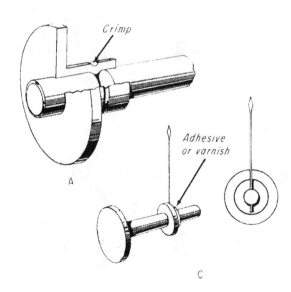

Crimp

A

Adhesive
or varnish

C

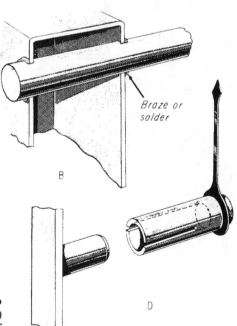

Braze or
solder

B

D

PERMANENT FASTENINGS of parts assembled to shafts are crimped (A); soldered, brazed or welded (B) or adhesive-held (C). Non-permanent fastening of small indicator-pointer is best achieved by providing simple push fit (D). If positive location is required here, dimple hub after assembly.

6 More Alternates for Doweled Fastenings

These simple but effective methods fasten or locate round and flat plates without dowels or other pins.

Federico Strasser

1 **TORQUE LIMITERS** are necessary in many cases where a doweled fastening would be useless. If shaft load becomes excessive a low-cost means of providing for its disengagement is to have a spring-loaded ball mounted externally (A) or internally (B).

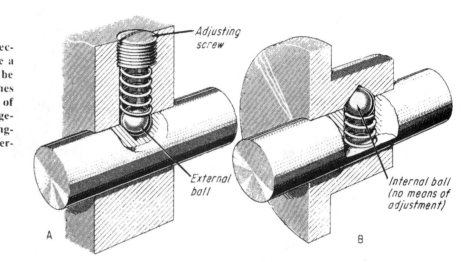

Adjusting screw

External ball

Internal ball (no means of adjustment)

A

B

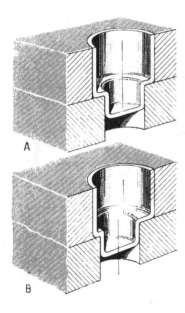

A

B

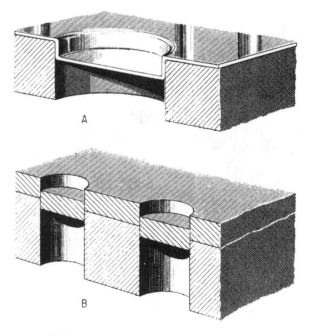

A

B

3 **SHEET METAL "DOWELS"** can be used where location of two parts is needed without precise hole location. Cup is drawn after assembly.

4 **SELF-DOWELING** part can be sheet metal (A) or other thin-material part (B). Merely emboss or punch slug halfway through to locate in hole.

BASIC FRICTION-CLUTCH can also provide for disengagement when torque exceeds a safe value. There is basically no difference between collared shaft (A) and ringed shaft (B), but adjustment of tension in collared shaft is limited by amount of threading on end of shaft.

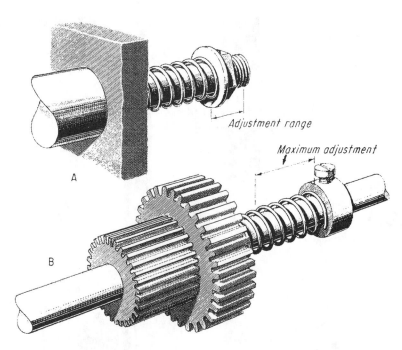

Adjustment range

Maximum adjustment

A

B

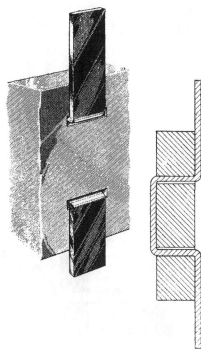

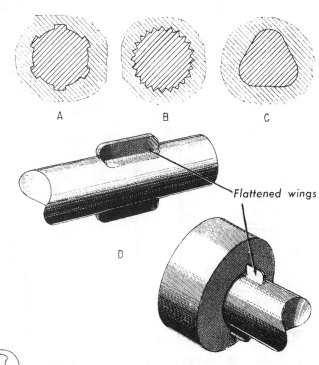

A   B   C

Flattened wings

D

FOLDED ASSEMBLIES eliminate the need for pins or other locating fasteners. Illustrated is a terminal mounted on insulating plates.

SPLINES such as square (A) and involute (B) are often the best way to locate and hold hubs. Don't overlook simpler methods shown at C and D.

29 Ways to Fasten Springs

Four pages of ingenious attachments for extension, compression, and torsion springs.

Federico Strasser

EXTENSION SPRINGS

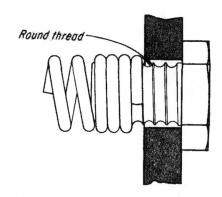

Round thread

Screw fits into spring end.

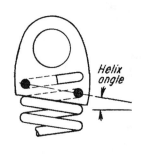
Helix angle

Tab with 3 holes engages 1½ spring-coils.

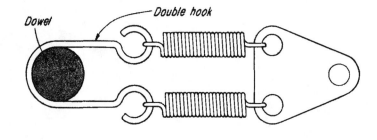
Dowel

Double hook

Twin-spring setup includes double hook and triangular tab.

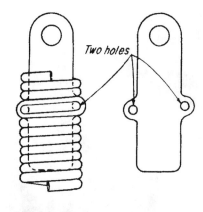
Two holes

Long tab with 2 holes in midsection provides ample adjustment.

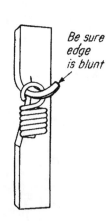
Be sure edge is blunt

Sheetmetal, slit and formed, suspends spring.

Tight fit for permanent support

Cross-pin holds spring deep in hole.

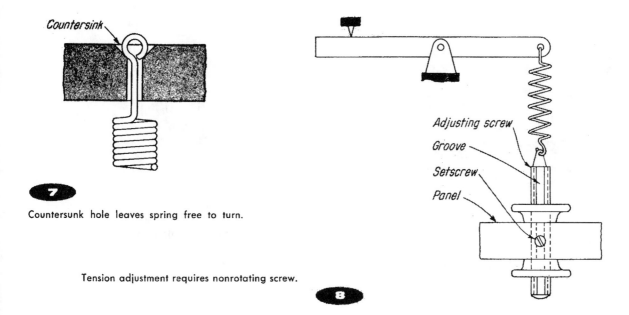

7

Countersunk hole leaves spring free to turn.

Tension adjustment requires nonrotating screw.

8

COMPRESSION SPRINGS

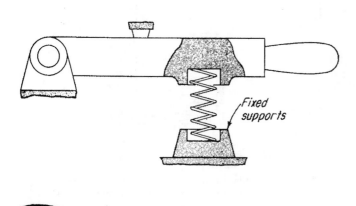

9

Unsupported spring-body must have somewhat more resistance to buckling in fixed supports (9) than pivoted ones (10).

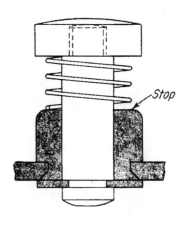

11

Supported springs are exemplified in push button (11) and friction clutch (12).

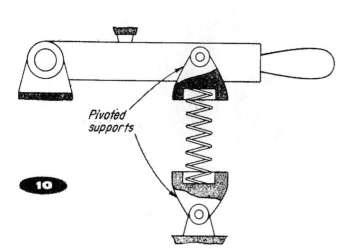

10

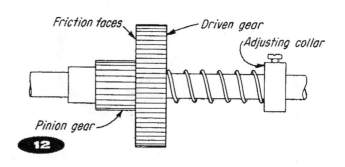

12

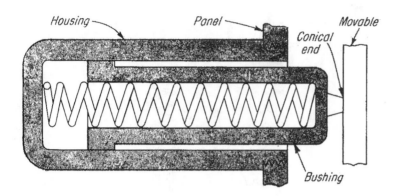

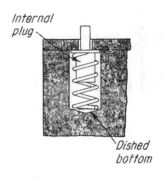

13

Concealed spring is supported externally by closed-end bushing, which also determines amount of compression.

14

Double guidance is exemplified by hole-bottom that centers spring end, and internal plug that supports spring body.

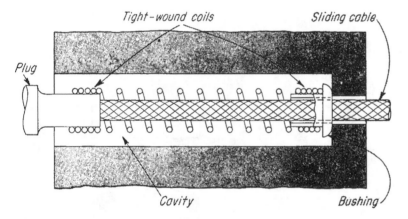

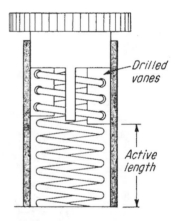

15

Tight-wound end-coils hold switchboard plug bushing—spring absorbs shock when weighted cable snaps entire assembly back into cavity after operator disconnects plug at end of message.

16

Adjustment vanes have holes that match spring pitch. Spring coils threaded through vanes become inactive, thus varying effective spring length.

FLAT TORSION-SPRINGS

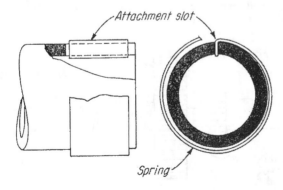

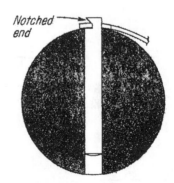

17

Saw-cut slot retains spring positively if slot end is peened over or otherwise closed.

18

Notched dowel provides hook for hole in spring end.

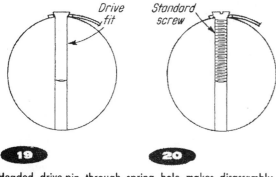

19   **20**

Headed drive-pin through spring hole makes disassembly difficult (19). Standard screw (20) eases disassembly.

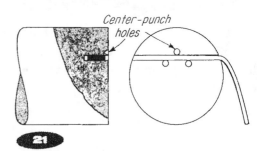

21

Chordal shot in shaft is closed tight by displacing metal with center punch for permanent spring retention.

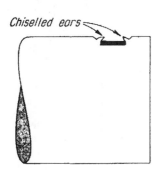

22

Chordal groove holds spring when ears are formed by chiselling or staking after assembly.

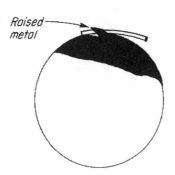

23

Raised metal, produced by staking, provides low-cost yet firm hook on shaft.

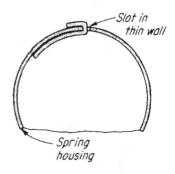

24

Slotted spring-housing is simple, but spring-end can be dangerous if housing revolves and is unprotected.

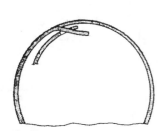

25

Lanced tab avoids hazard of external spring-end but dirt can enter housing.

26

Shoulder rivet provides dustproof fastening.

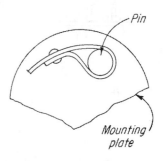

27

Mounting pin on plate may be plain, or headed—for more positive spring retention.

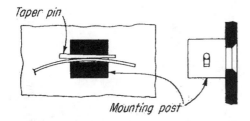

28

Taper pin allows end adjustment of precision, low-torque springs.

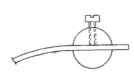

29

Setscrew and slotted post also provides adjustment feature but may be inefficient if spring-end becomes dimpled.

20 Tamper-Proof Fasteners

Ways to prevent or indicate unauthorized removal of fasteners in vending machines, instruments, radios, TV sets, and other units. Included are positively retained fasteners to prevent loss where retieval would be difficult.

Federico Strasser

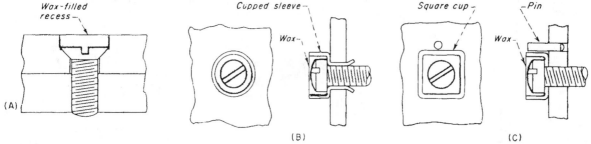

Fig. 1—(A) Wax or other suitable material fills recess above screw. Wax flush with plate hides screw position if surface is painted. (B) Cupped sleeve riveted in screw hole provides cavity for wax when plate is too thin for recessing. (C) Pin prevents rotation of square cup which would allow screw to be removed without disturbing wax.

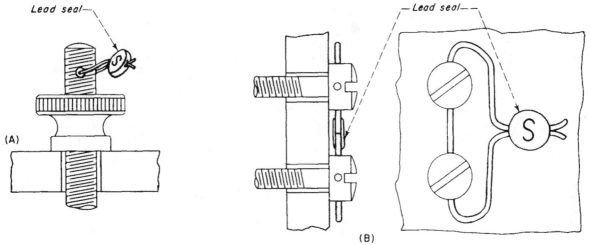

Fig. 2—(A) Lead seal crimped over twisted ends of wire passing through screw allows only limited slackening of nut. (B) Two or more screws strung through heads with wire are protected against unauthorized removal by only one seal. Code or other signet can be embossed on seals during crimping.

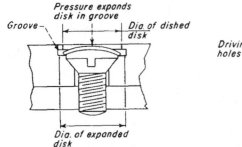

Fig. 3—Sheet-metal disk pressed into groove can only be removed with difficulty and discourages tampering.

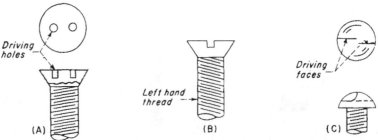

Fig. 4—(A) Spanner-head screws are available in all standard heads and sizes from U.S. manufacturers. Special driver is required for each screw size except ¼-in. dia and above. (B) Left-hand screw thread is sometimes sufficient to prevent unauthorized loosening, or (C) special head lets screw be driven but not unscrewed.

POSITIVELY RETAINED FASTENERS

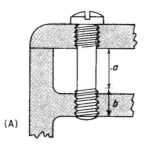

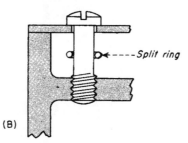

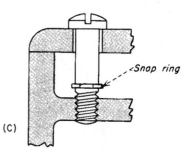

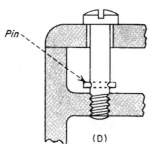

Fig. 5—(A) Tapped cover and casing allows screw (a > b) with reduced shank diameter to be completely unscrewed from casing yet retained positively in cover. For thin sheet-metal covers, split ring on reduced shank (B) is preferable. Snap ring in groove (C) or transverse pin (D) are effective on unreduced shank. Simple and cheap method (E) is fiber washer pushed over thread.

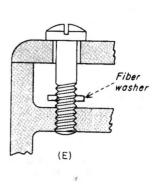

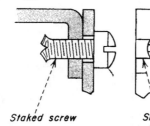

Fig. 6—Open-ended slot in sliding cover allows screw end to be staked or burred so screw cannot be removed, once assembled.

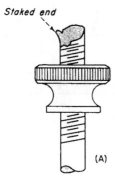

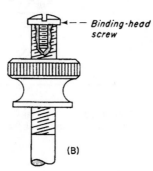

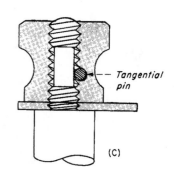

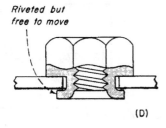

Fig. 7—(A) Nut is retained on screw by staking or similar method but, if removal of nut is occasionally necessary, coaxial binding-head screw (B) can be used. Where screw end must be flush with nut, pin through nut tangential to undercut screw (C) limits nut movement. Rotatable nut (D) or screw (E) should have sufficient lateral freedom to accommodate slight differences in location when two or more screws are used.

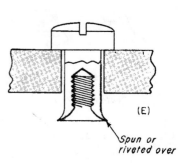

Lanced Metal Eliminates Separate Fasteners

15 ways in which sheetmetal tabs, ears, and lugs can serve to fasten and locate.

Federico Strasser

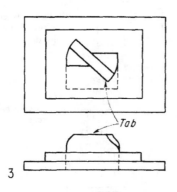

1

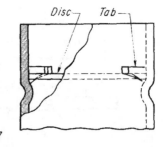

2

BENT-OVER TAB holds together up to four layers of sheetmetal. Designing tabs to stress in shear increases holding strength.

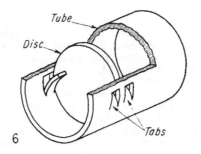

3

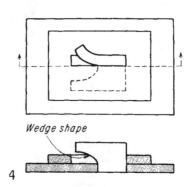

4

TWISTED TAB is less common than bent one. Shaped tab wedges tightly when twisted.

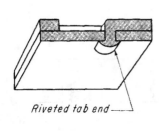

5

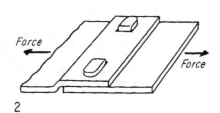

6

7

WITH THICK STOCK, end of tab can be riveted.

THESE TABS both locate and hold disk in tubing. When necking locates disk, tab only holds—again by wedge action.

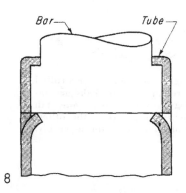

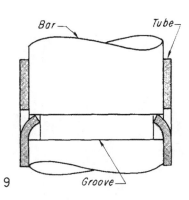

8

9

LIP AND TABS combine to join round bar and tubing. For longer bars, tabs fit into grooves. Bar can rotate inside tube if tabs are pressed lightly into groove.

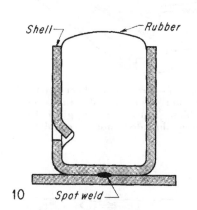

10

LANCED SHELL secures rubber in bumper or instrument foot.

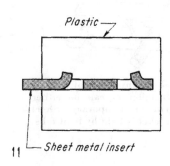

11

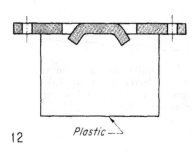

12

METAL REINFORCEMENTS and mounting pads grip plastic better if lanced.

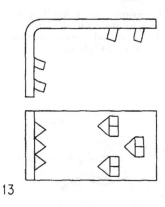

13

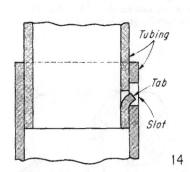

14

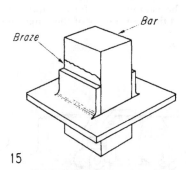

15

CORNER REINFORCEMENT grips wood, plastic or fiber with lanced teeth. Similarly, lanced nameplates or labels attach easily to equipment or instrument panels by pressing into the surface.

INTERLOCKING SLOT and tab connects two pieces of tubing. Joint is permanent if inner tubing has thick wall. If inner tube has thin wall, tab can be depressed and tubes pulled apart if desired.

LANCED FLAPS provide large contact areas for brazing or soldering sheet-metal fins to bars and pipes. An alternate method for round bars or pipe is an embossed collar around the hole. However, for angular shapes, flaps are easier to make.

Joining Circular Parts without Fasteners

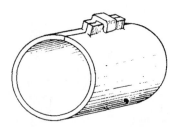

Fig. 1—Fastening for a rolled circular section. Tabs are integral with sheet; one tab being longer than the other, and bent over on assembly.

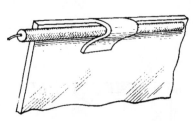

Fig. 5—Similar to Fig. 4 for supporting electrical wires. Tab is integral with plate and crimped over on assembly.

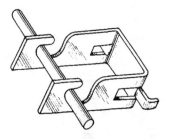

Fig. 6—For supporting of rods or tubes. Installation can be either permanent or temporary. Sheet metal bracket is held by bent tabs.

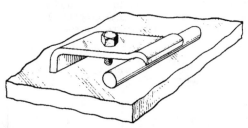

Fig. 7—Embossed sheet metal bracket to hold rods, tubes or cables. Tension is supplied by screw threaded into lower plate.

Fig. 10—Plate is embossed and tabs bent over on assembly. If two plates are used having tab edges (B) a piano-type hinge is formed. (A) and (B) can be combined to form a quick release door mechanism. A cable is passed through the eye of the hinge bolt, and a handle attached to the cable.

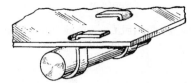

Fig. 11—Rods and tubes can be supported by sheet metal tabs. Tab is wrapped around circular section and bent through plate.

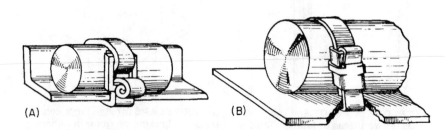

Fig. 14—Strap fastener to hold a circular section tight against a structural shape. Lock can be made from square bar stock (A) or from sheet metal (B) tabbed as shown. Strap is bent over for additional locking. Slotted holes in sheet should be spaced equal to rod dia to prevent tearing.

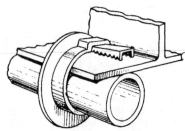

Fig. 15—C clamp support usually used for tubing. Serrated wedge is hammered tight; serrations keep wedge from unlocking.

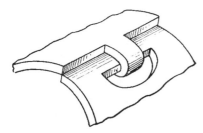

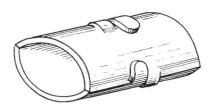

Fig. 2—Similar to Fig. 1 except tube is formed with a lap joint. Tab is bent over and inserted into cut-out on assembly. Joint tension is needed to maintain lock.

Fig. 3—Tab fastener for elliptical section. Tabs are formed integral with sheet. For best results tabs should be adjacent to each other as shown in sketch above.

Fig. 4—For supporting rod on plate. Tab is formed and bent over rod on assembly. Wedging action holds rod in place. Rod is free to move unless restrained.

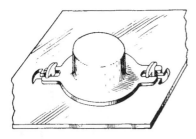

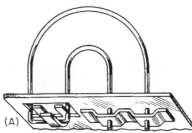

Fig. 8—Fastening of rod to plate. Rod is welded to plate with slotted holes. Tabs in bottom plate are bent on assembly.

Fig. 9—Tabs and bracket (A) used to support rod at right angle to plate. Bracket can be welded to plate. (B) has rod slotted into place. For mass production, the tabs and slots can be stamped into the sheet. For limited production, the tabs and slots can be hand formed.

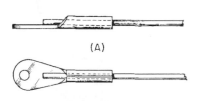

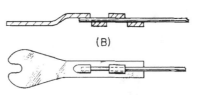

Fig. 12—For connecting wire ends to terminals. Sheet is crimped or tabbed to hold wire in place. Variety of terminal endings can be used. If additional fastening is required, in that parting of the wire and terminal end might create a safety or fire hazard, a drop of solder can be added.

Fig. 13—Spring joins two rods or tubes. Members are not limited in axial motion or rotation except by spring strength.

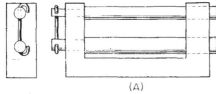

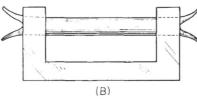

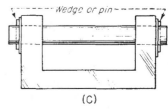

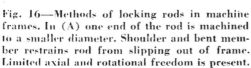

Fig. 16—Methods of locking rods in machine frames. In (A) one end of the rod is machined to a smaller diameter. Shoulder and bent member restrains rod from slipping out of frame. Limited axial and rotational freedom is present.

Split rod in (B) limits axial motion but permits rotation. Rod is split on assembly. Wedge or pin in (C) bear against washers. Axial motion can be restricted but rotation is possible. If rod is to be a roller, bearings can be inserted.

SECTION 5

LOCKING DEVICES & METHODS

Friction Clamping Devices

Bernard J. Wolfe

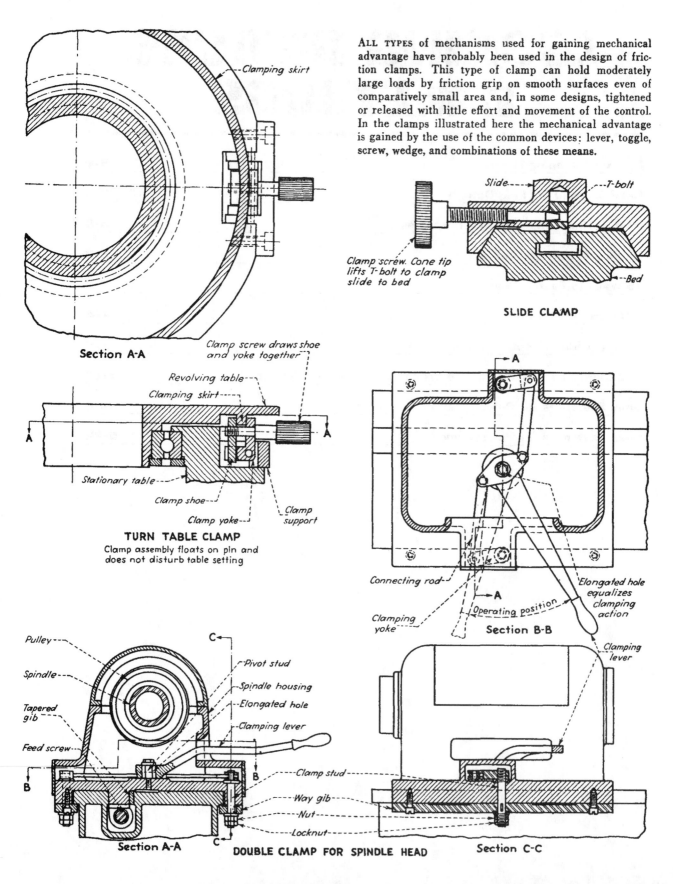

ALL TYPES of mechanisms used for gaining mechanical advantage have probably been used in the design of friction clamps. This type of clamp can hold moderately large loads by friction grip on smooth surfaces even of comparatively small area and, in some designs, tightened or released with little effort and movement of the control. In the clamps illustrated here the mechanical advantage is gained by the use of the common devices: lever, toggle, screw, wedge, and combinations of these means.

Clamping skirt

Section A-A

Clamp screw draws shoe and yoke together

Revolving table

Clamping skirt

Stationary table

Clamp shoe

Clamp yoke

Clamp support

TURN TABLE CLAMP
Clamp assembly floats on pin and does not disturb table setting

Slide *T-bolt*

Clamp screw. Cone tip lifts T-bolt to clamp slide to bed

Bed

SLIDE CLAMP

Connecting rod

Clamping yoke

Operating position

Elongated hole equalizes clamping action

Section B-B

Clamping lever

Pulley

Spindle

Tapered gib

Feed screw

Pivot stud

Spindle housing

Elongated hole

Clamping lever

Clamp stud

Way gib

Nut

Locknut

Section A-A

DOUBLE CLAMP FOR SPINDLE HEAD

Section C-C

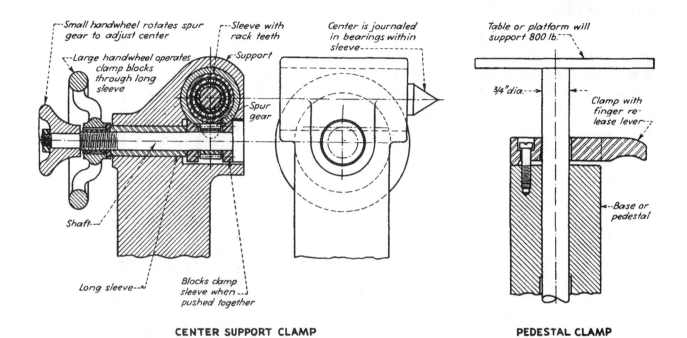

Small handwheel rotates spur gear to adjust center

Large handwheel operates clamp blocks through long sleeve

Sleeve with rack teeth

Support

Spur gear

Center is journaled in bearings within sleeve

Shaft

Long sleeve

Blocks clamp sleeve when pushed together

CENTER SUPPORT CLAMP

Table or platform will support 800 lb.

3/4" dia.

Clamp with finger release lever

Base or pedestal

PEDESTAL CLAMP

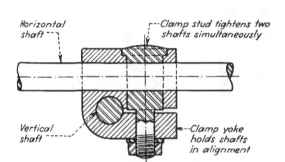

Horizontal shaft

Clamp stud tightens two shafts simultaneously

Vertical shaft

Clamp yoke holds shafts in alignment

RIGHT ANGLE CLAMP

Slide bed

Slide shaft

Clamp spring Lock

Slide

SLIDE CLAMP

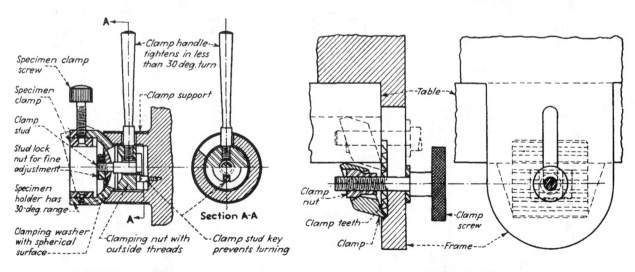

Clamp handle tightens in less than 30 deg. turn

Clamp support

Specimen clamp screw

Specimen clamp

Clamp stud

Stud lock nut for fine adjustment

Specimen holder has 30-deg. range

Clamping washer with spherical surface

Clamping nut with outside threads

Clamp stud key prevents turning

Section A-A

SPECIMEN HOLDER CLAMP

Table

Clamp nut

Clamp teeth

Clamp

Clamp screw

Frame

TABLE CLAMP

Retaining and Locking Detents

Many forms of detents are used for positioning
gears, levers, belts, covers, and similar parts.
Most of these embody some form of spring in
varying degrees of tension, the working end
of the detent being hardened to prevent wear.

Adam Fredericks

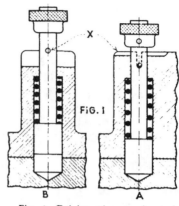

Fig. 1—Driving plunger, shown in
engagement at *A* is pulled out, and
given a 90-deg. turn, pin *X* slipping
into the shallow groove as shown at *B*,
thus disengaging both members.

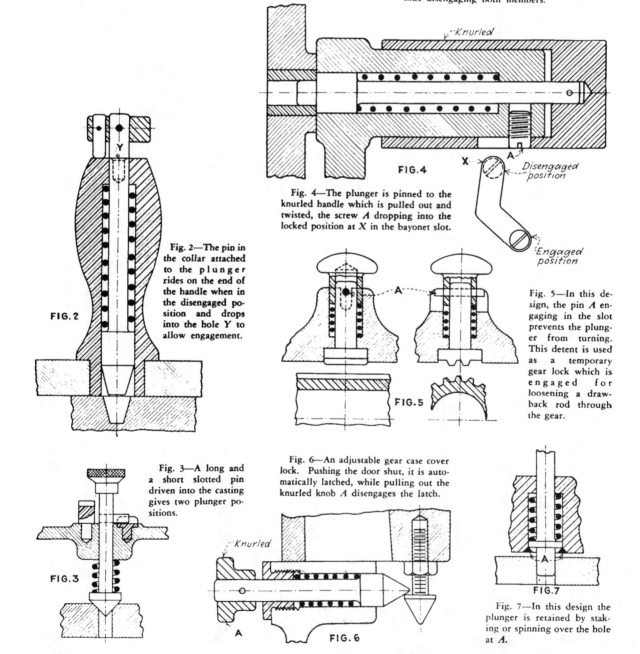

Fig. 4—The plunger is pinned to the
knurled handle which is pulled out and
twisted, the screw *A* dropping into the
locked position at *X* in the bayonet slot.

Fig. 2—The pin in
the collar attached
to the plunger
rides on the end of
the handle when in
the disengaged po-
sition and drops
into the hole *Y* to
allow engagement.

Fig. 5—In this de-
sign, the pin *A* en-
gaging in the slot
prevents the plung-
er from turning.
This detent is used
as a temporary
gear lock which is
engaged for
loosening a draw-
back rod through
the gear.

Fig. 3—A long and
a short slotted pin
driven into the casting
gives two plunger po-
sitions.

Fig. 6—An adjustable gear case cover
lock. Pushing the door shut, it is automatically latched, while pulling out the
knurled knob *A* disengages the latch.

Fig. 7—In this design the
plunger is retained by staking or spinning over the hole
at *A*.

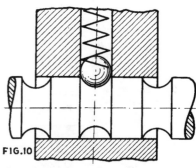

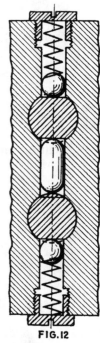

Fig. 10—Another form in which the grooves are cut all around the rod, which is then free to turn to any position.

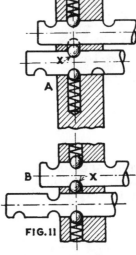

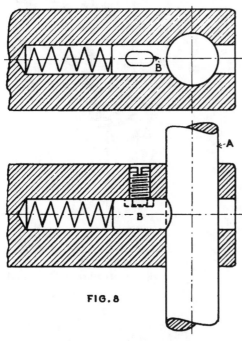

FIG. 8

Fig. 8—End of the plunger *B* bearing against the hand lever *A* is concaved and prevented from turning by the dog point setscrew engaging the splined slot. Friction is the only thing that holds the adjustable hand lever *A* in position.

Figs. 11 and 12—Above is shown a double-locking device for gear shift yoke rods. At *A* the neutral position is shown with ball *X* free in the hole. At *B* the lower rod is shifted, forcing ball *X* upwards, retaining the upper rod in a neutral position. The lower rod must also be in neutral position before the upper rod can be moved. To the right is shown a similar design wherein a rod with hemispherical ends is used in place of ball *X*.

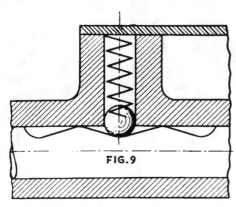

FIG. 9

Fig. 9—A spring-backed steel ball makes a cheap but efficient detent, the grooves in the rod having a long, easy riding angle. For economy, rejected or undersized balls can be purchased from manufacturers.

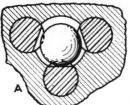

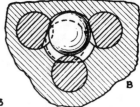

FIG. 13

Fig. 13—Without using a spring of any kind, three gear-shifting rods are locked by a large steel ball. At *A*, the neutral position is shown. At *B*, the lower rod has been shifted, forcing the ball upwards, thereby locking the other two rods. The dashed circle shows the position of the ball when the right-hand rod has been shifted.

How Spring Clamps Hold Workpieces

Here's a review of ways in which spring clamp devices can help you get a grip on things.

Federico Strasser

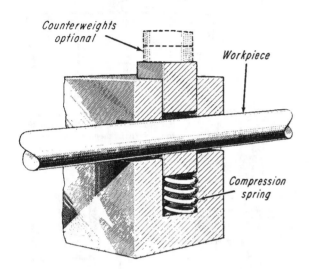

1 **RODS OF DIFFERENT SECTION** can be easily held by this device. Strength of grip can be varied if necessary.

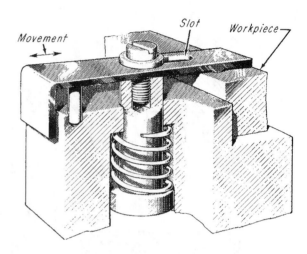

2 **SECOND-CLASS LEVER** gives low clamping forces for parts that are easily marked or require gentle handling.

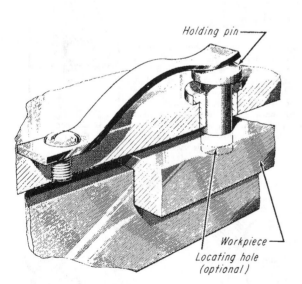

5 **FLAT SPRING ACTS THROUGH PIN** that holds the workpiece in the fixture. This device also positively locates parts.

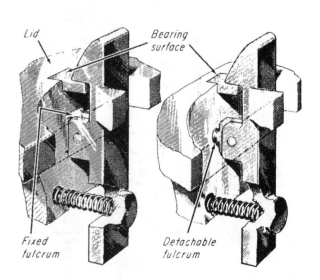

6 **COVER LATCH** is an ideal application for spring and notched lever. Make the fulcrum detachable for ease of repair.

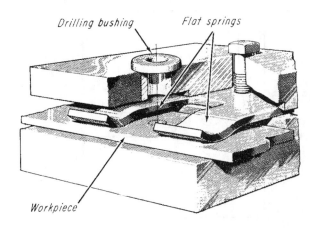

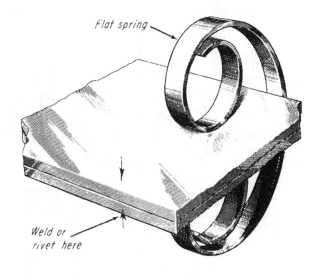

3 **FLAT WORKPIECES** of constant thickness are held with a couple of flat springs attached to the jig table.

4 **SIMPLE CLAMPING FIXTURE** is ideal for holding two flat pieces of material together for either welding or riveting.

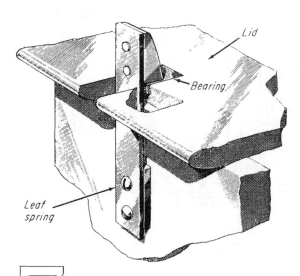

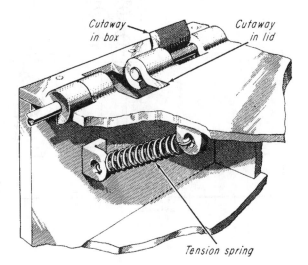

7 **LEAF-SPRING** latch can be fashioned as shown, or the spring itself can be formed to provide its own latching notch.

8 **POSITIVE OPEN-OR-SHUT** lid relies upon a spring. Over-center spring action makes the lid a simple toggle.

Holding Fixture for Workpiece

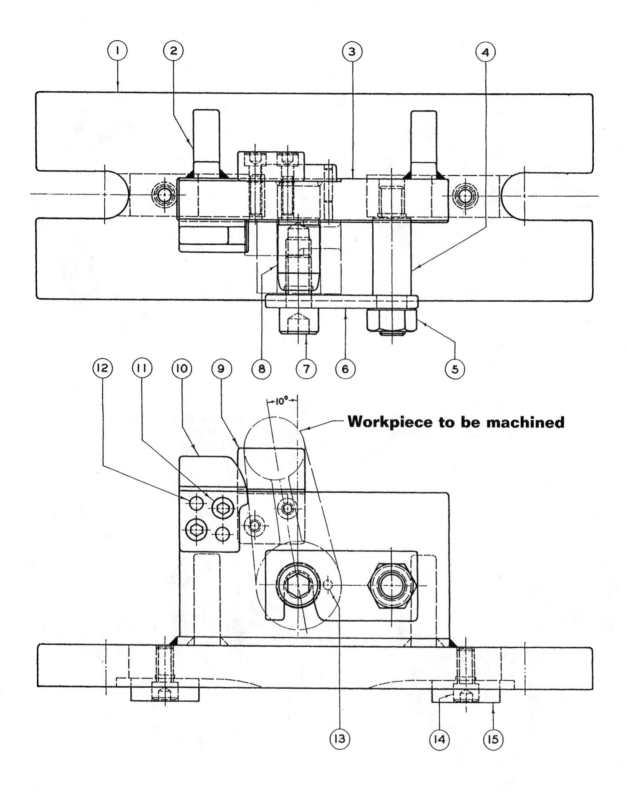

Workpiece to be machined

15 Ways to Fasten Gears to Shafts

So you've designed or selected a good set of gears for your unit–now how do you fasten them to their shafts? Here's a roundup of methods–some old, some new–with a comparison table to help make the choice.

L. M. Rich

1 PINNING

Pinning of gears to shafts is still considered one of the most positive methods. Various types can be used: dowel, taper, grooved, roll pin or spiral pin. These pins cross through shaft (A) or are parallel (B). Latter method requires shoulder and retaining ring to prevent end play, but allows quick removal. Pin can be designed to shear when gear is overloaded.

Main drawbacks to pinning are: Pinning reduces the shaft cross-section; difficulty in reorienting the gear once it is pinned; problem of drilling the pin holes if gears are hardened.

Recommended practices are:
• For good concentricity keep a maximum clearance of 0.0002 to 0.0003 in. between bore and shaft.
• Use steel pins regardless of gear material. Hold gear in place on shaft by a setscrew during machining.
• Pin dia should never be larger than $\frac{1}{3}$ the shaft—recommended size is 0.20 D to 0.25 D.
• Simplified formula for torque capacity T of a pinned gear is:

$$T = 0.787 \, Sd^2D$$

where S is safe shear stress and d is pin mean diameter.

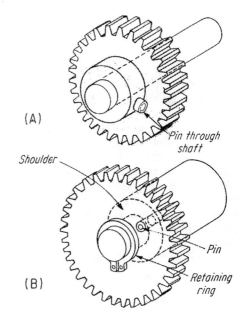

(A)   Pin through shaft

Shoulder

(B)   Pin

Retaining ring

2 CLAMPS AND COLLETS

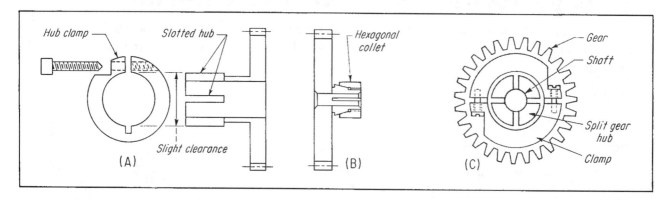

Clamping is popular with instrument-gear users because these gears can be purchased or manufactured with clamp-type hubs that are: machined integrally as part of the gear (A), or pressed into the gear bore. Gears are also available with a collet-hub assembly (B). Clamps can be obtained as a separate item.

Clamps of one-piece construction can break under excessive clamping pressure; hence the preference for the two-piece clamp (C). This places the stress onto the screw threads which hold the clamp together, avoiding possible fracture of the clamp itself. Hub of the gear should be slotted into three or four equal segments, with a thin wall section to reduce the size of the clamp. Hard-ened gears can be suitably fastened with clamps, but hub of the gear should be slotted prior to hardening.

Other recommendations are: Make gear hub approximately same length as for a pinned gear; slot through to the gear face at approximately 90° spacing. While clamps can fasten a gear on a splined shaft, results are best if both shaft and bore are smooth. If both splined, clamp then keeps gear from moving laterally.

Material of clamp should be same as for the gear, especially in military equipment because of specifications on dissimilarity of metals. However, if weight is a factor, aluminum-alloy clamps are effective. Cost of the clamp and slitting the gear hub are relatively low.

3 PRESS FITS

Press-fit gears to shafts when shafts are too small for keyways and where torque transmission is relatively low. Method is inexpensive but impractical where adjustments or disassemblies are expected.

Torque capacity is:

$$T = 0.785\, f D_1\, L e E \left[1 - \left(\frac{D_1}{D_2}\right)^2\right]$$

Resulting tensile stress in the gear bore is:

$$S = e E / D_1$$

where f = coefficient of friction (generally varies between 0.1 and 0.2 for small metal assemblies), D_1 is shaft dia, D_2 is OD of gear, L is gear width, e is press fit (difference in dimension between bore and shaft), and E is modulus of elasticity.

Similar metals (usually stainless steel when used in instruments) are recommended to avoid difficulties arising from changes in temperature. Press-fit pressures between steel hub and shaft are shown in chart at right (from Marks' Handbook). Curves are also applicable to hollow shafts, providing d is not over 0.25 D.

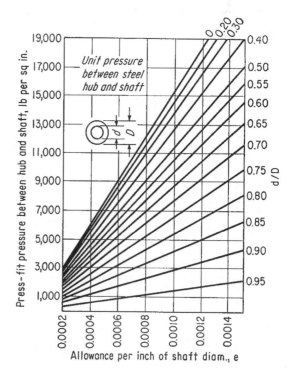

Comparison of Gear-Fastering Methods

Method	Torque Capacity	Ease of Replacing Gear	Reliability Under Operation	Versatility in Applications	Ability to Meet Environment Specs	Machining Requirements	Ability to Use Prehardened Parts	Relative Cost
—Pinning	Excellent	Poor	Excellent	Excellent	Excellent	High	Poor	High
—Clamping	Good	Excellent	Fair	Fair	Good	Moderate	Excellent	Medium
—Press fits	Fair	Fair	Good	Fair	Good	Moderate	Excellent	Medium
—Loctite	Good	Good	Good	Excellent	Excellent	Little	Excellent	Low
—Setscrews	Fair	Excellent	Poor	Good	Fair	Moderate	Good	Low
—Splining	Excellent	Excellent	Excellent	Fair	Excellent	High	Excellent	High
—Integral shaft	Excellent	Poor	Excellent	Good	Excellent	High	Excellent	High
—Knurling	Good	Poor	Good	Poor	Good	Moderate	Poor	Medium
—Keying	Excellent	Excellent	Excellent	Poor	Excellent	High	Excellent	High
—Staking	Poor	Fair	Poor	Poor	Good	Moderate	Poor	Low
—Spring washer	Poor	Excellent	Good	Fair	Good	Moderate	Excellent	Medium
—Tapered shaft	Excellent	Excellent	Excellent	Good	Excellent	High	Excellent	High
—Tapered rings	Good	Excellent	Good	Excellent	Good	Moderate	Excellent	Medium
—Tapered bushing	Excellent	Excellent	Excellent	Good	Good	Moderate	Excellent	High
—Die-cast assembly	Good	Poor	Good	Excellent	Good	Little	Fair	Low

4 RETAINING COMPOUNDS

Several different compounds can fasten the gear onto the shaft—one in particular is "Loctite," manufactured by American Sealants Co. This material remains liquid as long as it is exposed to air, but hardens when confined between closely fitting metal parts, such as with close fits of bolts threaded into nuts. (Military spec MIL-S-40083 approves the use of retaining compounds).

Loctite sealant is supplied in several grades of shear strength. The grade, coupled with the contact area, determines the torque that can be transmitted. For example: with a gear ⅜ in. long on a ⅛-in.-dia shaft, the bonded area is 0.22 in.² Using Loctite A with a shear strength of 1000 psi, the retaining force is 20 in.-lb.

Loctite will wick into a space 0.0001 in. or less and fill a clearance up to 0.010 in. It requires about 6 hr to harden, 10 min. with activator or 2 min. if heat is applied. Sometimes a setscrew in the hub is needed to position the gear accurately and permanently until the sealant has been completely cured.

Gears can be easily removed from a shaft or adjusted on the shaft by forcibly breaking the bond and then reapplying the sealant after the new position is determined. It will hold any metal to any other metal. Cost is low in comparison to other methods because extra machining and tolerances can be eased.

5 Setscrews

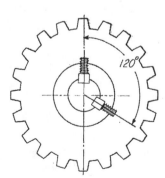

Two setscrews at 90° or 120° to each other are usually sufficient to hold a gear firmly to a shaft. More security results with a flat on the shaft, which prevents the shaft from being marred. Flats give added torque capacity and are helpful for frequent disassembly. Sealants applied on setscrews prevent loosening during vibration.

6 GEARS INTEGRAL WITH SHAFT

Fabricating a gear and shaft from the same material is sometimes economical with small gears where cost of machining shaft from OD of gear is not prohibitive. Method is also used when die-cast blanks are feasible or when space limitations are severe and there is no room for gear hubs. No limit to the amount of torque which can be resisted—usually gear teeth will shear before any other damage takes place.

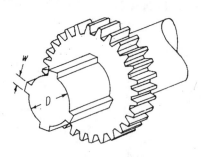

	4 - spline	6 - spline
D	w	w
1/2	0.120	0.125
3/4	0.181	0.188
7/8	0.211	0.219
1	0.241	0.250
1 - 1/4	0.301	0.313

7 SPLINED SHAFTS

Ideal where gear must slide in lateral direction during rotation. Square splines often used, but involute splines are self-centering and stronger. Non-sliding gears are pinned or held by threaded nut or retaining ring.

Torque strength is high and dependent on number of splines employed. Use these recommended dimensions for width of square tooth for 4-spline and 6-spline systems; although other spline systems are some times used. Stainless steel shafts and gears are recommended. Avoid dissimilar metals or aluminum. Relative cost is high.

8 KNURLING

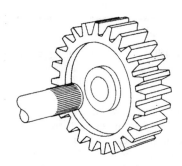

A knurled shaft can be pressed into the gear bore, to do its own broaching, thus keying itself into a close-fitting hole. This avoids need for supplementary locking device such as lock rings and threaded nuts.

The method is applied to shafts 1/4 in. or under and does not weaken or distort parts by the machining of groove or holes. It is inexpensive and requires no extra parts.

Knurling increases shaft dia by 0.002 to 0.005 in. It is recommended that a chip groove be cut at the trailing edge of the knurl. Tight tolerances on shaft and bore dia are not needed unless good concentricity is a requirement The unit can be designed to slip under a specific load—hence acting as a safety device.

9 KEYING

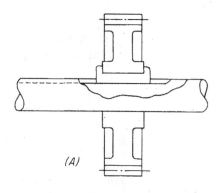

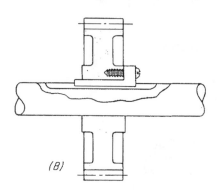

(A)   (B)

Generally employed with large gears, but occasionally considered for small gears in instruments. Feather key (A) allows axial movement but keying must be milled to end of shaft. For blind keyway (B), use setscrew against the key, but method permits locating the gear anywhere along length of shaft.

Keyed gears can withstand high torque, much more than the pinned or knurled shaft and, at times, more than the splined shafts because the key extends well into both the shaft and gear bore. Torque capacity is comparable with that of the integral gear and shaft. Maintenance is easy because the key can be removed while the gear remains in the system.

Materials for gear, shaft and key should be similar preferably steel. Larger gears can be either cast or forged and the key either hot- or cold-rolled steel. However, in instrument gears, stainless steel is required for most applications. Avoid aluminum gears and keys.

10 STAKING

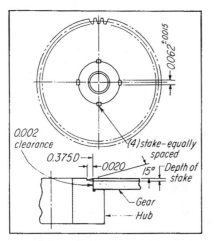

No. of stakes	Depth	Clearance	Torque in.-lb.
4	0.015	0.0020	27
4	0.015	0.0025	20
4	0.020	0.0020	28
4	0.020	0.0020	30
8	0.020	0	52

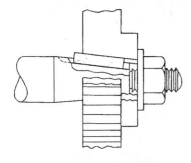

It is difficult to predict the strength of a staked joint—but it is a quick and economical method when the gear is positioned at the end of the shaft.

Results from five tests we made on gears staked on 0.375-in. hubs are shown here with typical notations for specifying staking on an assembly drawing. Staking was done with a 0.062-in. punch having a 15° bevel. Variables in the test were: depth of stake, number of stakes, and clearance between hub and gear. Breakaway torque ranged from 20 to 52 in.-lb.

Replacing a gear is not simple with this method because the shaft is mutirated by the staking. But production costs are low.

11 SPRING WASHER

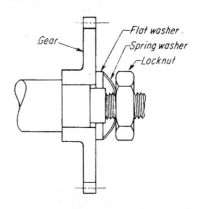

Assembly consists of locknut, spring washer, flat washer and gear. The locknut is adjusted to apply a predetermined retaining force to the gear. This permits the gear to slip when overloaded—hence avoiding gear breakage or protecting the drive motor from overheating.

Construction is simple and costs less than if a slip clutch is employed. Popular in breadboard models.

12 TAPERED SHAFT

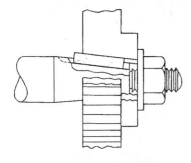

Tapered shaft and matching taper in gear bore need key to provide high torque resistance, and threaded nut to tighten gear onto taper. Expensive but suitable for larger gear applications where rigidity, concentricity and easy disassembly are important. A larger dia shaft is needed than with other methods. Space can be problem because of protruding threaded end. Keep nut tight.

13 TAPERED RINGS

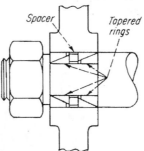

These interlock and expand when tightened to lock gear on shaft. A purchased item, the rings are quick and easy to use, and do not need close tolerance on bore or shaft. No special machining is required and torque capacity is fairly high. If lock washer is employed, the gear can be adjusted to slip at predetermined torque.

14 TAPERED BUSHINGS

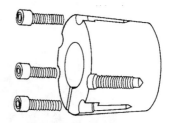

This, too, is a purchased item—but generally restricted to shaft diameters ½ in. and over. Adapters available for untapered bores of gears. Unthreaded half-holes in bushing align with threaded half-holes in gear bore. Screw pulls bushing into bore, also prevents rotational slippage of gear under load.

15 DIE-CAST HUB

Die-casting machines are available, which automatically assemble and position gear on shaft, then die-cast a metal hub on both sides of gear for retention. Method can replace staked assembly. Gears are fed by hopper, shafts by magazine. Method maintains good tolerances on gear wobble, concentricity and location. For high-production applications. Costs are low once dies are made.

8 Control Mountings

When designing control panels follow this 8-point guide and check for...

Frank William Wood, Jr.

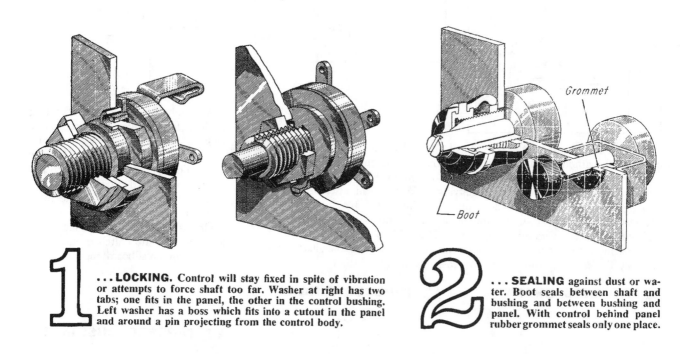

1 ...**LOCKING.** Control will stay fixed in spite of vibration or attempts to force shaft too far. Washer at right has two tabs; one fits in the panel, the other in the control bushing. Left washer has a boss which fits into a cutout in the panel and around a pin projecting from the control body.

2 ...**SEALING** against dust or water. Boot seals between shaft and bushing and between bushing and panel. With control behind panel rubber grommet seals only one place.

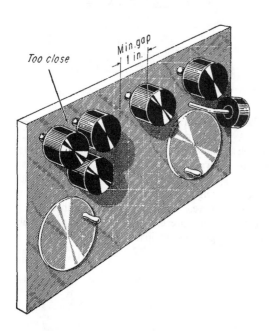

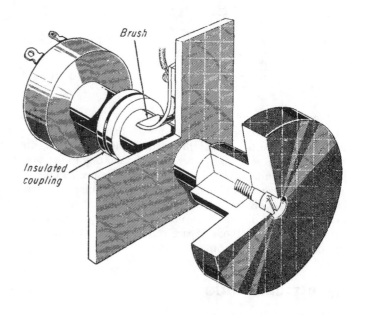

5 ...**HAND-ROOM** at front of the panel. Space knobs at least one inch apart. Extending knob to save space puts it where the operator can bump into it and bend the shaft. Best rule is to keep shaft as short as possible.

6 ...**"HOT" CONTROL KNOBS.** One approach is to ground them by installing a brush against the shaft. Another solution is to isolate the control by an insulated coupling or a plastic knob having recessed holding screws.

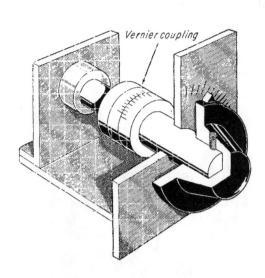

Vernier coupling

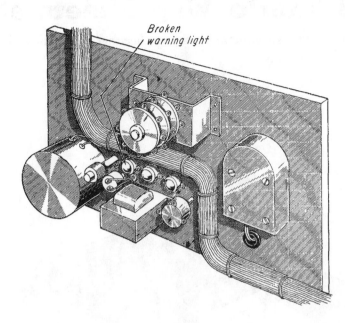

Broken warning light

3 ...**RESETTING** to match controls to panel markings. For crude adjustments a set-screw is enough. Where matching is critical a three-piece vernier coupling permits more accurate calibration.

4 ...**ACCESSIBILITY** behind the panel. Easy access reduces down time and maintenance costs especially if one man can do most jobs alone. Here, technician can't replace a warning light without dismantling other parts.

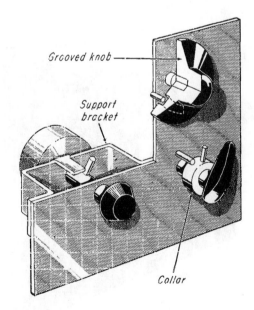

Grooved knob

Support bracket

Collar

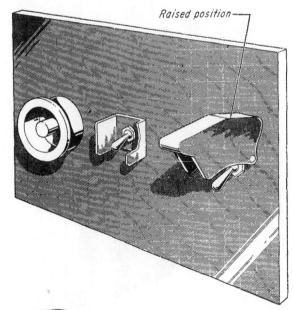

Raised position

7 ...**LIMIT STOPS** that are strong enough not to bend under heavy-handed use. Otherwise setting will change when stop moves. Collar and grooved knob permit adjustment; tab on bracket doesn't.

8 ...**GUARDS** to prevent accidental actuation of switches. Bell-shape guard for push-buttons is just finger-size. U-shape guard separates closely spaced toggle switches, and a swinging guard holds down special ones.

8 Interlocking Sheetmetal Fasteners

These eight sheetmetal parts join sheetmetal quickly with the simplest of tools, few screws or bolts.

L. Kasper

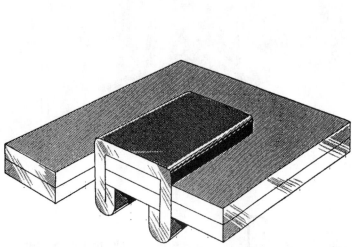

1 SQUEEZE CLIP holds two overlapping sheets together. The ends of the clip are pushed through parallel slots, then bent over much like a staple.

2 ALIGNING PIECE slides up out of the way in long slot while butting sheets are being positioned. Afterwards it slips down over lower sheet.

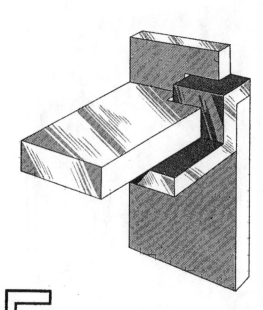

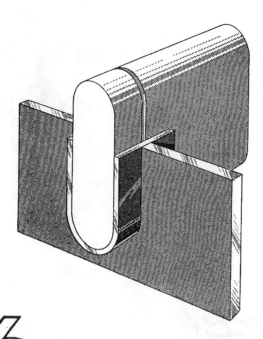

5 ESS supports shelf between uprights. By mating with notched edge it acts as a key to keep shelf from sliding back and forth, and provides positive location.

6 CUP carries a bar on both sides of divider. Here bars stick up above the top, but deeper cutout will lower them until they are flush or sunk.

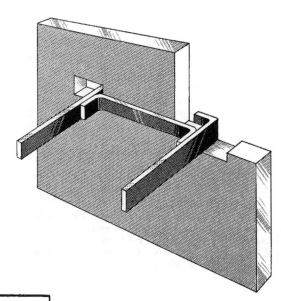

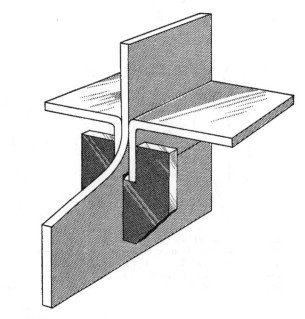

BRACKETS provide instantly mobile rack space for boxes. To install or remove, squeeze sides together and push hooked ends through slots.

FLANGE HOLDER does double duty by holding up shelves on both sides of a partition. Angular corners allow it to fit through small slit when tilted.

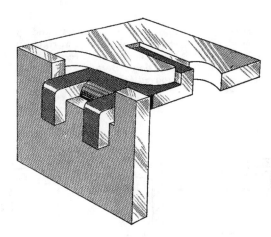

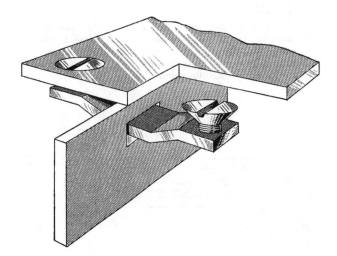

CLAW holds top sheet between two end pieces. Tail snaps into slot, then claw is hammered over edge. With notched edge, top is even with sides.

BAR clamps divider in place. Extruded holes provide a recess for screws so that they stay flush with upper surface of horizontal sheet.

Fastening Sheet-Metal Parts by Tongues, Snaps, or Clinching

Detachable and permanent assembly of sheet metal parts without using rivets, bolts, or screws.

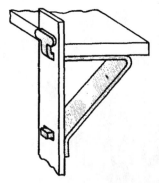

Fig. 1—Supporting bracket formed from sheet metal and having integral tabs. Upper tab is inserted into structure and bent. Ledge weight holds lower tab.

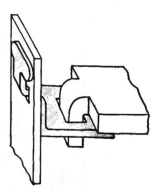

Fig. 2—Supporting bracket similar to Fig. 1 but offering restraint to shelf or ledge. Tabs are integral with sheet metal part and are bent on assembly.

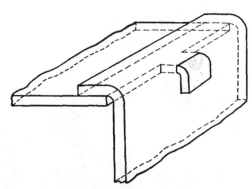

Fig. 3—Supporting ledge or shelf by direct attachment. Tab is integral and bent on assembly. Additional support is possible if sheet is placed on flange and tabbed.

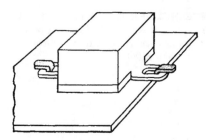

Fig. 7—Box section joined to a flat sheet or plate. Elongated holes are integral with box section and tabs are integral with plate. Design is not limited to edge location.

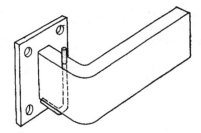

Fig. 8—Bar is joined to sheet metal bracket by a pin or rod. Right angle bends in pin restrict sidewise or rocking motion or bar. Bracket end of pin is peened.

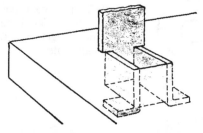

Fig. 9—To support and join sheet metal support at right angle to plate. Motion is restricted in all directions. Bottom surface can be grooved for tabs.

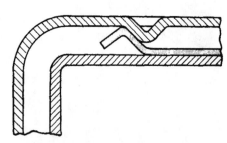

Fig. 13—A spacing method that can be used for circular sections. Formed sheet metal member support outer structure at set distance. Bead centers structure.

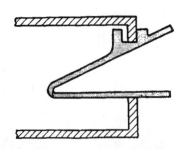

Fig. 14—A removable section held in place by elasticity of material. Design shown is a temporary or a removable cover for an elongated slotted hole in a sheet metal part.

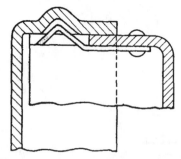

Fig. 15—A cover held in position by bead and formed sheet. Cover is restrained from motion but can be rotated. Used for covers that must be removable.

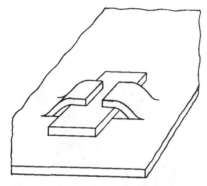

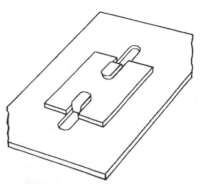

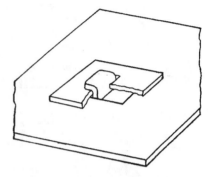

Fig. 4—To support or join a flat sheet metal form on a large plate. Tabs are integral with plate and bent over on assembly. Only sidewise motion is restricted.

Fig. 5—Similar to Fig. 4 but motion is restricted in all directions. Upper sheet is slotted, and tabs are bent over and into slots on assembly.

Fig. 6—Single tab design for complete restriction of motion. Upper plate has an elongated hole that matches width and thickness of integral lower plate tab.

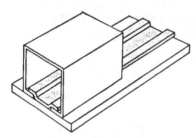

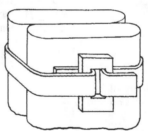

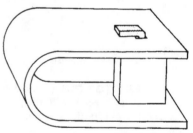

Fig. 10—Channel section spot welded to plate forms bottom surface and joins box section to plate. Channel edges can be crimped or spot welded to restrict motion.

Fig. 11—Sheet metal strap used to join two flat surfaces. Edges of plate are rounded to allow strap to follow contour and prevent cutting of plate by the metal strap.

Fig. 12—Sheet metal structures can be spaced and joined by use of a tabbed block. Formed sheet metal U section is held to form by the block as shown.

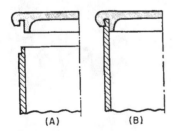

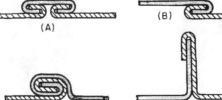

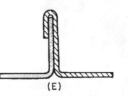

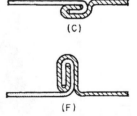

Fig. 16—A non-removable cover design. The vessel is notched as shown in A, and the cover crimped over, B, on assembly. This is a permanent cover assembly.

Fig. 17—Six methods of joining two sheet metal parts. These can be temporary or permanent joints. If necessary, joints can be riveted, bolted, screwed or welded for added strength and support. Such joints can also be used to make right angle corner joints on sheet metal boxes, or for attaching top and bottom covers on sheet metal containers.

Snap Fasteners for Polyethylene

It's difficult to cement polyethylene parts together, so eliminate extra cost of separate fasteners with these snap-together designs.

Edger Burns

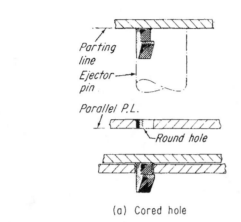

(a) Cored hole

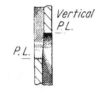

(b) "Shut-off" hole for female snap for different P.L.

1 **EJECTOR-PIN** of mold is cut to shape of snap. Ejected with the pin, the part is slid off the pin by the operator.

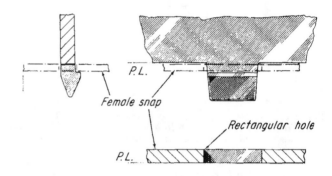

2 **WALL-END SNAP** is easier to remove from the mold than the ejector-pin snap. The best length for this snap is ¼ to ½ in.

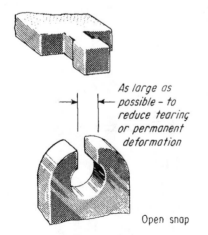

Open snap

5 **OPEN SNAP** relies on an undercut in the mold and on the ability of the polyethylene to deform and then spring back on ejection.

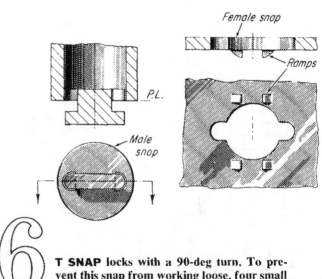

6 **T SNAP** locks with a 90-deg turn. To prevent this snap from working loose, four small ramps are added to the female part.

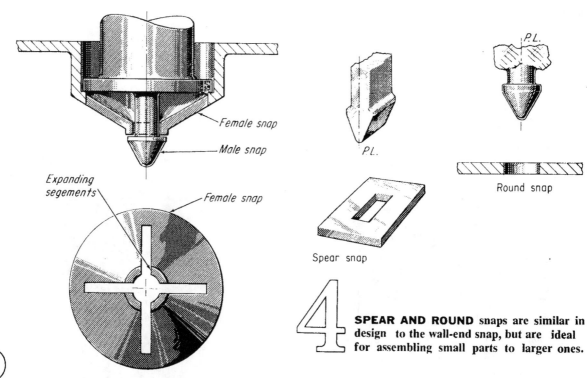

Female snap

Male snap

Expanding segments

Female snap

P.L.

P.L.

Round snap

Spear snap

3 **SEGMENTED WALL** of female snap allows a large-headed male snap to enter easily. The snap can not be pulled apart with light loads.

4 **SPEAR AND ROUND** snaps are similar in design to the wall-end snap, but are ideal for assembling small parts to larger ones.

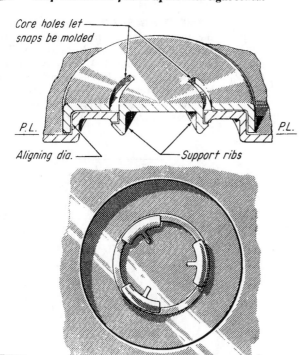

Core holes let snaps be molded

P.L.

Aligning dia.

Support ribs

P.L.

7 **ROTATING** parts can be snapped securely together with three (shown) or two snaps. Mostly for linear polyethylene, it's strong.

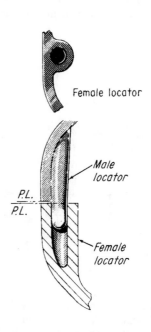

Female locator

Male locator

P.L.
P.L.

Female locator

8 **LOCATORS** are not really snaps, but align parts for subsequent eyeleting or riveting or in conjunction with other snaps.

Snap Fasteners for Polystyrene

Here's a low-cost way to join injection-molded polystyrene parts without the use of separate fasteners or solvents.

Edger Burns

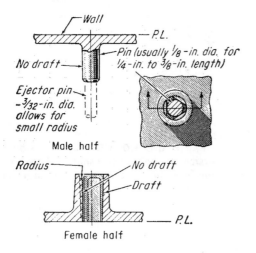

Male half

Female half

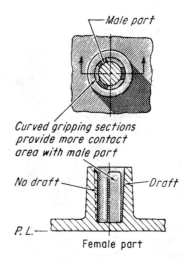

Male part

Curved gripping sections provide more contact area with male part

Female part

TRIANGULAR SNAPS actually depend upon friction, but are strong and easy to assemble. Space several around the parts to be joined. Grip can also be adjusted to suit.

CEMENTING SNAP will allow solvent cementing between the male and female members if required, although the snap will hold well without cement. Usually two or more such snaps are positioned around the parts to be joined. Male part is virtually the same as for the triangular snap. Blind, cored hole requires no shutoff.

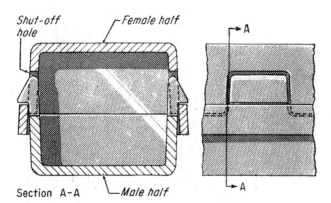

Section A-A

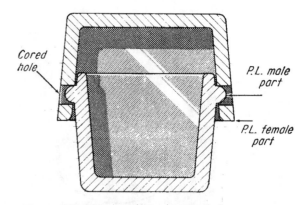

BOX SNAP requires a mold shutoff in the female half, large enough to accommodate the male part, which slips behind a shoulder and locks, as shown in the diagram.

DETENT SNAPS are ideal whenever a snap has to be frequently undone, and where a tight hold is not required. The detent itself can be a hemispherical bump, or a more elongated shape.

NOTE: *In the following illustrations, the parting line is marked "P.L." It is important in deciding which snap to use and how it should be molded.*

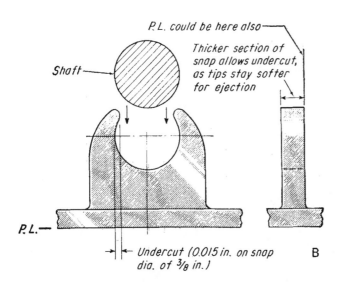

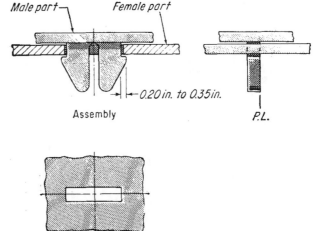

Female part

<table>
<tr><td>

3 **OPEN SNAPS** have undercuts of about 0.015 in. on a total snap dia of ⅜ in. Despite small undercuts, stiff polystyrene gives good grip. This snap is not suitable for the regular nonimpact polystyrene. If the parting line can be arranged to lie in the other plane, as shown in B, ejection from the mold would be trouble free, thus avoiding excessive scrap.

</td><td>

4 **PRONG SNAPS** are ideal for snapping small parts onto a larger assembly. The male member is usually on the small part. Slot length in the female part must be designed for maximum holding power, without cracking the prongs on the male member.

</td></tr>
</table>

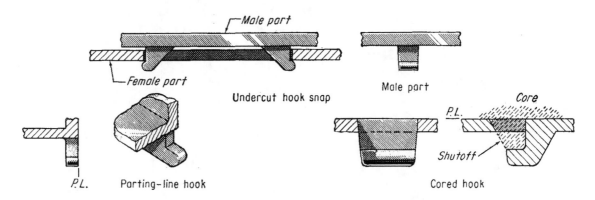

7 **HOOK SNAPS.** *Undercut hook* snap relies on an undercut in the mold. To prevent polystyrene breakage on ejection from the mold, make the hooks thicker than the other parts — they then retain more heat, stay softer. Since large undercuts cannot be made this way, however, this snap loosens quite easily.

Parting-line hook is much simpler to apply and is easier to design. Choose this snap whenever the part-ing line can be arranged to be in the plane shown on the drawing. It can be almost any strength and shape desired, and is simple to cut into the mold.

Cored hook requires a core to come down from the other half of the mold, which then produces the inside of the snap. This will leave a hole in the wall to which the hook is attached. Three shutoff surfaces are also required in the mold construction.

SECTION 6

GEARS & GEARING

Nomenclature of Gears

SPUR & HELICAL GEARS

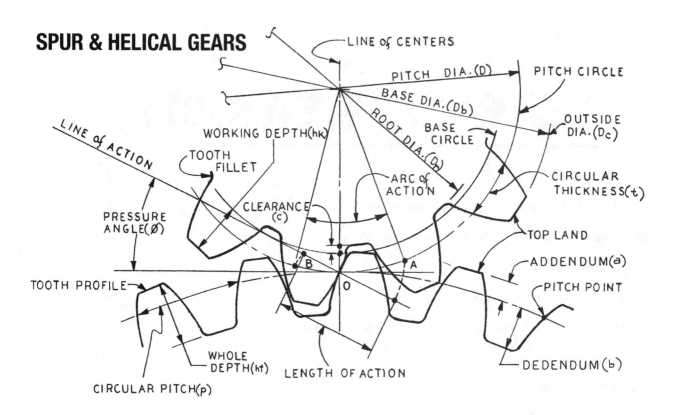

BEVEL GEARS

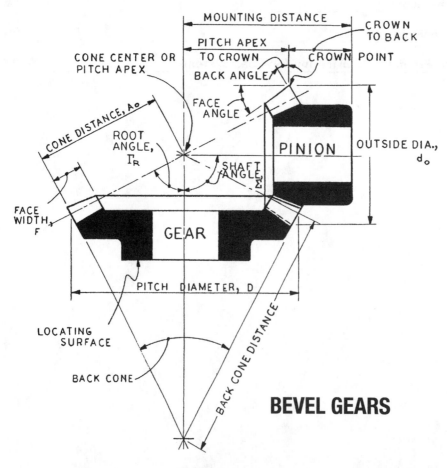

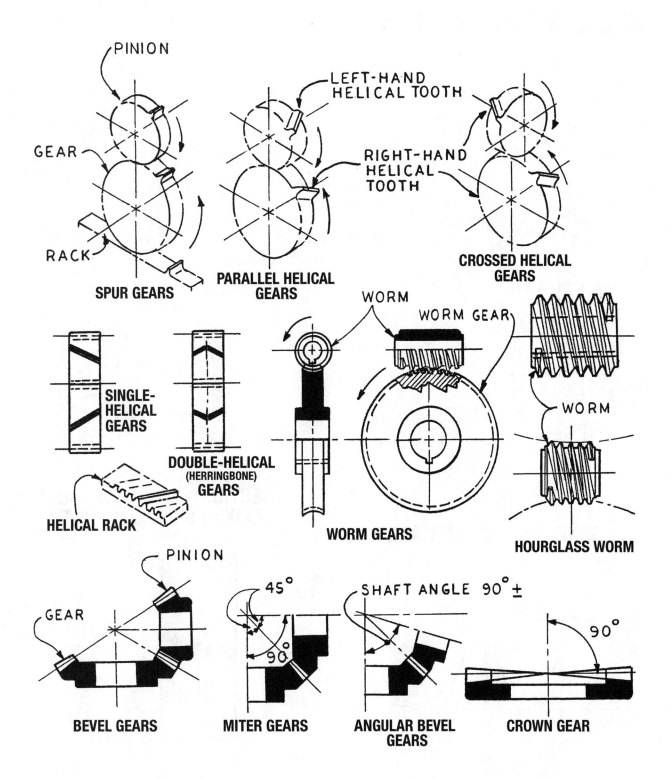

Adapted from: *The New American Machinist's Handbook*, © 1955 McGraw-Hill

Graphical Representation of Gear Dimensions

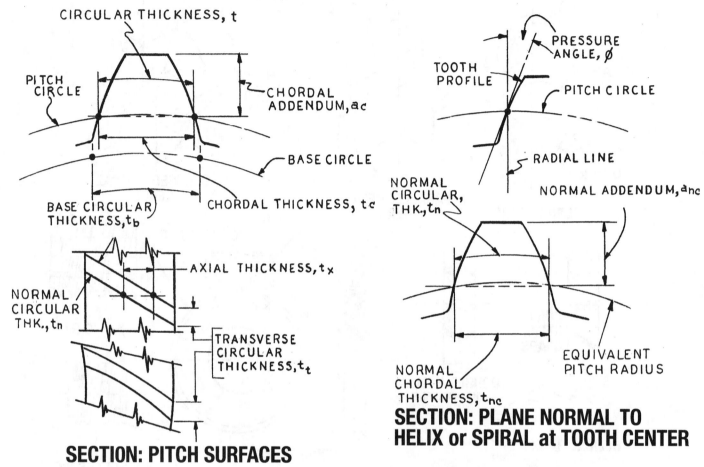

SECTION: PITCH SURFACES

SECTION: PLANE NORMAL TO HELIX or SPIRAL at TOOTH CENTER

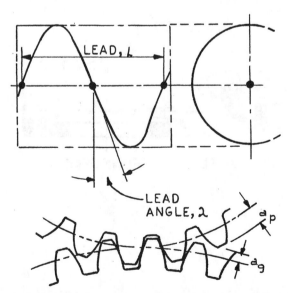

LONG & SHORT ADDENDUM TEETH

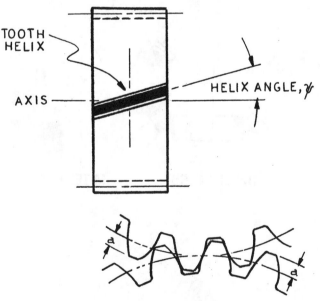

EQUAL ADDENDUM TEETH

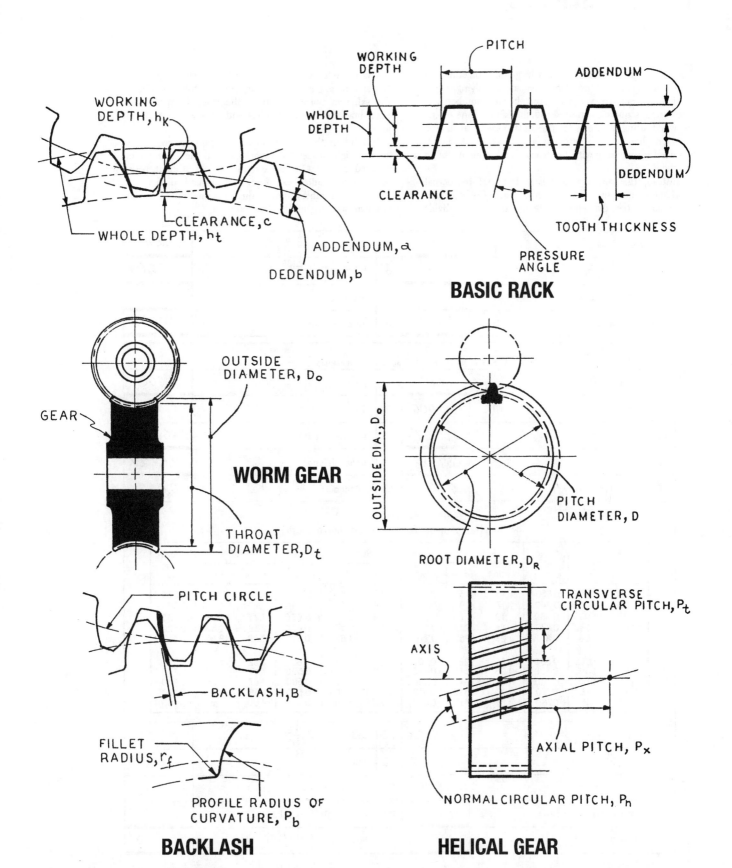

WORKING DEPTH, h$_K$

WHOLE DEPTH, h$_t$

CLEARANCE, c

ADDENDUM, a

DEDENDUM, b

WORKING DEPTH

WHOLE DEPTH

CLEARANCE

PITCH

ADDENDUM

DEDENDUM

TOOTH THICKNESS

PRESSURE ANGLE

BASIC RACK

GEAR

OUTSIDE DIAMETER, D$_o$

THROAT DIAMETER, D$_t$

WORM GEAR

OUTSIDE DIA., D$_o$

ROOT DIAMETER, D$_R$

PITCH DIAMETER, D

PITCH CIRCLE

BACKLASH, B

FILLET RADIUS, r$_f$

PROFILE RADIUS OF CURVATURE, P$_b$

BACKLASH

TRANSVERSE CIRCULAR PITCH, P$_t$

AXIS

AXIAL PITCH, P$_x$

NORMAL CIRCULAR PITCH, P$_h$

HELICAL GEAR

Adapted from: *The New American Machinist's Handbook,* © 1955 McGraw-Hill

Worksheet Streamlines Bevel-Gear Calculations

B. J. Mumken

The following worksheet neatly gathers together the many mathematical problems that need solving when designing straight bevel-gears. And they are numbered in the correct sequence—no need to hunt "all over the place" as when using formulas in the usual bevel-gear tables. In fact, there are no formulas as such—and, therefore, no need for working with the many Greek symbols found in them.

Instead, the language here is in terms of the actual working operations. For example, space (9) tells you to obtain pitch diameter of the pinion—simply divide

the value in space (1) by the value in space (3). And to get root angle for the gear, you are told to subtract the value in space (24) from the value in space (14). Each bracketed number refers you to a value previously filled in.

Just fill in the known values for pinion and gear in the first eight spaces, then work through the sheet, which is based on the Gleason system for 90° straight bevel-gears. Final result (next page) is gear-blank dimensions.

Colored numbers show values obtained in a sample problem worked out by this method.

1	No. of teeth, pinion	40	5	Working depth = $\frac{2.000}{(3)}$		0.200
2	No. of teeth, gear	80	6	Whole depth = $\frac{2.188}{(3)}$ + 0.002 ("D + F")		0.2208
3	Diametral pitch	10	7	Pressure angle		20°
4	Face width	0.750	8	Total backlash		0.003

	PINION			GEAR	
	(Thick underlining indicates working dimensions)				
9	Pitch dia. $\frac{(1)}{(3)}$	4.000	10	Pitch dia. $\frac{(2)}{(3)}$	8.000
11	Tan $\frac{(1)}{(2)}$	0.5000	12	Tan $\frac{(2)}{(1)}$	2.0000
13	Pitch angle (11), in deg.	26° 34'	14	Pitch angle (12)	63° 26'
15	2 X cos (13)	1.7888	16	Cone distance $\frac{(10)}{(15)}$	4.4722
17	Addendum (5) − (17)	0.135	18	Addendum = $\frac{\text{(see table)}}{(3)}$	0.065

Gear Addendum for 1 D.P.
Ratio = (No. of gear teeth)/(No. of pinion teeth)

Ratios		Addendum, in.	Ratios		Addendum, in.	Ratios		Addendum, in.	Ratios		Addendum, in.
From	To		From	To		From	To		From	To	
1.00	1.00	0.850	1.15	1.17	0.750	1.41	1.44	0.650	1.99	2.10	0.550
1.00	1.02	0.840	1.17	1.19	0.740	1.44	1.48	0.640	2.10	2.23	0.540
1.02	1.03	0.830	1.19	1.21	0.730	1.48	1.52	0.630	2.23	2.38	0.530
1.03	1.05	0.820	1.21	1.23	0.720	1.52	1.57	0.620	2.38	2.58	0.520
1.05	1.06	0.810	1.23	1.26	0.710	1.57	1.63	0.610	2.58	2.82	0.510
1.06	1.08	0.800	1.26	1.28	0.700	1.63	1.68	0.600	2.82	3.17	0.500
1.08	1.09	0.790	1.28	1.31	0.690	1.68	1.75	0.590	3.17	3.67	0.490
1.09	1.11	0.780	1.31	1.34	0.680	1.75	1.82	0.580	3.67	4.56	0.480
1.11	1.13	0.770	1.34	1.37	0.670	1.82	1.90	0.570	4.56	7.00	0.470
1.13	1.15	0.760	1.37	1.41	0.660	1.90	1.99	0.560	7.00	α	0.460

19	Dedendum = $\frac{2.188}{(3)}$ − (18)	0.0838	20	Dedendum = $\frac{2.188}{(3)}$ − (17)	0.1538
21	Tan $\frac{(19)}{(16)}$	0.0187	22	Tan $\frac{(20)}{(16)}$	0.0343
23	Ded angle (21)	1° 4'	24	Ded angle (22)	1° 58'
25	Face angle (13) + (24)	28° 32'	26	Face angle (14) + (23)	64° 30'
27	Root angle (13) − (23)	25° 30'	28	Root angle (14) − (24)	61° 28'

No.	Formula	Value	No.	Formula	Value
29	cos (13)	0.8944	30	cos (14)	0.4472
31	[2 x (18)] x (29)	0.2414	32	[2 x (17)] x (30)	0.0581
33	OD = (9) + (31)	4.2415	34	OD = (10) + (32)	8.0581
35	(18) x (30)	0.0603	36	(17) x (29)	0.0581
37	Pitch-apex to crown = [0.5 x (10)] – (35)	3.9396	38	Pitch-apex to crown = [0.5 x (91)] – (36)	1.9419
39	Circular pitch = $\frac{3.1416}{(3)}$	0.3141	40	(18) – (17)	0.0700
41	0.5 x (39)	0.1570	42	(41) x tan (7)	0.0254
43	Circular tooth thickness = (39)–(43)	0.1825	44	Circular tooth thickness = (40)–(42)	0.1316
45	(44)^3	0.0060	46	(44)^3	0.0022
47	(9)^2	16.0000	48	(10)^2	64.0000
49	6 x (47)	96.0000	50	6 x (48)	384.000
51	$\frac{(45)}{(49)}$	0.00006	52	$\frac{(46)}{(50)}$	0.0000
53	Chordal tooth thickness = (44) – (51) – [0.5 x (8)]	0.181	54	Chordal tooth thickness = (43) – (52) – [0.5 x (8)]	0.1301
55	(44)^2 x (29)	0.0298	56	(43)^2 x (30)	0.0077
57	4 x (9)	16.0000	58	4 x (10)	32.0000
59	$\frac{(55)}{(57)}$	0.0019	60	$\frac{(56)}{(58)}$	0.0002
61	Chordal addendum (18) + (59)	0.1369	62	Chordal addendum (17) + (60)	0.0652
63	sin (28)	0.8785	64	sin (27)	0.4771
65	cos (28)	0.4776	66	cos (27)	0.8788

PINION   G E A R

67	(4) x (63)	0.6589	68	(4) x (64)	0.3579
69	(4) x (65)	0.3583	70	(4) x (66)	0.6591
71	$\frac{(16) – (4)}{(16)}$	0.8323			
72	(18) x (71)	0.1124	73	(17) x (71)	0.0541
74	(19) x (71)	0.0697	75	(20) x (71)	0.1280
76	[(72) + (74)] x (30)	0.0815	77	[(73) + (75)] x (29)	0.1629
78	(33) – [2 x (69)]	3.5249	79	(34) – [2 x (70)]	6.7399
80	(76) + mfg. std.	0.125	81	(77) + mfg. std.	0.250

Alignment Chart for Face Gears

B. Bloomfield

THE MAXIMUM PRACTICAL DIAMETER for face gears is that diameter at which the teeth become pointed. The limiting inside diameter is the value at which tooth trimming occurs. This is always larger than the diameter for which the operating pressure angle is zero. The two alignment charts that follow can be used to find the maximum OD and the minimum ID if the numbers of teeth in the face gear and pinion are known. They eliminate lengthy calculations.

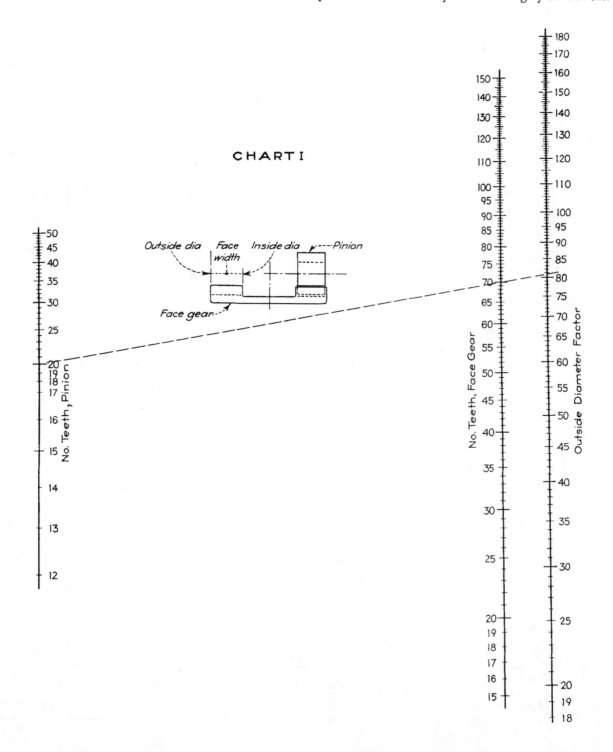

CHART I

FOR BOTH CHARTS the pinions are assumed to be spur gears of standard AGMA proportions, and the axes of the face gear and pinion are assumed to intersect at right angles. Both should be used only for tooth ratios of 1.5 to 1 or larger. Smaller ratios require pinion modifications not allowed for in these data. For both charts, the appropriate face gear diameter is found by dividing the factor from the chart by the diametral pitch of the pinion.

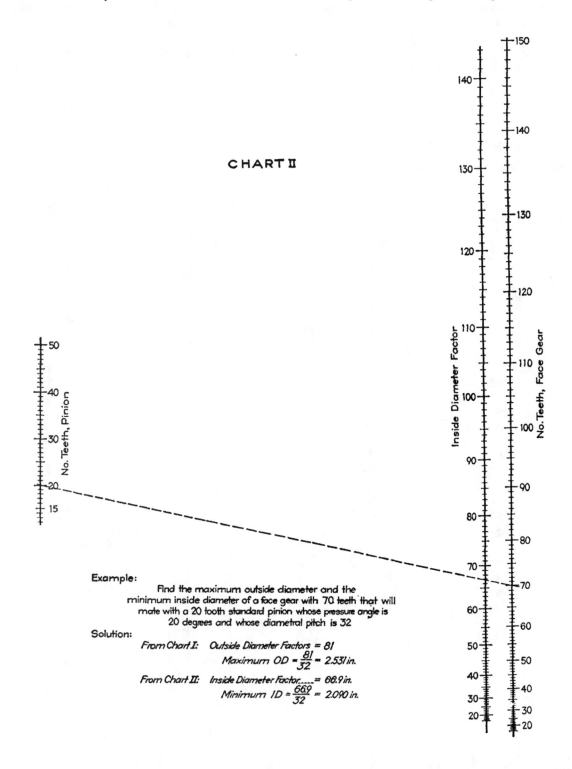

CHART II

Example:
Find the maximum outside diameter and the minimum inside diameter of a face gear with 70 teeth that will mate with a 20 tooth standard pinion whose pressure angle is 20 degrees and whose diametral pitch is 32

Solution:

From Chart I: Outside Diameter Factors = 81

$$Maximum\ OD = \frac{81}{32} = 2.531\ in.$$

From Chart II: Inside Diameter Factor......= 66.9 in.

$$Minimum\ ID = \frac{66.9}{32} = 2.090\ in.$$

No. Teeth, Pinion

Inside Diameter Factor

No. Teeth, Face Gear

Power Capacity of Spur Gears

Charles Tiplitz

Maximum rated horsepower that can safely be transmitted by a gear depends upon whether it runs for short periods or continuously. Capacity may be based on tooth strength if the gear is run only periodically; durability or wear governs rated horsepower for continuous running.

Checking strength and surface durability of gears can be a lengthy procedure. The following charts simplify the work and give values accurate to 5 to 10%. They are based on AGMA standards for strength and durability of spur gears.

Strength Nomograph is used first. Apart from the

Strength of Spur Gears.
Based on AGMA 220.01

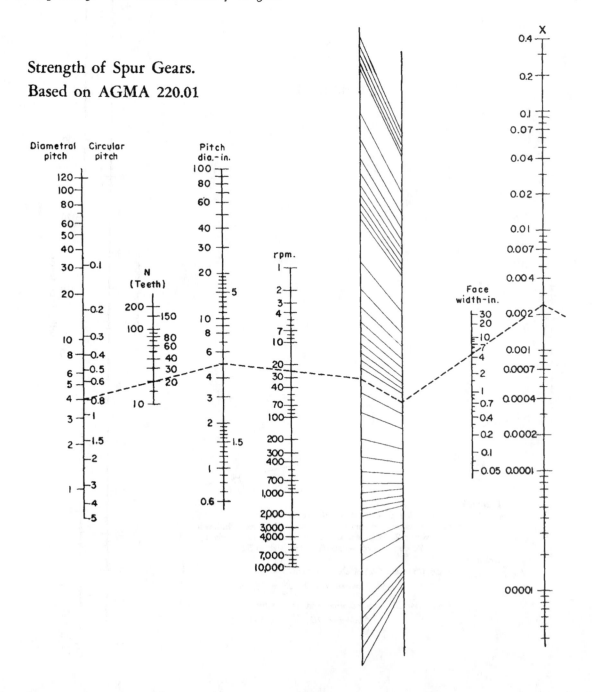

usual design constants only two of the following three need be known: pitch, number of teeth and pitch diameter. To use the charts connect the two known factors by a straight line, cutting the third scale. From this point on the scale continue drawing straight lines through known factors, cutting the pivot scales. Between the double pivot scales the line should be drawn parallel to the adjacent lines.

Durability nomograph must be entered on scale **X** at the same value that was cut on the **X** scale on the strength chart. Both pinion and gear should be checked if made of different materials and the smaller of the values obtained should be used.

Strength of Spur Gears (cont.)

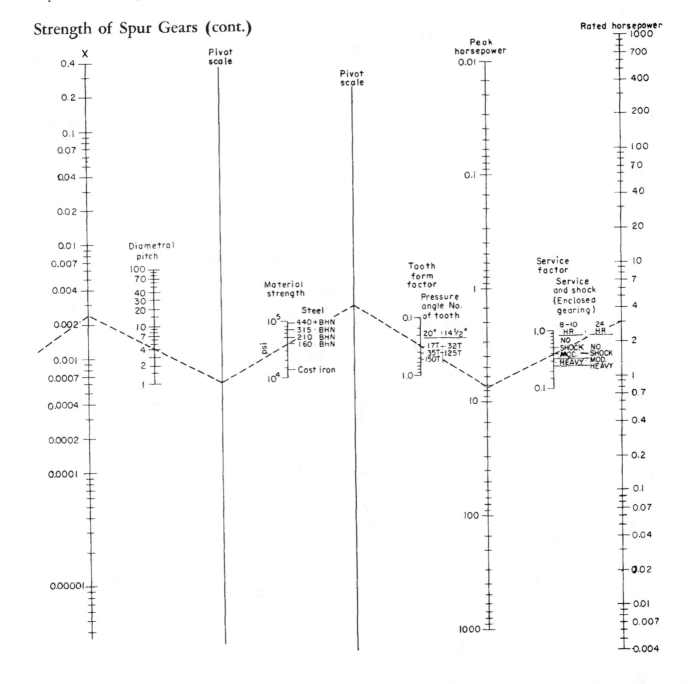

Surface Durability of Spur Gears. Based on AGMA 210.01

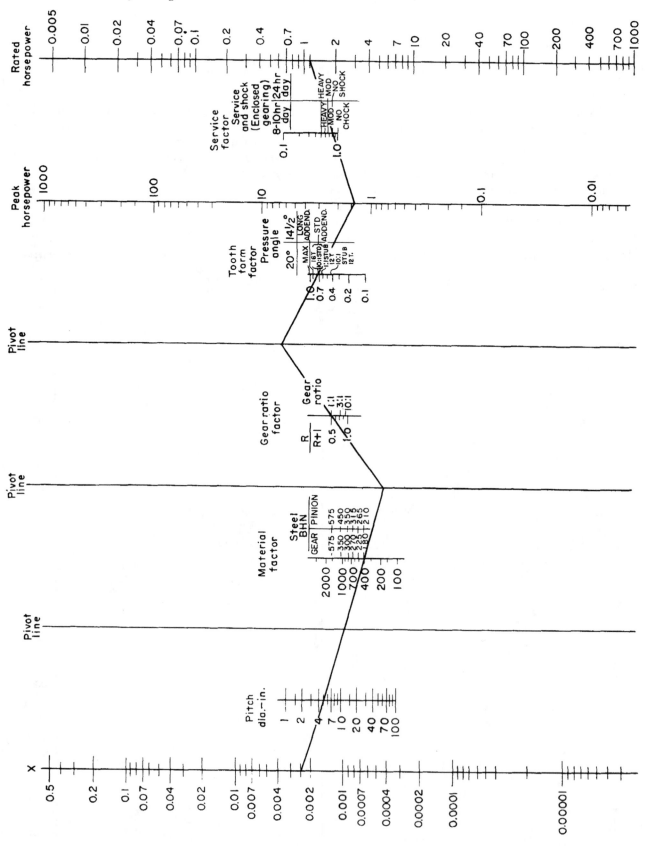

Linear to Angular Conversion of Gear-Tooth Index Error

For pitch diameters up to 200 inch, chart quickly converts index error from ten-thousandths of an inch to seconds or arc.

Harold R. Ronan, Jr.

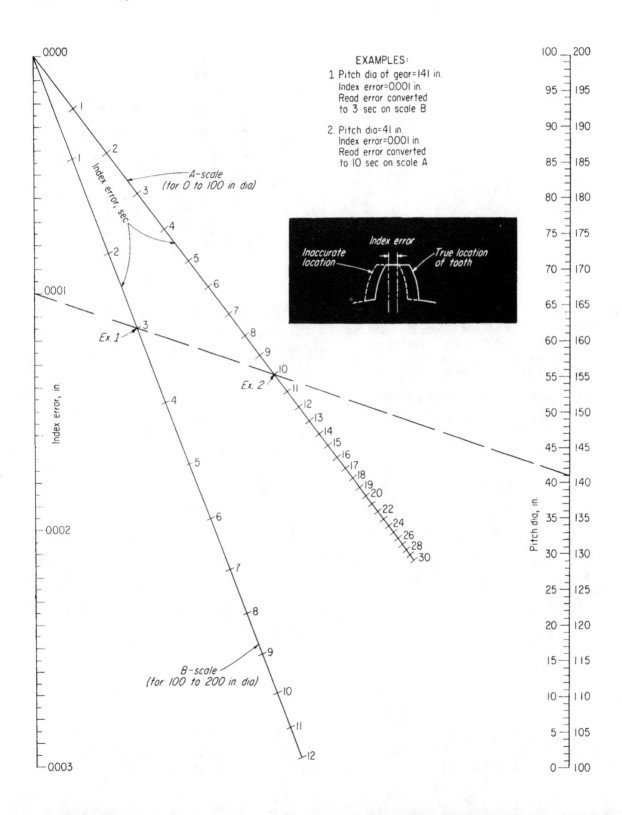

EXAMPLES:

1. Pitch dia of gear=141 in.
   Index error=0.001 in.
   Read error converted
   to 3 sec on scale B

2. Pitch dia=41 in.
   Index error=0.001 in.
   Read error converted
   to 10 sec on scale A

A-scale
(for 0 to 100 in dia)

Index error, sec

Index error, in.

Ex. 1

Ex. 2

Pitch dia, in.

B-scale
(for 100 to 200 in. dia)

Index error

Inaccurate location

True location of tooth

Checklist for Planetary-Gear Sets

These five tests quickly tell whether the gears will mesh, and whether there is room for them to fit together.

Hugh P. Hubbard

You have decided to design a planetary-gear system with a certain gear ratio, and have chosen the number of teeth for each gear to get that ratio. Will it work? Will the gears fit together to make a workable system?

If they can pass the following five tests, they will.

1—Do all gears have the same circular pitch?

If they do not, the gears will not mesh.

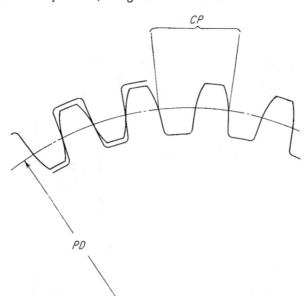

Circular pitch $CP = \pi/DP = PD/N$

Circular pitch and number of teeth determine pitch diameter, which leads to the next test:

2—Will the gears mate at the pitch diameters?

This equation shows whether the planet gear will fill the space between the sun gear and the ring gear:

$$N_p = \frac{N_r - N_s}{2}$$

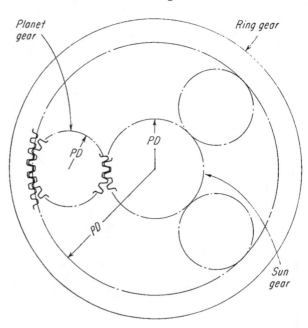

3—Will the teeth mesh?

Gears that pass the first two tests will not necessarily pass this one. If the gears have the wrong number of teeth, the planet gear will not mesh with the sun gear and the ring gear at the same time. Gears with numbers of teeth divisible by three will mesh. There are two other possible cases.

Case I—The number of teeth on the sun gear divides evenly into the number of teeth on the ring gear. This set will mesh, if allowance is made by spacing the planet gears unevenly around the sun gear.

EXAMPLE: In a set of planetary gears the ring gear has 70 teeth, the sun gear 14 teeth and each of the three planet gears 28 teeth. Even spacing would place the planet gears every 120°, but in this case they must be placed slightly to one side of the 120° point to mesh. Since N_s divides evenly into N_r, there is a tooth on the ring gear opposite every tooth on the sun gear. Therefore, it is possible to fit a planet gear opposite any tooth on the sun gear. Tooth 6, five circular pitches from tooth 1, is the choice because it is closest to being one-third of the way around. It is opposite tooth 26 on the ring gear, because $N_r/N_s = 70/14 = 25/5$.

Case II—The number of teeth on the sun gear does not divide evenly into the number of teeth on the ring gear. This set may or may not mesh; the following example shows how to tell.

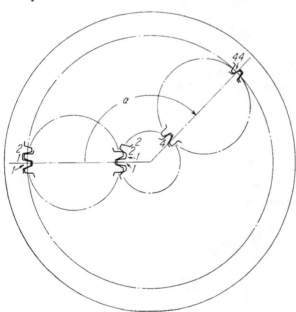

EXAMPLE: In a set of planetary gears with three planets, the ring gear has 134 teeth, the sun gear 14 and the planet gears 60 each. $N_s/3 = 14/3 = 4.67$, so the whole number $x = 4$. $N_r/3 = 134/3 = 44.67$, so the whole number $y = 44$.

Plug these numbers into the locating equation

$$(x+z) N_r/N_s = y+(1-z) = (4+z) 134/14 = 44+(1-z)$$
$$10.57 z = 6.72$$
$$z = 0.636$$

Location of the planet gear as a fractional part of the circular distance around the set is $(x + z)/N_s = 4.636/14 = 0.3311$, and $y + (1 - z)/N_r = 44.364/$

134 = 0.3311. The answers agree to four places, so the gears will mesh. If the answers don't agree to four places, there will be interference.

Angle $a = 0.3311 \times 360 = 119.2°$

4—Can three planets fit around the sun gear?

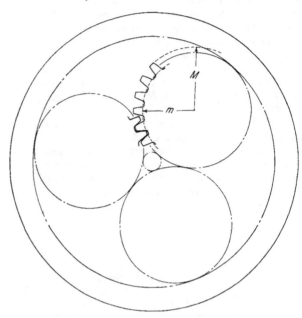

They will if the major diameters adhere to the limitation $M_p + m_s/2 < m_r$ by a safety clearance of $\frac{1}{32}$ in. more than maximum tolerances.

5—Will irregularly spaced planets hit each other?

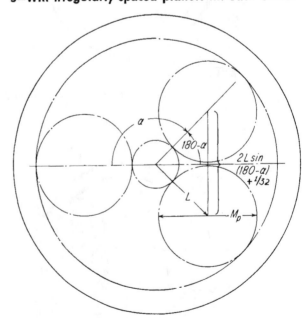

Two adjacent planets will not hit each other if $2L \sin (180 - a) > M_p + \frac{1}{32}$ in. safety clearance. Sun-to-planet center-to-center distance $L = (PD_s + PD_p)/2$.

■

Epicyclic Gear Trains

M. F. Spotts

EPICYCLIC GEAR TRAIN shown in Fig. 1 has Arm A integral with the right hand shaft. Gears C and D are keyed to a short length of shaft which is mounted in a bearing in Arm A. Gear C meshes with the fixed internal gear B. Gear D meshes with internal gear E which is keyed to the left hand shaft.

To find the ratio of the speed of shaft E to the speed of shaft A proceed as follows. Let N_b be the number of teeth in gear B, N_c the number in gear C, and so on. Let arm A, which was originally in a vertical position, be given an angular displacement θ. In so doing gear C will traverse through arc ab on gear B. Arc bc of gear C must be equal to arc ab of gear B. Since angles are inversely proportional to radii, or to the number of teeth, gears C and D will have turned through angle $\theta N_b/N_c$.

While the foregoing was taking place gears D and E were rotating on each other through the equal arcs ed and ef. Gear E will have been turned in the reverse direction through angle

$$\theta N_b/N_c \times N_d/N_e$$

The net effect of these two operations is to move the point of gear E, which was originally vertical at g, over to location f. Gear E has thus been rotated through angle

$$(1 - N_bN_d/N_cN_e)\,\theta$$

This latter value when divided by θ, the angular movement of shaft A, gives the ratio of the rotations of shafts E and A respectively.

This method of analysis gives a graphical representation of the movement of all the parts. It may be easily applied to all types of epicyclic systems including those containing bevel gears. Additional examples are shown in Figs. 2 to 6 inclusive. Either of shafts A or E may be used as the driver.

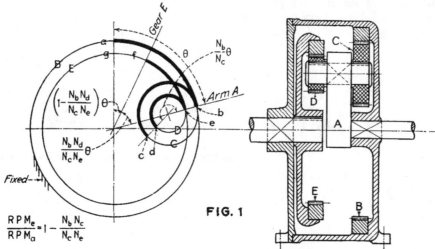

$$\frac{RPM_e}{RPM_a} = 1 - \frac{N_b\,N_c}{N_c\,N_e}$$

FIG. 1

Drive and driven shafts rotate in **same** direction

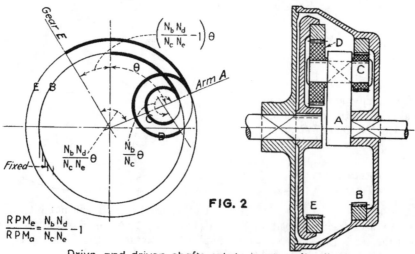

$$\frac{RPM_e}{RPM_a} = \frac{N_b\,N_d}{N_c\,N_e} - 1$$

FIG. 2

Drive and driven shafts rotate in **opposite** directions

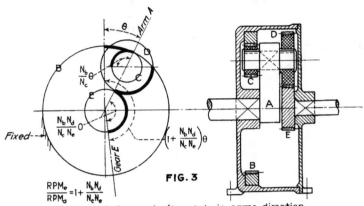

$$\frac{RPM_e}{RPM_a} = 1 + \frac{N_b N_d}{N_c N_e}$$

FIG. 3

Drive and driven shafts rotate in **same** direction
Equation is valid for $N_c = N_d$, and for $N_c > N_d$

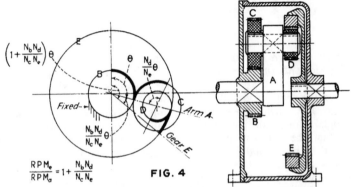

$$\frac{RPM_e}{RPM_a} = 1 + \frac{N_b N_d}{N_c N_e}$$

FIG. 4

Drive and driven shafts rotate in **same** direction
Equation is valid for $N_c = N_d$, and for $N_c > N_d$

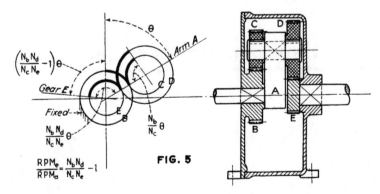

$$\frac{RPM_e}{RPM_a} = \frac{N_b N_d}{N_c N_e} - 1$$

FIG. 5

Drive and driven shafts rotate in **opposite** directions

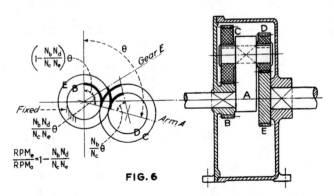

$$\frac{RPM_e}{RPM_a} = 1 - \frac{N_b N_d}{N_c N_e}$$

FIG. 6

Drive and driven shafts rotate in **same** direction

Cycloid Gear Mechanisms

Cycloidal motion is becoming popular for mechanisms in feeders and automatic machines. Here are arrangements, formulas, and layout methods.

Preben W. Jensen

THE appeal of cycloidal mechanisms is that they can easily be tailored to provide one of these three common motion requirements:

- **Intermittent motion**—with either short or long dwells
- **Rotary motion with progressive oscillation**—where the output undergoes a cycloidal motion during which the forward motion is greater than the return motion
- **Rotary-to-linear motion with a dwell period**

All the cycloidal mechanisms covered in this article are geared; this results in compact positive devices capable of operating at relatively high speeds with little backlash or "slop." The mechanisms can also be classified into three groups:

Hypocycloid—where the points tracing the cycloidal curves are located on an external gear rolling inside an internal ring gear. This ring gear is usually stationary and fixed to the frame.

Epicycloid—where the tracing points are on an external gear which rolls in another external (stationary) gear

Pericycloid—where the tracing points are located on an internal gear which rolls on a stationary external gear.

Hypocycloid Mechanisms

1. Basic hypocycloid curves

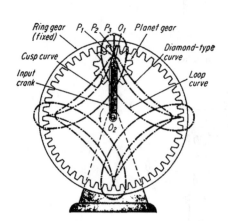

2. Double-dwell mechanism

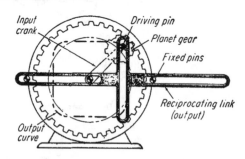

Coupling the output pin to a slotted member produces a prolonged dwell in each of the extreme positions. This is another application of the diamond-type hypocycloidal curve.

Input drives a planet in mesh with a stationary ring gear. Point P_1 on the planet gear describes a diamond-shape curve, point P_2 on the pitch line of the planet describes the familiar cusp curve, and point P_3, which is on an extension rod fixed to the planet gear, describes a loop-type curve. In one application, an end miller located at P_1 was employed in production for machining a diamond-shape profile.

3. Long-dwell geneva drive

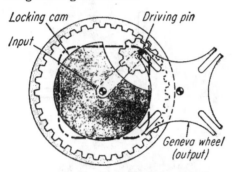

As with standard four-station genevas, each rotation of the input indexes the slotted geneva 90 deg. By employing a pin fastened to the planet gear to obtain a rectangular-shape cycloidal curve, a smoother indexing motion is obtained because the driving pin moves on a noncircular path.

4. Internal-geneva drive

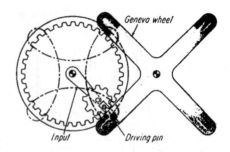

Loop-type curve permits driving pin to enter slot in a direction that is radially outward from the center, and then loop over to rapidly index the cross member. As with the previous geneva, the output rotates 90 deg, then goes into a long dwell period during each 270-deg rotation of the input.

5. Cycloidal parallelogram

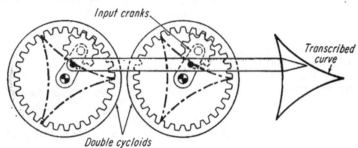

Two identical hypocycloid mechanisms guide the point of the bar along the triangularly shaped path. They are useful also in cases where there is limited space in the area where the curve must be described. Such double-cycloid mechanisms can be designed to produce other types of curves.

7. Cycloidal rocker

6. Short-dwell rotary

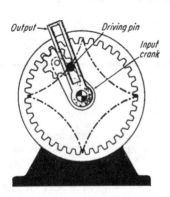

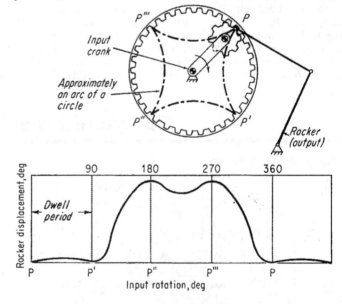

Here the pitch circle of the planet gear is exactly one-quarter that of the ring gear. A pin on the planet will cause the slotted output member to have four instantaneous dwells for each revolution of the input shaft.

The curvature of the cusp is approximately that of an arc of a circle. Hence the rocker comes to a long dwell at the right extreme position while point *P* moves to *P′*. There is then a quick return from *P′* to *P″*, with a momentary dwell at the end of this phase. The rocker then undergoes a slight oscillation from point *P″* to *P‴*, as shown in the displacement diagram.

8. Cycloidal reciprocator

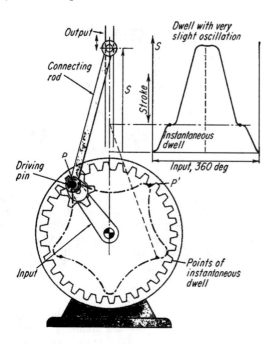

Portion of curve, *P-P'*, produces the long dwell (as in previous mechanism), but the five-lobe cycloidal curve avoids a marked oscillation at the end of the stroke. There are also two points of instantaneous dwell where the curve is perpendicular to the connecting rod.

By making the pitch diameter of the planet equal to half that of the ring gear, every point on the planet gear (such as points P_2 and P_3) will describe elliptical curves which get flatter as the points are selected closer to the pitch circle. Point P_1, at the center of the planet, describes a circle; point P_4 at the pitch circle describes a straight line. When a cutting tool is placed at P_3, it will cut almost-flat sections from round stock, as when machining a bolt. The other two sides of the bolt can be cut by rotating the bolt, or the cutting device, 90 deg. (Reference: H. Zeile, *Unrund- und Mehrkantdrehen*, VDI-Berichte, Nr. 77,1965.)

9. Adjustable harmonic drive

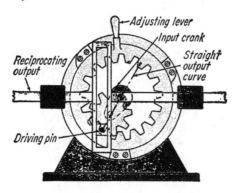

By making the planet-gear half that of the internal gear, a straight-line output curve is produced by the driving pin which is fastened to the planet gear. The pin engages the slotted member to cause the output to reciprocate back and forth with harmonic (sinusoidal) motion. The position of the fixed ring gear can be changed by adjusting the lever, which in turn rotates the straight-line output-curve. When the curve is horizontal, the stroke is at a maximum; when the curve is vertical, the stroke is zero.

10. Elliptical-motion drive

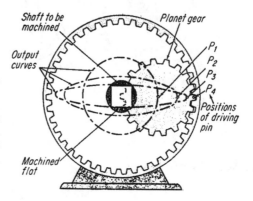

Epicycloid Mechanisms

11. Epicycloid reciprocator

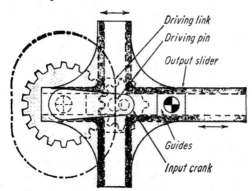

Here the sun gear is fixed and the planet gear driven around it by means of the input link. There is no internal ring gear as with the hypocycloid mechanisms. Driving pin *P* on the planet describes the curve shown which contains two almost-flat portions. By having the pin ride in the slotted yoke, a short dwell is produced at both the extreme positions of the output member. The horizontal slots in the yoke ride the end-guides, as shown.

12. Progressive oscillating drive

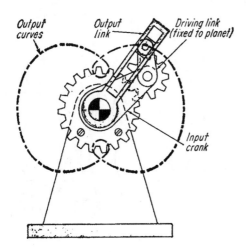

By fixing a crank to the planet gear, a point P can be made to describe the double loop curve illustrated. The slotted output crank oscillates briefly at the vertical portions.

13. Parallel-guidance mechanisms

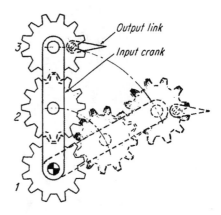

The input crank contains two planet gears. The center sun-gear is fixed as in the previous epicycloid mechanisms. By making the three gears equal in diameter and having gear *2* serve as an idler, any member fixed to gear *3* will remain parallel to its previous positions throughout the rotation of the input ring crank.

Motion Equations

14. Equations for epicycloid drives

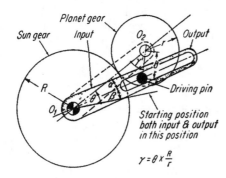

$$\gamma = \theta \times \frac{R}{r}$$

The equations for angular displacement, velocity and acceleration for basic epicyclic drive are given below. (Reference: Schmidt, E. H., "Cycloidal Cranks," *Transactions of the 5th Conference on Mechanisms*, 1958, pp 164-180):

Angular displacement

$$\tan \beta = \frac{(R + r) \sin \theta - b \sin (\theta + \gamma)}{(R + r) \cos \theta - b \cos (\theta + \gamma)} \qquad (1)$$

Angular velocity

$$V = \omega \frac{1 + \dfrac{b^2}{r(R + r)} - \left(\dfrac{2r + R}{r}\right)\left(\dfrac{b}{R + r}\right)\left(\cos \dfrac{R}{r} \theta\right)}{1 + \left(\dfrac{b}{R + r}\right)^2 - \left(\dfrac{2b}{R + r}\right)\left(\cos \dfrac{R}{r} \theta\right)} \qquad (2)$$

Angular acceleration

$$A = \omega^2 \frac{\left(1 - \dfrac{b^2}{(R + r)^2}\right)\left(\dfrac{R^2}{r^2}\right)\left(\dfrac{b}{R + r}\right)\left(\sin \dfrac{R}{r} \theta\right)}{\left[1 + \dfrac{b^2}{(R + r)^2} - \left(\dfrac{2b}{R + r}\right)\left(\cos \dfrac{R}{r} \theta\right)\right]^2} \qquad (3)$$

Symbols

A = angular acceleration of output, deg/sec^2

b = radius of driving pin from center of planet gear

r = pitch radius of planet gear

R = pitch radius of fixed sun gear

V = angular velocity of output, deg/sec

β = angular displacement of output, deg

γ = $\theta R/r$

θ = input displacement, deg

ω = angular velocity of input, deg/sec

15. Equations for hypocycloid drives

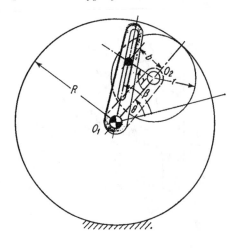

$$\tan \beta = \frac{\sin \theta - \left(\dfrac{b}{R - r} \right)\left(\sin \dfrac{R - r}{r}\,\theta \right)}{\cos \theta + \left(\dfrac{b}{R - r} \right)\left(\cos \dfrac{R - r}{r}\,\theta \right)} \tag{4}$$

$$V = \omega\, \frac{1 - \left(\dfrac{R-r}{r} \right)\left(\dfrac{b^2}{(R+r)^2} \right) + \left(\dfrac{2r-R}{r} \right)\left(\cos \dfrac{R}{r}\,\theta \right)}{1 + \dfrac{b^2}{(R+r)^2} + \left(\dfrac{2b}{R+r} \right)\left(\cos \dfrac{R}{r}\,\theta \right)} \tag{5}$$

$$A = \omega^2\, \frac{\left(1 - \dfrac{b^2}{(R+r)^2} \right)\left(\dfrac{b}{R+r} \right)\left(\dfrac{R^2}{r^2} \right)\left(\sin \dfrac{R}{r}\,\theta \right)}{\left[1 + \dfrac{b^2}{(R+r)^2} + \left(\dfrac{2b}{R+r} \right)\left(\cos \dfrac{R}{r}\,\theta \right) \right]^2} \tag{6}$$

Describing Approximate Straight Lines

16. Gear rolling on a gear—flatten curves

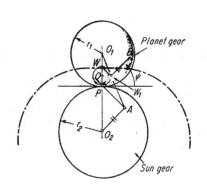

It is frequently desirable to find points on the planet gear that will describe approximately straight lines for portions of the output curve. Such points will yield dwell mechanisms, as shown in Fig 2 and 11. Construction is as follows (shown at left):

1. Draw an arbitrary line PB.
2. Draw its parallel O_2A.
3. Draw its perpendicular PA at P. Locate point A.
4. Draw O_1A. Locate W_1.
5. Draw perpendicular to PW_1 at W_1 to locate W.
6. Draw a circle with PW as the diameter.

All points on this circle describe curves with portions that are approximately straight. This circle is also called the inflection circle because all points describe curves which have a point of inflection at the position illustrated. (Shown is the curve passing through point W.)

17. Gear rolling on a rack—vee curves

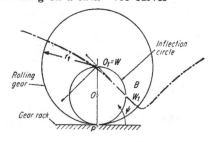

This is a special case. Draw a circle with a diameter half that of the gear (diameter O_1P). This is the inflection circle. Any point, such as point W_1, will describe a curve that is almost straight in the vicinity selected. Tangents to the curves will always pass through the center of the gear, O_1 (as shown).

18. Gear rolling inside a gear—zig-zag

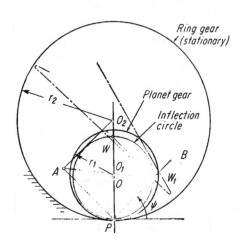

To find the inflection circle for a gear rolling inside a gear:

1. Draw arbitrary line PB from the contact point P.
2. Draw its parallel O_2A, and its perpendicular, PA. Locate A.
3. Draw line AO_1 through the center of the rolling gear. Locate W_1.
4. Draw a perpendicular through W_1. Obtain W. Line WP is the diameter of the inflection circle. Point W_1, which is an arbitrary point on the circle, will trace a curve of repeated almost-straight lines, as shown.

19. Center of curvature—gear rolling on gear

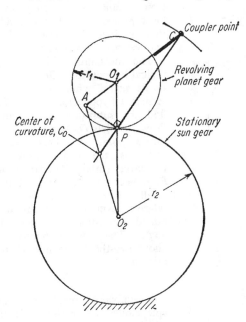

20. Center of curvature—gear rolling on a rack

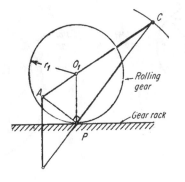

Construction is similar to that of the previous case.
1. Draw an extension of line *CP*.
2. Draw a perpendicular at *P* to locate *A*.
3. Draw a perpendicular from *A* to the straight suface to locate *C₀*.

By locating the centers of curvature at various points, one can then determine the proper length of the rocking or reciprocating arm to provide long dwells (as required for the mechanisms in Fig 7 and 8), or proper entry conditions (as for the drive pin in the mechanism in Fig 3).

In the case of a gear with an extended point, point *C*, rolling on another gear, the graphical method for locating the center of curvature is given by these steps:
1. Draw a line through points *C* and *P*.
2. Draw a line through points *C* and *O₁*.
3. Draw a perpendicular to *CP* at *P*. This locates point *A*.
4. Draw line *AO₂*, to locate *C₀*, the center of curvature.

21. Center of curvature—gear rolling inside a gear

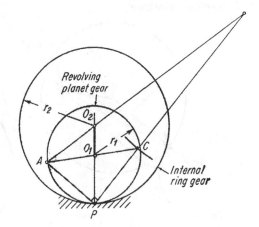

1. Draw extensions of *CP* and *CO₁*.
2. Draw a perpendicular of *PC* at *P* to locate *A*.
3. Draw *AO₂* to locate *C₀*.

22. Analytical solutions

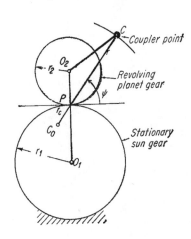

The centure of curvature of a gear rolling on a external gear can be computed directly from the Euler-Savary equation:

$$\left(\frac{1}{r} - \frac{1}{r_c}\right) \sin \psi = \text{constant} \quad (7)$$

where angle ψ and r locate the position of *C*.

By applying this equation twice, specifically to point O_1 and O_2 which have their own centers of rotation, the following equation is obtained:

$$\left(\frac{1}{r_2} + \frac{1}{r_1}\right) \sin 90° =$$

$$\left(\frac{1}{r} + \frac{1}{r_c}\right) \sin \psi$$

or

$$\frac{1}{r_2} + \frac{1}{r_1} = \left(\frac{1}{r} + \frac{1}{r_c}\right) \sin \psi$$

This is the final design equation. All factors except r_c are known; hence solving for r_c leads to the location of C_0.

For a gear rolling inside an internal gear, the Euler-Savary equation is

$$\left(\frac{1}{r} + \frac{1}{r_c}\right) \sin \psi = \text{constant}$$

which leads to

$$\frac{1}{r_2} - \frac{1}{r_1} = \left(\frac{1}{r} - \frac{1}{r_c}\right) \sin \psi$$

23. Hypocycloid substitute

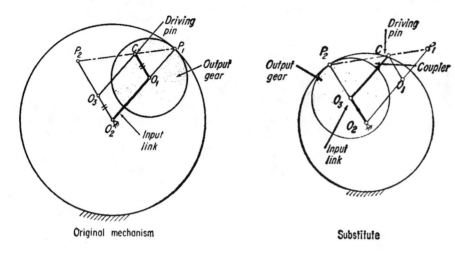

Original mechanism

Substitute

It is not always realized that cycloid mechanisms can frequently be replaced by other cycloids that produce the same motion and yet are more compact.

The mechanism (right) is a typical hypocycloid. Gear *1* rolls inside gear *2* while point *C* describes a hypocycloid curve. To find the substitute mechanism, draw parallels O_3O_2 and O_3C to locate point P_2. Then select O_2P_2 as the new radius of the large (internal) gear. Line P_2O_3 becomes the radius of the small gear. Point *C* has the same relative position and can be obtained by completing the triangles. The new mechanism is about two-thirds the size of the original.

24. Epicycloid substitute

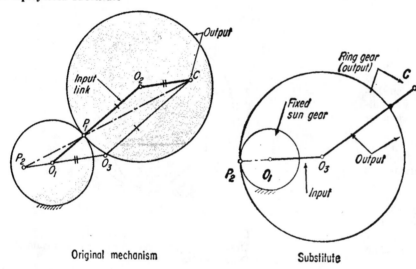

Original mechanism

Substitute

The equivalent mechanisms of epicycloids are pericycloids in which the planetary gear is stationary and the output is taken from the ring gear. Such arrangements usually lead to a more-compact design.

In the above mechanism, point *C* traces an epicycloidal curve. Draw the proper parallels to find P_2, then use P_2O_3 to construct the compact substitute mechanism shown at right of original.

25. Multigear substitute

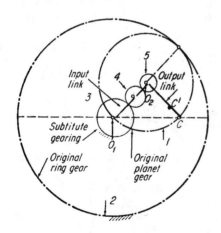

This is another way of producing a compact substitute for a hypocycloid mechanism. The original mechanism is shown in dashed lines—gear *1* rolls inside gear *2* and point *C* describes the curve. The three external gears (gears *3*, *4*, and *5*) replace gears *1* and *2* with a remarkable savings in space. The only criterion is that gear *5* must be one-half the size of gear *3*; gear *4* is only an idler. The new mechanism thus has been reduced to approximately one-half that of the original in size.

Cardan-Gear Mechanisms

These gearing arrangements convert rotation into straight-line motion, without need for slideways.

Sigmund Rappaport

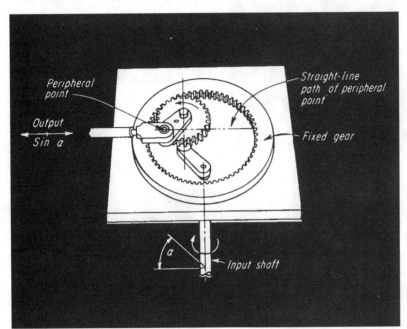

1

Cardan gearing . . .

works on the principle that any point on the periphery of a circle rolling on the inside of another circle describes, in general, a hypocyloid. This curve degenerates into a true straight line (diameter of the larger circle) if diameters of both circles are in the ratio of 1:2. Rotation of input shaft causes small gear to roll around the inside of the fixed gear. A pin located on pitch circle of the small gear describes a straight line. Its linear displacement is proportional to the theoretically true sine or cosine of the angle through which the input shaft is rotated. Among other applications, Cardan gearing is used in computers, as a component solver (angle resolver).

2

Cardan gearing and Scotch yoke . . .

in combination provide an adjustable stroke. Angular position of outer gear is adjustable. Adjusted stroke equals the projection of the large dia, along which the drive pin travels, upon the Scotch-yoke centerline. Yoke motion is simple harmonic.

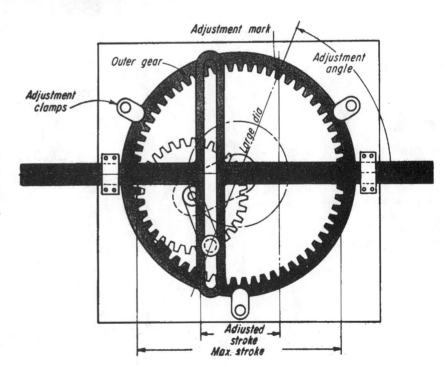

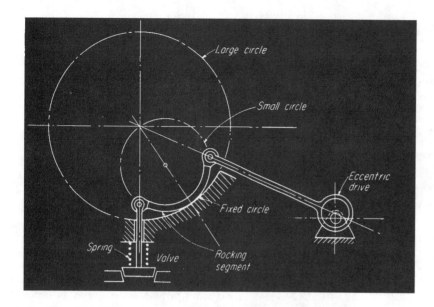

Valve drive . . .
exemplifies how Cardan principle may
be applied. A segment of the smaller
circle rocks to and fro on a circular
segment whose radius is twice as large.
Input and output rods are each attached to points on the small circle.
Both these points describe straight lines.
Guide of the valve rod prevents the
rocking member from slipping.

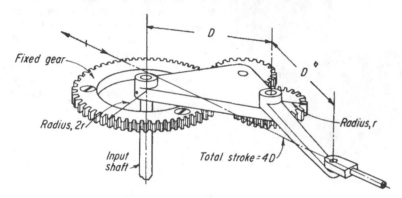

Simplified Cardan principle . . .
does away with need for the relatively expensive internal gear. Here,
only spur gears may be used and the
basic requirements should be met, i.e.
the 1:2 ratio and the proper direction
of rotation. Latter requirement is
easily achieved by introducing an
idler gear, whose size is immaterial. In
addition to cheapness, this drive delivers a far larger stroke for the com-
parative size of its gears.

Rearrangement of gearing . . .
in (4) results in another useful motion. If
the fixed sun-gear and planet pinion are
in the ratio of 1:1, then an arm fixed to
the planet shaft will stay parallel to itself
during rotation, while any point on it
describes a circle of radius R. An example of application: in conjugate pairs
for punching holes on moving webs of
paper.

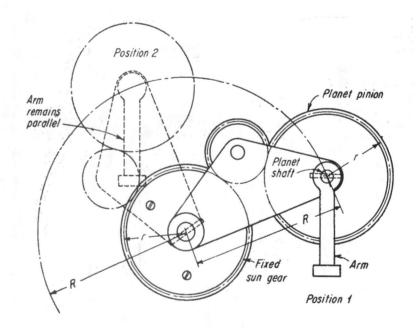

Typical Methods of Providing Lubrication for Gear Systems

Below are shown various lubricating systems that can serve as guides when designing for successful, efficient gear systems.

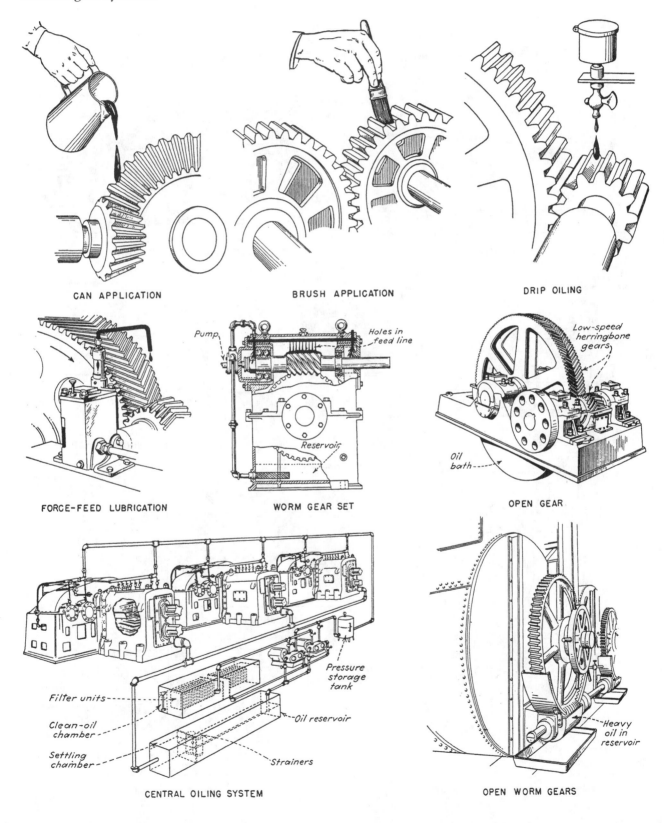

CAN APPLICATION

BRUSH APPLICATION

DRIP OILING

FORCE-FEED LUBRICATION

WORM GEAR SET

OPEN GEAR

CENTRAL OILING SYSTEM

OPEN WORM GEARS

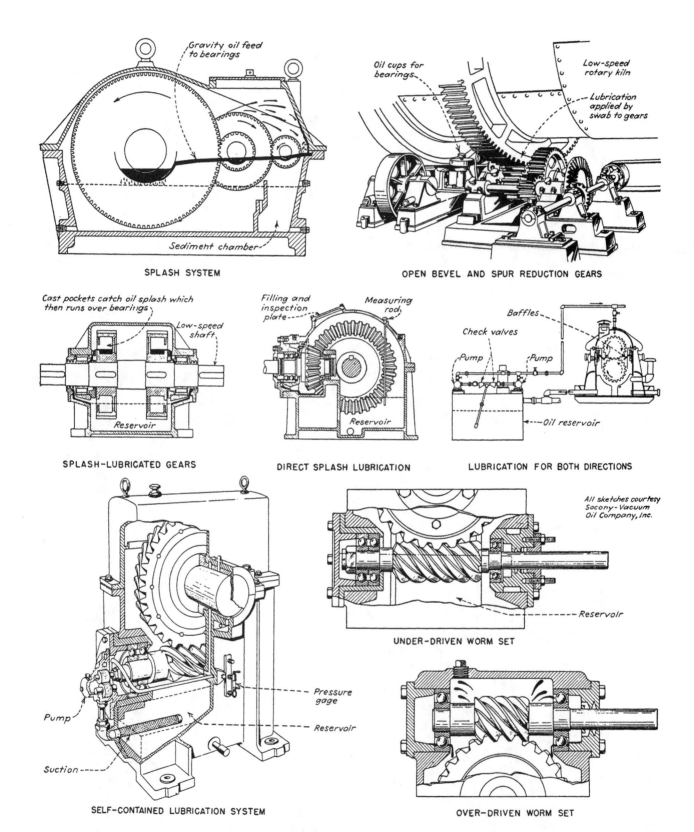

SPLASH SYSTEM

OPEN BEVEL AND SPUR REDUCTION GEARS

SPLASH-LUBRICATED GEARS

DIRECT SPLASH LUBRICATION

LUBRICATION FOR BOTH DIRECTIONS

UNDER-DRIVEN WORM SET

SELF-CONTAINED LUBRICATION SYSTEM

OVER-DRIVEN WORM SET

All sketches courtesy
Socony-Vacuum
Oil Company, Inc.

SECTION 7

CLUTCHES

Basic Types of Mechanical Clutches

Sketches include both friction and positive types. Figs. 1-7 are classified as externally controlled; Figs. 8-12 are internally controlled. The latter are further divided into overload relief, over-riding, and centrifugal types.

Marvin Taylor

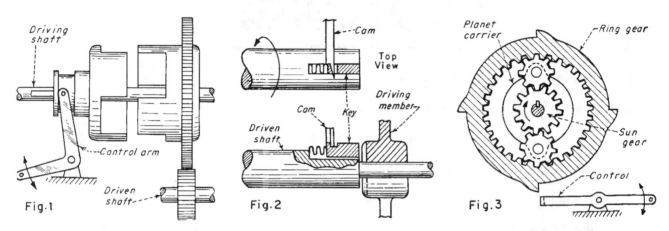

Fig. 1   Fig. 2   Fig. 3

1. JAW CLUTCH. Left sliding half is feathered to the driving shaft while right half rotates freely. Control arm activates the sliding half to engage or disengage the drive. This clutch, though strong and simple, suffers from disadvantages of high shock during engagement, high inertia of the sliding half, and considerable axial motion required for engagement.

2. SLIDING KEY CLUTCH. Driven shaft with a keyway carries freely-rotating member which has radial slots along its hub; sliding key is spring loaded but is normally restrained from engaging slots by the control cam. To engage the clutch, control cam is raised and key

enters one of the slots. To disengage, cam is lowered into the path of the key; rotation of driven shaft forces key out of slot in driving member. Step on control cam limits axial movement of the key.

3. PLANETARY TRANSMISSION CLUTCH. In disengaged position shown, driving sun gear will merely cause the free-wheeling ring gear to idle counter-clockwise, while the driven member, the planet carrier, remains motionless. If motion of the ring gear is blocked by the control arm, a positive clockwise drive is established to the driven planet carrier.

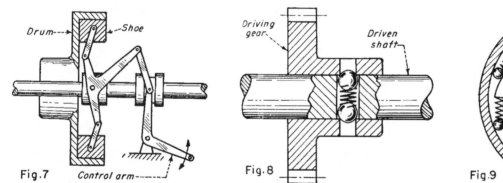

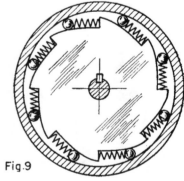

Fig. 7   Fig. 8   Fig. 9

7. EXPANDING SHOE CLUTCH. In sketch above, engagement is obtained by motion of control arm which operates linkages to force friction shoes radially outward into contact with inside surface of drum.

8. SPRING AND BALL RADIAL DETENT CLUTCH. This design will positively hold the driving gear and driven shaft in a given timing relationship until the torque becomes excessive. At this point the balls will be forced inward against their spring pressure and out of engagement with the holes in the hub, thus permitting the driving gear to continue rotating while the driven shaft is stationary.

9. CAM AND ROLLER CLUTCH. This over-running clutch is suited for higher speed free-wheeling than the pawl and ratchet types. The inner driving member has camming surfaces at its outer rim and carries light springs that force rollers to wedge between these surfaces and the inner cylindrical face of the driven member. During driving, self-energizing friction rather than the springs forces the roller to tightly wedge between the members and give essentially positive drive in a clockwise direction. The springs insure fast clutching action. If the driven member should attempt to run ahead of the driver, friction will force the rollers out of a tight wedging position and break the connection.

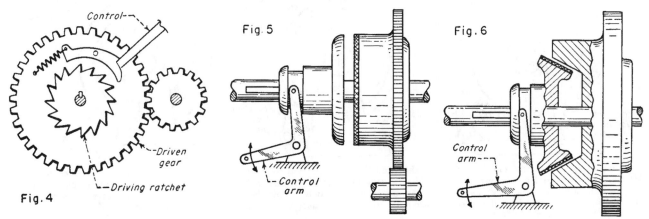

Fig. 5

Fig. 6

Fig. 4

Control

Driven gear

Driving ratchet

Control arm

Control arm

4. PAWL AND RATCHET CLUTCH. (External Control). Ratchet is keyed to the driving shaft; pawl is carried by driven gear which rotates freely on the driving shaft. Raising the control member permits the spring to pull the pawl into engagement with the ratchet and drive the gear. Engagement continues until control member is lowered into the path of a camming surface on the pawl. The motion of the driven gear will then force the pawl out of engagement and bring the driven assembly to a solid stop against the control member. This clutch can be converted into an internally controlled type of unit by removing the external control arm and replacing it with a slideable member on the driving shaft.

5. PLATE CLUTCH. Available in many variations, with single and multiple plates, this unit transmits power through friction force developed between the faces of the left sliding half which is fitted with a feather key and the right half which is free to rotate on the shaft. Torque capacity depends upon the axial force exerted by the control member when it activates the sliding half.

6. CONE CLUTCH. This type also requires axial movement for engagement, but the axial force required is less than that required with plate clutches. Friction material is usually applied to only one of the mating surfaces. Free member is mounted to resist axial thrust.

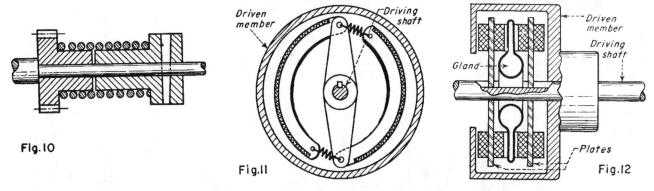

Fig. 10

Driven member

Driving shaft

Gland

Driven member

Driving shaft

Plates

Fig. 11

Fig. 12

10. WRAPPED SPRING CLUTCH. Makes a simple and inexpensive uni-directional clutch consisting of two rotating members connected by a coil spring which fits snugly over both hubs. In the driving direction the spring tightens about the hubs producing a self energizing friction grip; in the opposite direction it unwinds and will slip.

11. EXPANDING SHOE CENTRIFUGAL CLUTCH. Similar in action to the unit shown in Fig. 7 with the exception that no external control is used. Two friction shoes, attached to the driving member, are held inward by springs until they reach the "clutch-in" speed, at which centrifugal force energizes the shoes outward into contact with the drum. As the driver rotates faster the pressure between the shoes and the drum increases thereby providing greater torque capacity.

12. MERCURY GLAND CLUTCH. Contains two friction plates and a mercury filled rubber gland, all keyed to the driving shaft. At rest, mercury fills a ring shaped cavity near the shaft; when revolved at sufficient speed, the mercury is forced outward by centrifugal force spreading the rubber gland axially and forcing the friction plates into driving contact with the faces of the driven housing. Axial thrust on driven member is negligible.

Construction Details of Overriding Clutches

A. DeFeo

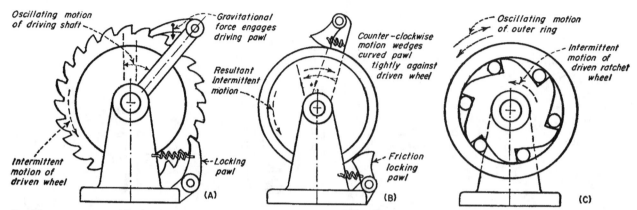

1 Elementary over-riding clutches: (A) Ratchet and Pawl mechanism is used to convert reciprocating or oscillating movement to intermittent rotary motion. This motion is positive but limited to a multiple of the tooth pitch. (B) Friction-type is quieter but requires a spring device to keep eccentric pawl in constant engagement. (C) Balls or rollers replace the pawls in this device. Motion of the outer race wedges rollers against the inclined surfaces of the ratchet wheel.

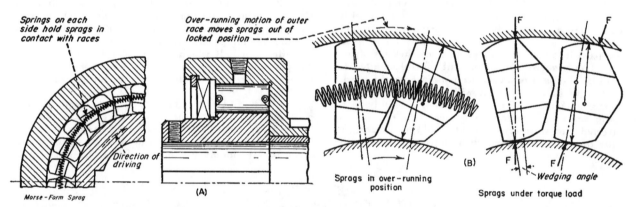

4 With cylindrical inner and outer races, sprags are used to transmit torque. Energizing springs serves as a cage to hold the sprags. (A) Compared to rollers, shape of sprag permits a greater number within a limited space; thus higher torque loads are possible. Not requiring special cam surfaces, this type can be installed inside gear or wheel hubs. (B) Rolling action wedges sprags tightly between driving and driven members. Relatively large wedging angle insures positive engagement.

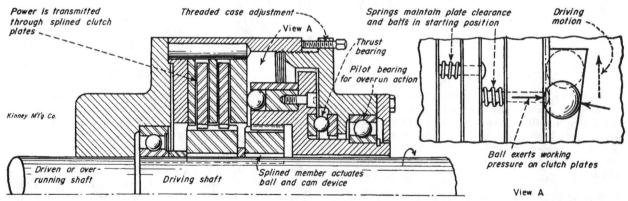

6 Multi-disk clutch is driven by means of several sintered-bronze friction surfaces. Pressure is exerted by a cam actuating device which forces a series of balls against a disk plate. Since a small part of the transmitted torque is carried by the actuating member, capacity is not limited by the localized deformation of the contacting balls. Slip of the friction surfaces determine the capacity and prevent rapid, shock loads. Slight pressure of disk springs insure uniform engagement.

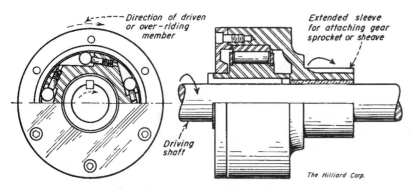

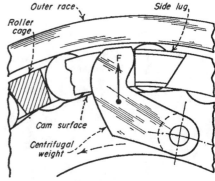

2 Commercial over-riding clutch has springs which hold rollers in continuous contact between cam surfaces and outer race; thus there is no backlash or lost motion. This simple design is positive and quiet. For operation in the opposite direction, the roller mechanism can easily be reversed in the housing.

3 Centrifugal force can be used to hold rollers in contact with cam and outer race. Force is exerted on lugs of the cage which controls the position of the rollers.

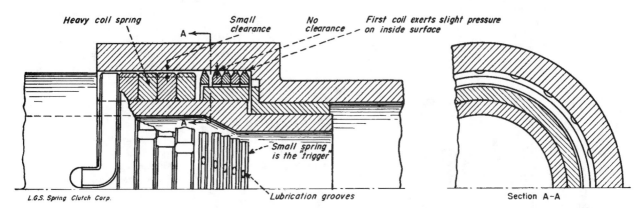

5 Engaging device consists of a helical spring which is made up of two sections: a light trigger spring and a heavy coil spring. It is attached to and driven by the inner shaft. Relative motion of outer member rubbing on trigger causes this spring to wind-up. This action expands the spring diameter which takes up the small clearance and exerts pressure against the inside surface until the entire spring is tightly engaged. Helix angle of spring can be changed to reverse the over-riding direction.

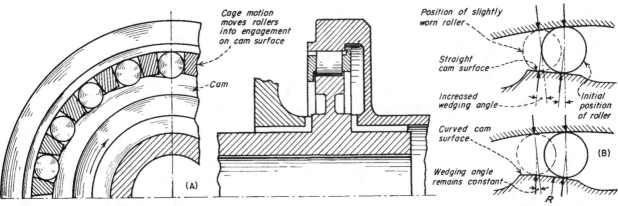

7 Free-wheeling clutch widely used in power transmission has a series of straight-sided cam surfaces. An engaging angle of about 3 deg is used; smaller angles tend to become locked and are difficult to disengage while larger ones are not as effective. (A) Inertia of floating cage wedges rollers between cam and outer race. (B) Continual operation causes wear of surfaces; 0.001 in. wear alters angle to 8.5 deg. on straight-sided cams. Curved cam surfaces maintain constant angle.

10 Ways to Apply Overrunning Clutches

These clutches allow freewheeling, indexing, and backstopping applicable to many design problems. Here are some clutch setups.

W. Edgar Mulholland & John L. King, Jr.

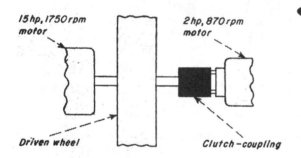

Precision Sprags . . .

act as wedges and are made of hardened alloy steel. In the Formsprag clutch, torque is transmitted from one race to another by wedging action of sprags between the races in one direction; in other direction the clutch freewheels.

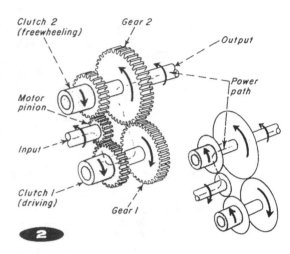

2-Speed Drive—I . . .

requires input rotation to be reversible. Counterclockwise input as shown in the diagram drives gear 1 through clutch 1; output is counterclockwise; clutch 2 over-runs. Clockwise input (schematic) drives gear 2 through clutch 2; output is still counterclockwise; clutch 1 over-runs.

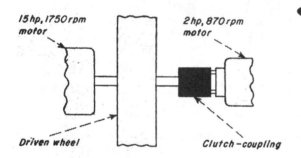

2-Speed Drive—II . . .

for grinding wheel can be simple, in-line design if over-running clutch couples two motors. Outer race of clutch is driven by gearmotor; inner race is keyed to grinding-wheel shaft. When gearmotor drives, clutch is engaged; when larger motor drives, inner race over-runs.

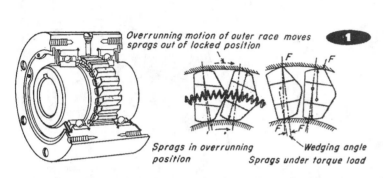

Indexing Table . . .

is keyed to clutch shaft. Table is rotated by forward stroke of rack, power being transmitted through clutch by its outer-ring gear only during this forward stroke. Indexing is slightly short of position required. Exact position is then located by spring-loaded pin, which draws table forward to final positioning. Pin now holds table until next power stroke of hydraulic cylinder

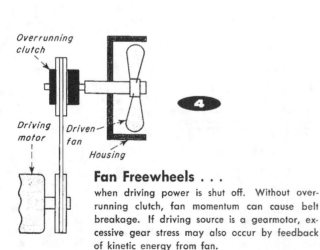

Fan Freewheels . . .

when driving power is shut off. Without overrunning clutch, fan momentum can cause belt breakage. If driving source is a gearmotor, excessive gear stress may also occur by feedback of kinetic energy from fan.

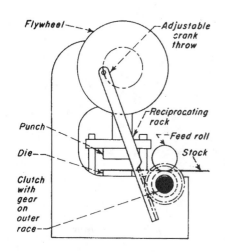

6

Punch Press Feed..

is so arranged that strip is stationary on down-stroke of punch (clutch freewheels); feed occurs during upstroke when clutch transmits torque. Feed mechanism can easily be adjusted to vary feed amount.

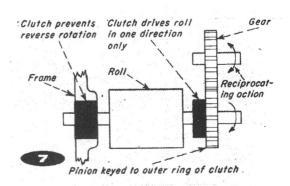

7

Indexing and Backstopping . . .

is done with two clutches so arranged that one drives while the other freewheels. Application here is for capsuling machine; gelatin is fed by the roll and stopped intermittently so blade can precisely shear material to form capsules.

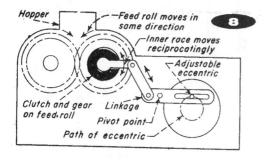

8

Intermittent Motion . . .

of candy machine is adjustable; function of clutch is to ratchet the feed rolls around. This keeps the material in the hopper agitated.

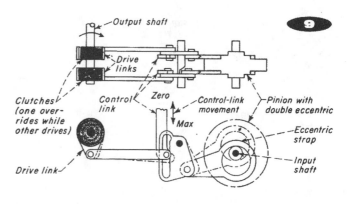

9

Double-impulse Drive . . .

employs double eccentrics and drive clutches. Each clutch is indexed 180° out of phase with the other. One revolution of eccentric produces two drive strokes. Stroke length, and thus the output rotation, can be adjusted from zero to max by the control link.

Anti-backlash Device . . .

uses over-running clutches to insure that no backlash is left in the unit. Gear A drives B and shaft II with the gear mesh and backlash as shown in (A). The over-running clutch in gear C permits gear D (driven by shaft II) to drive gear C and results in the mesh and backlash shown in (B). The over-running clutches never actually over-run. They provide flexible connections (something like split and sprung gears) between shaft I and gears A, C to allow absorption of all backlash.

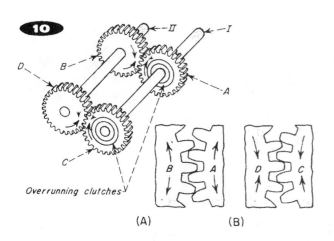

10

(A)   (B)

Low-Cost Designs for Overrunning Clutches

All are simple devices that can be constructed inexpensively in the laboratory workshop.

James F. Machen

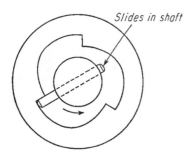

Lawnmower type

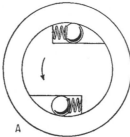

Wedging balls or rollers: internal (A); external (B)

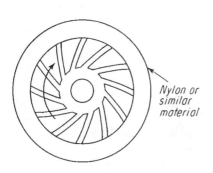

Molded sprags
(for light duty)

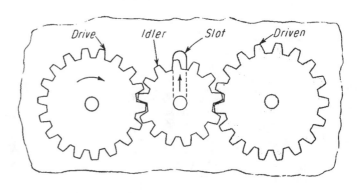

Disengaging idler rises in slot when drive direction
is reversed

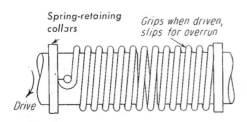

Slip-spring coupling

Internal ratchet and
spring-loaded pawls

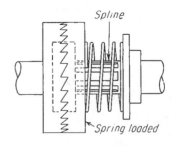

One-way dog clutch

(FOR ANGULAR AND AXIAL ADJUSTMENT)

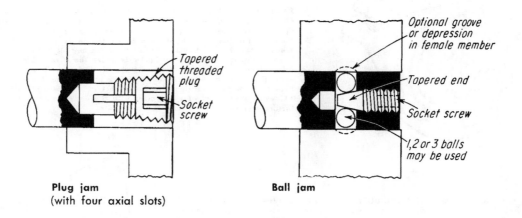

Plug jam
(with four axial slots)

Tapered
threaded
plug

Socket
screw

Ball jam

Optional groove
or depression
in female member

Tapered end

Socket screw

1,2 or 3 balls
may be used

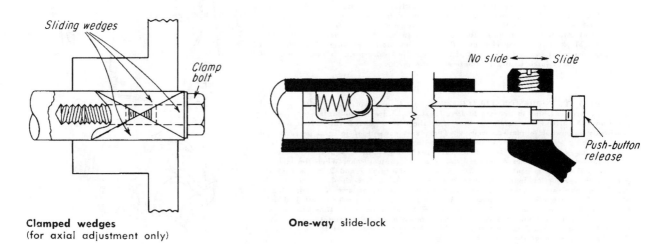

Sliding wedges

Clamp
bolt

Clamped wedges
(for axial adjustment only)

No slide — Slide

One-way slide-lock

Push-button
release

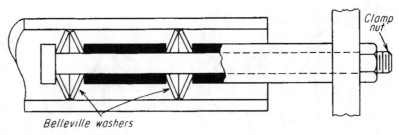

Clamp
nut

Belleville washers

Belleville-washer clamp

Small Mechanical Clutches for Precise Service

Clutches used in calculating machines must have: (1) Quick response–lightweight moving parts; (2) Flexibility–permit multiple members to control operation; (3) Compactness–for equivalent capacity positive clutches are smaller than friction; (4) Dependability; and (5) Durability.

Marvin Taylor

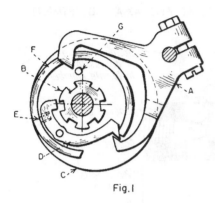

Fig. 1

PAWL AND RATCHET SINGLE CYCLE CLUTCH (Fig. 1). Known as Dennis Clutch, parts B, C and D, are primary components, B, being the driving ratchet, C, the driven cam plate and, D, the connecting pawl carryied by the cam plate. Normally the pawl is held disengaged by the lower portion of clutch arm A. When activated, arm A rocks counter-clockwise until it is out of the path of rim F on cam plate C and permits pawl D under the effect of spring E to engage with ratchet B. Cam plate C then turns clockwise until, near the end of one cycle, pin G on the plate strikes the upper part of arm A camming it clockwise back to its normal position. The lower part of A then performs two functions: (1) cams pawl D out of engagement with the driving ratchet B and (2) blocks further motion of rim F and the cam plate.

PAWL AND RATCHET SINGLE CYCLE DUAL CONTROL CLUTCH—(Fig. 2). Principal parts are: driving ratchet, B, directly connected to the motor and rotating freely on rod A; driven crank, C, directly connected to the main shaft of the machine and also free on A; and spring loaded ratchet pawl, D, which is carried by crank, C, and is normally held disengaged by latch E. To activate the clutch, arm F is raised, permitting latch E to trip and pawl D to engage with ratchet B. The left arm of clutch latch G, which is in the path of the lug on pawl D, is normally permitted to move out of interference by the rotation of the camming edge of crank C. For certain operations block H is temporarily lowered, preventing motion of latch G, resulting in disengagement of the clutch after part of the cycle until subsequent raising of block H permits motion of latch G and resumption of the cycle.

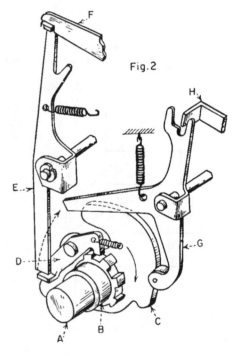

Fig. 2

PLANETARY TRANSMISSION CLUTCH (Fig. 3). A positive clutch with external control, two gear trains to provide bi-directional drive to a calculator for cycling the machine and shifting the carriage. Gear A is the driver, gear L the driven member is directly connected to planet carrier F. The planet consists of integral gears B and C; B meshing with sun gear A and free-wheeling ring gear G, and C meshing with free-wheeling gear D. Gears D and G carry projecting lugs, E and H respectively, which can contact formings on arms J and K of the control yoke. When the machine is at rest, the yoke is centrally positioned so that the arms J and K are out of the path of the projecting lugs permitting both D and G to free-wheel. To engage the drive, the yoke rocks clockwise as shown, until the forming on arm K engages lug H blocking further motion of ring gear G. A solid gear train is thereby established driving F and L in the same direction as the drive A and at the same time altering the speed of D as it continues counter-clockwise. A reversing signal rotates the yoke counter-clockwise until arm J encounters lug E blocking further motion of D. This actuates the other gear train of the same ratio.

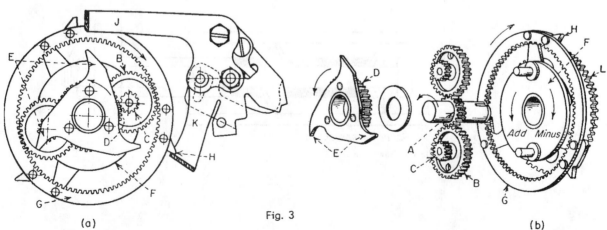

(a)   Fig. 3   (b)

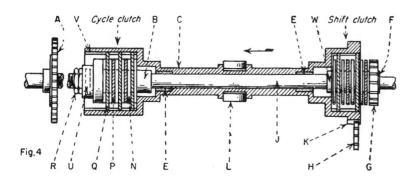

Fig. 4

MULTIPLE DISK FRICTION CLUTCH (Fig. 4).
Two multiple disk friction clutches are combined in a single two-position unit which is shown shifted to the left. A stepped cylindrical housing C enclosing both clutches is carried by self-lubricated bearing E on shaft J and is driven by the transmission gear H meshing with the housing gear teeth K. At either end, the housing carries multiple metal disks Q that engage keyways V and can make frictional contact with formica disks N which, in turn, can contact a set of metal disks P which have slotted openings for coupling with flats on sleeves B and W. In the position shown, pressure is exerted through rollers L forcing the housing to the left making the left clutch compact against adjusting nuts R, thereby driving gear A via sleeve B which is connected to jack shaft J by pin U. When the carriage is to be shifted, rollers L force the housing to the right, first relieving the pressure between the adjoining disks on the left clutch then passing through a neutral position in which both clutches are disengaged and finally making the right clutch compact against thrust bearing F, thereby driving gear G through sleeve W which rotates freely on the jack shaft.

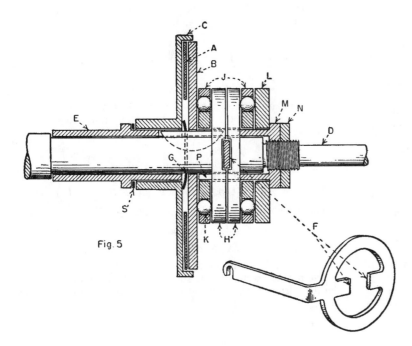

Fig. 5

SINGLE PLATE FRICTION CLUTCH (Fig. 5).
The basic clutch elements, formica disk A, steel plate B and drum C, are normally kept separated by spring washer G. To engage the drive, the left end of a control arm is raised, causing ears F, which sit in slots in plates H, to rock clockwise spreading the plates axially along sleeve P. Sleeves E and P and plate B are keyed to the drive shaft; all other members can rotate freely. The axial motion loads the assembly to the right through the thrust ball bearings K against plate L and adjusting nut M, and to the left through friction surfaces on A, B and C to thrust washer S, sleeve E and against a shoulder on shaft D, thus enabling plate A to drive the drum C.

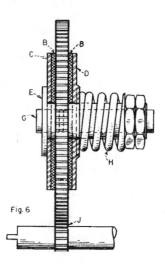

Fig. 6

OVERLOAD RELIEF CLUTCH (Fig. 6). This is a simply constructed, double-plate, spring loaded, friction coupling. Shaft G drives collar E which drives slotted plates C and D and formica disks B. Spring H is forced by the adjusting nuts, which are screwed on to collar E, to maintain the unit under axial pressure against the shoulder at the left end of the collar. This enables the formica disks B to drive through friction against both faces of the gear which is free to turn on the collar, causing output pinion J to rotate. If the machine should jam and pinion J prevented from turning, the motor can continue running without overloading

Centrifugal Clutches

These simple devices provide low-cost clutching for machines operating under fast-changing load conditions.

M. F. Spotts

IF you want a practical way of connecting a motor— or any other type of prime mover—to a load when frequent stopping and starting are involved, centrifugal clutches are a good bet. Low in initial cost, centrifugal clutches can save you the expense of buying another form of electric motor, not to mention auxiliary starting equipment. Centrifugal clutches are built in a wide range of sizes and types—all the way from the little gasoline cars run on tracks at funfairs (here the motor disengages at low speed, and as the child presses on the gas pedal the car starts moving) to 500-hp diesel engines.

Other advantages: These clutches are good starters for high-inertia loads. They tolerate a considerable amount of manufacturing variations and are well suited to drives that undergo vibrations and heavy shock loads. Delayed engagements are possible by varying the clutch-spring force, and installation and service costs are low.

A typical centrifugal clutch has a set of shoes that are forced out against the output drum by centrifugal force. The shoes may be loosely held within the drum (Fig 1, next page), but in the more refined designs the shoes are connected to the input member by means of a floating link (Fig 2) or a fixed pivot (Fig 5).

Both attached-shoe designs (Fig 2 and 5) are analyzed here. Until now, however, the design procedure has been basically a graphical one. The design formulas derived here obviate the need for a graphical layout (or the graphical solution can serve as a check).

Floating-link design

The shoe of this type (Fig 2) is supported at the free end H of the floating link BH. End B of the link is attached to the hub of the driving member. The lining contacts the drum of the driven member and covers an angle ϕ, with support H at the midpoint which determines the v and w axes for the shoe. Angle ϕ is large so pressure p between the lining and drum is not constant but varies according to the equation

$$p = p_o \cos(\psi - \theta) \qquad (1)$$

where p_o is the maximum value of the lining pressure located at angle θ to the v-axis. Angle ψ is the angle from the v-axis to the element under consideration.

The pressure on the lining p, when multiplied by the area $br\,d\psi$ of an element of lining, gives the normal force between lining and drum. The component of force parallel to the v-axis is found by multiplying by $\cos \psi$. Eq 1 is substituted for p and the result integrated over the length of the lining, $-\phi/2$ to $\phi/2$, to obtain N_v, the component of normal force parallel to v-axis:

$$N_v = -\int_{-\phi/2}^{\phi/2} pbr \cos \psi \, d\psi$$
$$N_v = -\tfrac{1}{2}brp_o \cos \theta \,(\phi + \sin \phi) \qquad (2)$$

In a similar manner, the total component of normal force N_w parallel to the w-axis is found by multiplying by $\sin \psi$. The integration gives

$$N_w = -\int_{-\phi/2}^{\phi/2} pbr \sin \psi \, d\psi$$
$$N_w = -\tfrac{1}{2}brp_o \sin \theta \,(\phi - \sin \phi) \qquad (3)$$

The accompanying friction forces are

$$F_v = \int \mu pbr \sin \psi \, d\psi = -\mu N_w \qquad (4)$$
$$F_w = -\int \mu pbr \cos \psi \, d\psi = \mu N_v \qquad (5)$$

where μ is the coefficient of friction.

The torque exerted by the shoe is found by multiplying the normal force on the element by μ, and multiplying again by r to give:

Torque equation

$$T = \int_{-\phi/2}^{\phi/2} \mu pbr^2 \, d\psi = 2\mu br^2 \, p_o \cos \theta \sin \frac{\phi}{2} \qquad (6)$$

(Eq 2 to 6 were obtained by G. A. G. Fazekas, "Graphical Shoe-Brake Analysis," Trans. ASME, vol 79, 1957, p 1322.)

The clutch shoe is in equilibrium from the following three forces:

1. An outward radial force consisting of the difference between the centrifugal force F_c on the shoe and the inward force F_s from the springs.

2. Force Q which is the resultant of the previously determined forces N_v, N_w, F_v and F_w.

3. Reaction R which must have a BH direction since this is a two-force member.

When a body with three forces is in equilibrium, the three forces must intersect at a single point. Because R and F_c-F_s intersect at H, force Q must pass through this point also. Fig 3 shows Q passing through H as the resultant of the other two forces.

Since Q is also the resultant of the forces arising from p and μp, the moment made by such forces about H must be zero. This fact is utilized for finding θ, the inclination of the line of maximum pressure. The moment arms about H for p and μp are marked in the figure. Then

$$\int pbrh \sin \psi \, d\psi - \int \mu pbr \, (r - h \cos \psi) \, d\psi = 0$$

The value of the variable p from Eq 1 is substituted, and the resulting expressions are integrated between the limits of $-\phi/2$ and $\phi/2$. The result can be solved for $\tan \theta$ to give:

Line of maximum pressure

$$\tan \theta = \frac{4\mu r \sin \dfrac{\phi}{2} - \mu h \, (\phi + \sin \phi)}{h \, (\phi - \sin \phi)} \qquad (7)$$

When θ is determined, the forces represented by Eq 2 to 5 inclusive can be found and added vectorially as shown in Fig 3. This gives:

Resultant of forces, Q

$$Q = \sqrt{(N_v + F_v)^2 + (N_w + F_w)^2} \qquad (8)$$

Inclination β of force Q

$$\tan \beta = \frac{N_w + F_w}{N_v + F_v} \qquad (9)$$

At low or idling speeds, the shoe is pulled inwardly by the spring force F_s. The outward force F_c on the shoe is equal to:

Centrifugal force

$$F_c = \frac{W}{g} \, r_o \, \omega^2 \qquad (10)$$

where W is the weight of the shoe assumed to be concentrated at radius r_o; ω is the angular velocity, rad/sec, and g is the gravitational constant, 386 in./sec².

As shown in Fig 2, link BH is inclined at angle α to the v-axis. The sine theorem gives equations for:

Forces R and F_c-F_s

$$R = \frac{Q \sin \beta}{\sin \alpha} \qquad (11)$$

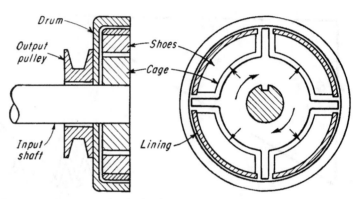

1. Free-shoe centrifugal clutch

SYMBOLS

b = width of lining

F_c = centrifugal force

F_s = spring force

F_v = component along v-axis of friction lining force

F_w = component along w-axis of friction lining force

g = gravitational constant, 386 in./sec²

h = distance from center of rotation to pivot of shoe

hp = transmittable horsepower

M_f = moment of friction forces about fixed pivot

M_n = moment of normal forces about fixed pivot

n = revolutions per minute

N_v = component along v-axis of forces normal to lining

N_w = component along w-axis of forces normal to lining

p = pressure between lining and drum

p_o = maximum lining pressure

Q = resultant of N_v, N_w, F_v and F_w

r = radius of drum

r_o = radius to center of gravity of shoe; outer radius of fixed pivot shoe

r_i = inner radius of fixed pivot shoe

r_p = distance from center of rotation to fixed pivot

R = reaction along floating link

T = torque about center of rotation

W = weight of shoe

α = inclination of floating link with v-axis

β = inclination of resultant Q with v-axis

γ = weight per in.³

θ = inclination of line of maximum pressure with v-axis

μ = coefficient of friction

ϕ = angular extent of lining

ω = angular velocity before slip occurs, radians/sec

$$F_c - F_s = \frac{Q \sin (\alpha - \beta)}{\sin \alpha} \qquad (12)$$

Numerical example

A centrifugal clutch shoe has a radius of 5.25 in. and a width of lining of 2.50 in. Lining pressure is not to exceed 100 psi. Angular length of lining is $\phi = 108$ deg, and the link is at an angle of $\alpha = 48$ deg. The shoe is pivoted at a distance h of 4 in. from the center. Total inward spring force of both springs is 15 lb. Weight of shoe is 3 lb with its center of gravity at a radius of 4.6 in.

As there are linings on the market with a different coefficient of friction, for values of μ of 0.1, 0.2, 0.3, 0.4 and 0.5 find the corresponding values of forces Q, R, and $F_c\text{-}F_s$. Find the torque T, the corresponding slipping rpm, and the horsepower per shoe. Plot the curve for hp vs μ to determine the best value for μ.

Solution

By Eq (7)

$$\tan \theta = \frac{\mu \,[(4)(5.25)(0.80902) - 4(1.88496 + 0.95106)]}{4\,(1.88496 - 0.95106)}$$

$$= 1.511\mu$$

By Eq (2)

$$N_v = -(\tfrac{1}{2})(2.5)(5.25)(100)(2.83602)\cos \theta$$
$$= -1861 \cos \theta$$

By Eq (3)

$$N_w = -612.9 \sin \theta$$

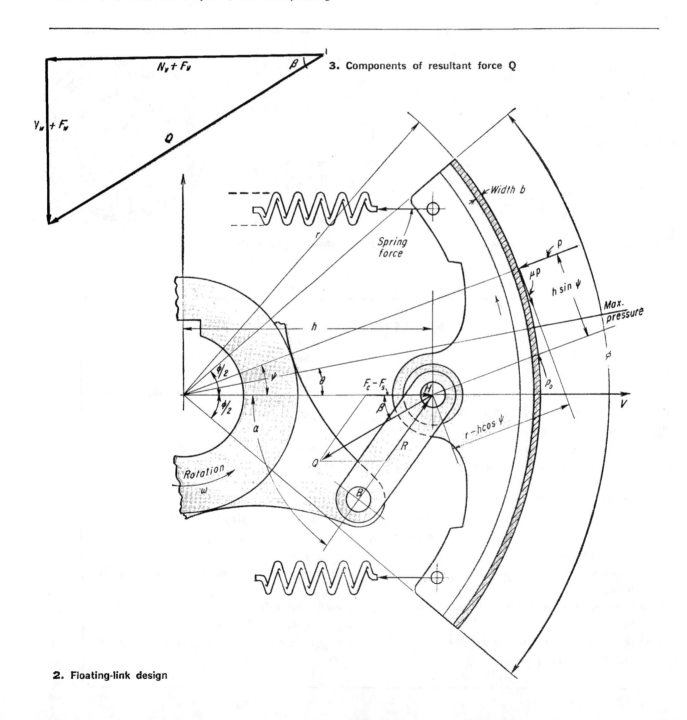

3. Components of resultant force Q

2. Floating-link design

By Eq (6)

$$T = 2\mu (2.5)(5.25)^2(100)(0.80902)\cos\theta$$
$$= 11{,}149\mu \cos\theta$$

From Eq (10)

$$\omega = \sqrt{\frac{g F_c}{W r_o}} = \sqrt{\frac{386 F_c}{(3) 4.6}} = 5.29 \sqrt{F_c}$$

Then

$$n = \frac{\omega}{2\pi} 60 = 9.55\omega = 50.5 \sqrt{F_c}$$

also

$$hp = \frac{Tn}{63{,}000}$$

The calculations are best carried out in tabular form (see table). Thus, for the case of $\mu = 0.1$

$$\tan\theta = 0.1511$$
$$\theta = 8°\,35.6'; \quad \cos\theta = 0.9888$$
$$N_v = -1861 (0.9888) = -1840.2 \text{ lb}$$
$$N_w = -612.9 (0.1494) = -91.6 \text{ lb}$$

By Eq 4, 5, and 8

$$F_v = -0.1 (-91.6) = 9.16 \text{ lb}$$
$$F_w = 0.1 (-1840.2) = -184.0 \text{ lb}$$
$$Q = \sqrt{(-1840.2+9.16)^2+(-91.6-184.0)^2}$$
$$= 1852 \text{ lb}$$

By Eq 9

$$\tan\beta = \frac{-275.6}{-1831} = 0.15052$$

$$\beta = 8°\,33.6'$$

By Eq 11 and 12

$$R = \frac{1852 (\sin 8°\,33.6')}{\sin 48°} = 371 \text{ lb}$$

$$F_c - F_s = \frac{1852 \sin (48° - 8°\,33.6')}{\sin 48°} = 1583 \text{ lb}$$

$$F_c = 1583 + 15 = 1598 \text{ lb}$$

Thus

$$T = 11{,}149 (0.1) (0.9888) = 1102 \text{ lb-in.}$$
$$n = 50.5 \sqrt{1598} = 2019 \text{ rpm}$$

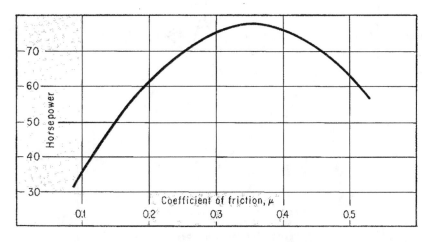

4. Variation of horsepower with coefficient of friction for the floating-link clutch analyzed in the numerical example. Note that best gripping power is obtained with shoe-linings having a coefficient of about 0.35%.

Calculations for various linings

μ	tan θ	θ	N_v	N_w	F_v	F_w	N_v+F_v	N_w+F_w	$(N_w+F_w)^2$	Q
0.1	0.15112	8°35.6′	−1840.2	−91.6	9.2	−184.0	−1831.0	−275.6	75,960	1852
0.2	0.30225	16°49.0′	−1774.1	−177.3	35.5	−354.8	−1738.6	−532.1	283,130	1818
0.3	0.45337	24°23.3′	−1695.1	−253.1	75.9	−508.5	−1619.2	−761.6	580,030	1789
0.4	0.60449	31° 9.2′	−1592.7	−317.1	126.8	−637.1	−1465.9	−954.2	910,500	1749
0.5	0.75562	37° 4.5′	−1484.9	−369.5	184.7	−742.4	−1300.2	−1111.9	1,236,320	1711

μ	tan β	β	R	α−β	sin (α−β)	F_c-F_s	T	F_c	n, rpm	hp
0.1	0.15052	8°33.6′	371	39°26.4′	0.63527	1583	1102	1598	2019	35.3
0.2	0.30605	17° 1.0′	716	30°59.0′	0.51478	1259	2126	1274	1803	60.8
0.3	0.47036	25°11.4′	1024	22°48.6′	0.38768	933	3046	948	1556	75.2
0.4	0.65093	33° 3.7′	1284	14°56.3′	0.25795	607	3817	622	1260	76.3
0.5	0.85518	40°32.2′	1496	7°27.8′	0.12990	299	4448	314	1011	63.2

Serrated Clutches and Detents

L. N. Canick

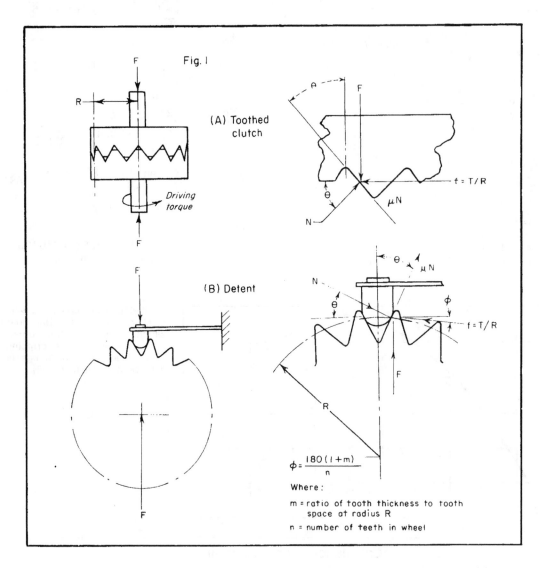

Fig. I

(A) Toothed clutch

(B) Detent

$f = T/R$

μN

$$\phi = \frac{180(1+m)}{n}$$

Where:

m = ratio of tooth thickness to tooth space at radius R

n = number of teeth in wheel

IN THE DESIGN OF straight toothed components such as serrated clutches, Fig. 1(A), and detent wheels, Fig. 1(B), the effective pitch radius is usually set by size considerations. The torque transmitting capacity of the clutch, or the torque resisting capacity of the detent wheel, is then obtained by assigning suitable values to the engaging force, tooth angle, and coefficient of friction.

The nomogram, Fig. 2, is designed to be a convenient means for considering the effect of variations in the values of tooth angle and coefficient of friction. For a given coefficient of friction, there is a tooth angle below which the clutch or detent is self-locking and will transmit torque limited only by its structural strength. Where

T = torque transmitted without clutch slip, or torque resisted by detent wheel, lb in.
R = effective clutch, or detent wheel, radius, in.
F = axial, or radial, force, lb
f = tangential force acting at radius R, lb
N = reaction force of driven tooth, or detent, acting normal to tooth face, lb

μ = coefficient of friction of tooth material
θ = see Fig. 1(B)
  = angle of tooth face, deg
K = $(1 + \mu \tan \theta)/(\tan \theta - \mu)$

a statement of the conditions of equilibrium for the forces acting on a clutch tooth will lead to the following equation

$$T = R F K \qquad (1$$

A similar statement of the conditions of equilibrium for the forces acting on a tooth of the detent wheel shown in Fig. 1(B) will lead to the following equation:

$$T = \frac{R F}{[(\cos \varphi)/K] - \sin \varphi} \qquad (2)$$

From Eqs (1) and (2), when all other terms have constant values, it is obvious that the required axial force, or the radial force, diminishes as the value of K increases. Dependent upon the values of θ and μ, the value of K can vary from zero to infinity.

The circular nomogram shown in Fig. 2 relates the values of the parameters K, θ, and μ that satisfy the basic equation

$$K = (1 + \mu \tan \theta)/(\tan \theta - \mu)$$

EXAMPLE I. Find the maximum tooth angle for a self-locking clutch, or for which K is infinity, taking the coefficient of friction as 0.4 minimum.

SOLUTION I. Line I through these values for K and μ on the nomogram gives a maximum tooth angle slightly less than 22 deg for the self-locking condition.

EXAMPLE II. Find the minimum value of K to be expected for a clutch having a tooth angle of 30 deg and a coefficient of friction of 0.2 minimum.

SOLUTION II. Line II through these values for θ and μ on the nomogram gives a value for K of 3 approximately.

EXAMPLE III. Find the value of K for a flat-face (θ equals 90 deg) friction clutch, the face material of which has a coefficient of friction of 0.2. Compare its torque transmitting capacity with that of the toothed clutch of Example II.

SOLUTION III. Line III through these values for θ and μ on the nomogram gives a value for K of 0.2.

Torque transmitting capacity of flat-face clutch:
$$T = 0.2\, R\, F$$

Torque transmitting capacity of toothed-clutch:
$$T = 3\, R\, F$$

Thus for equal effective radii and engaging forces, the torque capacity of the toothed-clutch is 3/0.2, or 15, times greater than that of the flat face clutch.

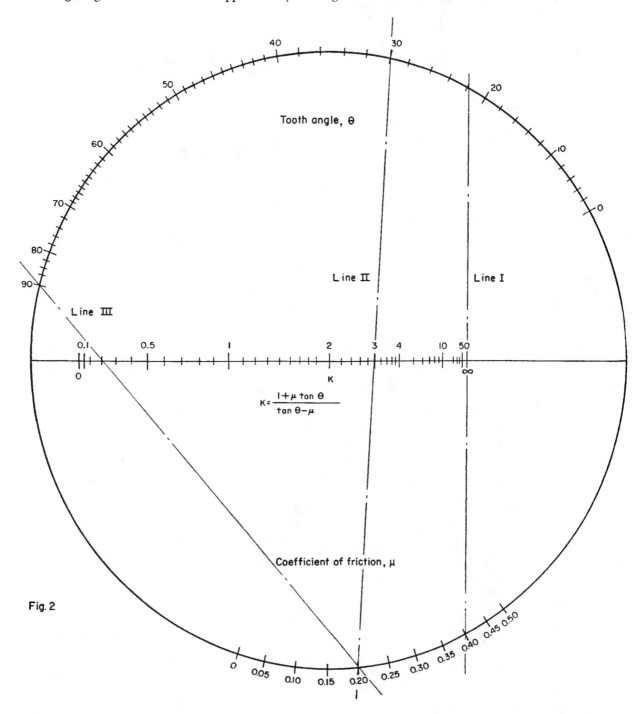

Fig. 2

Tooth angle, θ

Line III   Line II   Line I

$$K = \frac{1 + \mu \tan \theta}{\tan \theta - \mu}$$

Coefficient of friction, μ

Spring Bands Grip Tightly to Drive Overrunning Clutch

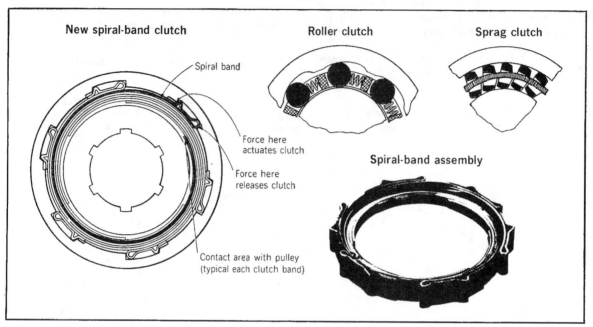

New spiral-band clutch

Spiral band

Force here actuates clutch

Force here releases clutch

Contact area with pulley (typical each clutch band)

Roller clutch

Sprag clutch

Spiral-band assembly

Spiral bands direct force inward as outer ring drives counterclockwise. Roller and sprag types direct force outward.

A new type of overrunning clutch that takes up only half the usual space employs a series of spiral-wound bands instead of the conventional rollers or sprags to transmit high torques. The new design (drawings, above) also simplifies the assembly, cutting costs as much as 40% by eliminating more than half the parts in conventional clutches.

The key to the savings in cost and bulk is the new design's freedom from the need for a hardened outer race. Roller and sprag types must have hardened races because they transmit power by a wedging action between the inner and outer races.

Role of spring bands. Overrunning clutches, including the spiral-band type—slip and overrun when reversed—in drawing above, when outer member is rotated clockwise and inner ring is the driven member.

The new clutch, developed by National Standard Co., Niles, Mich., contains a set of high-carbon spring-steel bands (six in the design illustrated) that grip the inner member when the clutch is driving. The outer member merely serves to retain the spring anchors and to play

a part in actuating the clutch. Since it isn't subject to wedging action, it can be made of almost any material, and this accounts for much of the cost saving. For example, in the automotive torque converter in the drawing at right, the bands fit into the aluminum die-cast reactor.

Reduced wear. The bands are spring-loaded over the inner member of the clutch, but they are held and rotated by the outer member. The centrifugal force on the bands thus releases much of the force on the inner member and considerably decreases the overrunning torque. Wear is, therefore, greatly reduced.

The inner portion of the bands fits into a V-groove in the inner member. When the outer member is reversed, the bands wrap, creating a wedging action in this V-groove. This action is similar to that of a spring clutch with a helical-coil spring, but the spiral-band type has very little unwind before it overruns, compared with the coil type. Thus it responds faster.

Edges of the clutch bands carry the entire load, and there is also a compound action of one band upon

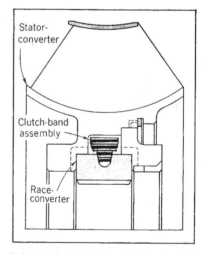

Stator-converter

Clutch-band assembly

Race-converter

Spiral clutch bands can be bought separately to fit in user's assembly.

another. As the torque builds up, each band pushes down on the band beneath it, so each tip is forced more firmly into the V-groove.

National Standard plans to sell the bands as separate components, without the inner and outer clutch members (which the user customarily builds as part of his product). The bands are rated for torque capacities from 85 to 400 ft.-lb. Applications include auto transmissions and starters and industrial machinery. □

Accurate Solution for Disk-Clutch Torque Capacity

Nils M. Sverdrup

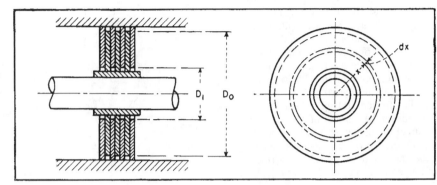

IN COMPUTING TORQUE CAPACITY, the mean radius R of the clutch disks is often used. The torque equation then assumes the following form:

$$T = P\mu Rn \qquad (1)$$

Where

T = torque, in.-lb
P = pressure, lb.
μ = coefficient of friction
R = mean radius of disks, in.
n = no. of friction surfaces

This formula, however, is not mathematically correct and should be used cautiously. The formula's accuracy varies with the ratio D_1/D_0. When D_1/D_0 approaches unity, the error is negligible; but as the value of this ratio decreases, the induced error will increase to a maximum of 33 percent.

By introducing a correction factor, ϕ, Eq (1) can be written

$$T = P\mu Rn\phi \qquad (2)$$

The value of the correction factor can be derived by the calculus derivation of Eq (2).

Sketch above represents a disk clutch with n friction surfaces, pressure between plates being p psi. Inside and outside diameters of effective friction areas are D_1 and D_0 in., respectively. Since the magnitude of pressure on an element of area, dA, at distance x from center is pdA, the friction force is $pdA\mu$ and the moment of this force around the center is $pdA\mu x$.

Integrating within limits $D_1/2$ and $D_0/2$ and multiplying by n friction surfaces, the expression for total torque in in.-lb is obtained.

Hence

$$T = \int_{D_1/2}^{D_0/2} pdA\mu xn \qquad (3)$$

but

$$dA = 2\pi x dx \qquad (4)$$

Substituting in Eq (3)

$$T = \int_{D_1/2}^{D_0/2} p\,(2\pi x dx)\,\mu x n$$

$$= (2/3)\pi p\mu n \left[\left(\frac{D_0}{2}\right)^3 - \left(\frac{D_1}{2}\right)^3 \right]$$

$$\text{or} \quad T = 0.262\, p\mu n\,(D_0{}^3 - D_1{}^3) \qquad (5)$$

If the total pressure acting on clutch disks be P lb, the expression for pressure per unit area is

$$p = \frac{P}{(\pi/4)\,(D_0{}^2 - D_1{}^2)}$$

Substituting this value for p in Eq (5)

$$T = 0.333\, P\mu n \frac{D_0{}^2 + D_0 D_1 + D_1{}^2}{D_0 + D_1} \qquad (6)$$

Now let

$$\frac{D_1}{D_0} = m \text{ so that, } D_1 = m D_0 \qquad (7)$$

Substituting in Eq (6)

$$T = 0.333\, P\mu n D_0 \frac{1 + m^2 + m}{1 + m} \qquad (8)$$

Similarly, by substituting value of D_1 from Eq (7) in Eq (2), and having

$$R = \frac{D_0 + D_1}{4},$$

$$T = P\mu \frac{D_0 + m D_0}{4} n \phi$$

or

$$T = 0.25\, P\mu n\phi\, D_0\,(1 + m) \qquad (9)$$

Equating expressions (8) and (9)

$$0.25\, P\mu n\phi\, D_0\,(1 + m) =$$

$$0.333\, P\mu n D_0 \frac{1 + m^2 + m}{1 + m}$$

and solving for ϕ, the result is

$$\phi = 1.333 \times \frac{1 + m^2 + m}{(1 + m)^2} \qquad (10)$$

With various diameter ratios, the values for ϕ were computed and represented in graph herewith. By using this graph and Eq (2), accurate values of torque can be easily determined.

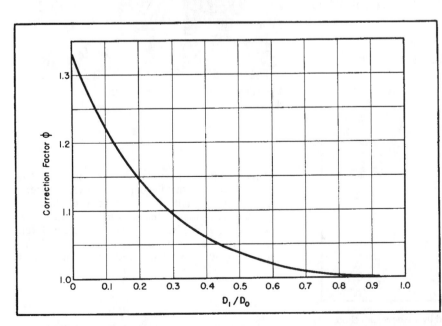

Spring-Loaded Pins Aid Sprags in One-Way Clutch

Sprags combined with cylindrical rollers in a bearing assembly can provide a simple, low-cost method for meeting the torque and bearing requirements of most machine applications. Designed and built by Est. Nicot of Paris, this unit gives one-direction-only torque transmission in an overrunning clutch. In addition, it also serves as a roller bearing.

The torque rating of the clutch depends on the number of sprags. A minimum of three, equally spaced around the circumference of the races, is generally necessary to get acceptable distribution of tangential forces on the races.

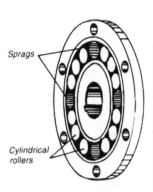 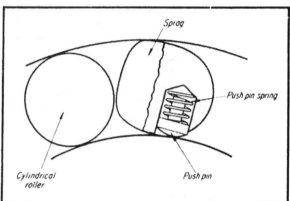

Races are concentric; a locking ramp is provided by the sprag profile, which is composed of two nonconcentric curves of different radius. A spring-loaded pin holds the sprag in the locked position until the torque is applied in the running direction. A stock roller bearing cannot be converted because the hard-steel races of the bearing are too brittle to handle the locking impact of the sprag. The sprags and rollers can be mixed to give any desired torque value.

Rolling-Type Clutch

This clutch can be adapted for either electrical or mechanical actuation, and will control ½ hp at 1500 rpm with only 7 W of power in the solenoid. The rollers are positioned by a cage (integral with the toothed control wheel —see diagram) between the ID of the driving housing and the cammed hub (integral with the output gear).

When the pawl is disengaged, the drag of the housing on the friction spring rotates the cage and wedges the rollers into engagement. This permits the housing to drive the gear through the cam.

When the pawl engages the control wheel while the housing is rotating, the friction spring slips inside the housing and the rollers are kicked back, out of engagement. Power is therefore interrupted.

According to the manufacturer, Tiltman Langley Ltd, Surrey, England, the unit operated over the full temperature range of –40° to 200°F.

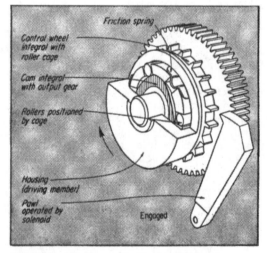

A positive drive is provided by this British roller clutch.

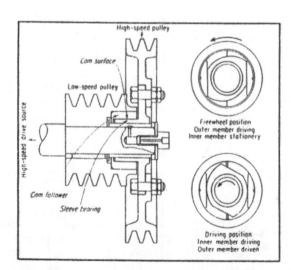

Two-speed operation is provided by the new cam clutch

This clutch consists of two rotary members (see diagrams), arranged so that the outer (follower) member acts on its pulley only when the inner member is driving. When the outer member is driving, the inner member idles. One application was in a dry-cleaning machine. The clutch functions as an intermediary between an ordinary and a high-speed motor to provide two output speeds that are used alternately.

Source: *Mechanisms and Mechanical Devices Sourcebook*, 3E, by Chironis & Sclater, © 2001, McGraw-Hill

SECTION 8

CHAINS, SPROCKETS & RATCHETS

History of Chains

William R. Edgerton

The fundamental concept of creating a strong, yet flexible, chain structure by joining together a consecutive series of individual links is an idea that dates back to the earliest human utilization of metals. The use of iron for this purpose probably dates to the eighth century B.C.

The second step in the development process was the fashioning of wheels adapted to interact with the flexible chains, by the provision of teeth and pockets on the circumference of the wheels. These specially adapted wheels, known as "sprocket wheels," but usually referred to simply as "sprockets," were first developed by the military engineers of Greece some 22 centuries ago. From the writings of Philo of Byzantium, c. 200 B.C., we learn of chain and sprocket drives being used to transmit power from early water wheels, of a pair of chains fitted with buckets to lift water to higher elevations, and of a pair of reciprocating chain drives which acted as a tension linkage to feed and cock a repeating catapult.

The first two of these instances of chain and sprocket interaction probably used simple round-link chain, but the third involved a flat-link chain concept designed by Dionysius of Alexandria while working at the Arsenal at Rhodes. The design conceived by Dionysius employed what is now known as the inverted-tooth chain-sprocket engagement principle, a major advance over the cruder round-link design.

Despite its very early origins, rather little practical use was made of chain and sprocket interaction for the transmission of power or the conveyance of materials until the advent of the Industrial Revolution, which took place largely during the nineteenth century. The development of machinery to mechanize textile manufacture, agricultural harvesting, and metalworking manufacturing brought with it a need for the positive transmission of power and accurate timing of motions that only a chain-and-sprocket drive could provide.

The earliest sprocket chains manufactured in the United States employed cast components, usually of malleable iron, and many configurations of detachable link chain and pintle chain were produced in large quantities. As the need for higher strength and improved wear resistance became evident, chains employing heat-treated steel components were introduced. The use of rolled or drawn steel as a raw material required manufacturing machinery which provided greater dimensional accuracy than was possible in foundry practice, with the result that certain of the new types of sprocket chains came to be known as "precision chain." This developed somewhat earlier in Great Britain than in the United States, starting with the Slater chain, patented in England in 1864. The Slater design was further refined by Hans Renold with the development of precision roller chain, patented in England in 1880.

Chain manufacture in the United States continued to be principally concerned with cast and detachable link designs until the American introduction of the "Safety Bicycle" in 1888. Drop-forged steel versions of the cast detachable chains were first used, then precision steel block chain for bicycle driving, and progressively larger sizes were manufactured in the U.S. as the horseless carriage craze swept the country in the 1890s.

Precision inverted-tooth chain, popularly known as "silent chain," was introduced in the late 1890s, with many proprietary styles being developed during the early part of the twentieth century.

The first efforts toward standardization of roller chain were begun in the 1920s, resulting in the publication of the first chain standard, American Standard B29a, on July 22, 1930. Since that time, eighteen B29 standards have been developed, covering inverted-tooth chain; detachable chain; pintle and offset-sidebar chains; cast, forged, and combination chains; mill and drag chains; and many other styles.

There are eighteen American National Standards which relate to the various types of sprocket chains in general use. This family of standards is the result of over 50 years of standardization activity, which had its beginning in the work that led to the publication of *American Standard B29a—Roller Chain, Sprockets, and Cutters* in 1930. The chain types covered by the current standards are as follows:

ANSI B29.1	Precision roller chain
ANSI B29.2	Inverted-tooth (or silent) chain
ANSI B29.3	Double-pitch roller chain for power transmission
ANSI B29.4	Double-pitch roller chain for conveyor usage
ANSI B29.6	Steel detachable chain
ANSI B29.7	Malleable iron detachable chain
ANSI B29.8	Leaf chain
ANSI B29.10	Heavy-duty offset-sidebar roller chain
ANSI B29.11	Combination chain
ANSI B29.12	Steel-bushed rollerless chain
ANSI B29.14	Mill chain (H type)
ANSI B29.15	Heavy-duty roller-type conveyor chain
ANSI B29.16	Mill chain (welded type)
ANSI B29.17	Hinge-type flat-top conveyor chain
ANSI B29.18	Drag chain (welded type)
ANSI B29.19	Agricultural roller chain (A and CA types)
ANSI B29.21	Chains for water and sewage treatment plants
ANSI B29.22	Drop-forged rivetless chain

The basic size dimension for all types of chain is pitch—the center-to-center distance between two consecutive joints. This dimension ranges from $3/16$ in (in the smallest inverted-tooth chain) to 30 in (the largest heavy-duty roller-type conveyor chain).

Chains and sprockets interact with each other to convert linear motion to rotary motion or vice versa, since the chain moves in an essentially straight line between sprockets and moves in a circular path while engaged with each sprocket. A number of tooth-form designs have evolved over the years, but the prerequisite of any tooth form is that it must provide:

1. Smooth engagement and disengagement with the moving chain
2. Distribution of the transmitted load over more than one tooth of the sprocket
3. Accommodation of changes in chain length as the chain elongates as a result of wear during its service life

The sprocket layout is based on the pitch circle, the diameter of which is such that the circle would pass through the center of each of the chain's joints when that joint is engaged with the sprocket. Since each chain link is rigid, the engaged chain forms a polygon whose sides are equal in length to the chain's pitch. The pitch circle of a sprocket, then, is a circle that passes through each corner, or vertex, of the pitch polygon. The calculation of the pitch diameter of a sprocket follows the basic rules of geometry as they apply to pitch and number of teeth. This relationship is simply

$$\text{Pitch diameter} = \frac{\text{pitch}}{\sin(180°/\text{number of teeth})}$$

The action of the moving chain as it engages with the rotating sprocket is one of consecutive engagement. Each link must articulate, or swing, through a specific angle to accommodate itself to the pitch polygon, and each link must be completely engaged, or seated, before the next in succession can begin its articulation.

Source: *Mechanical Components Handbook* by Robert O. Parmley ©1985

Ingenious Jobs for Roller Chain

How this low-cost industrial workhorse can be harnessed in a variety of
ways to perform tasks other than simply transmitting power.

Peter C. Noy

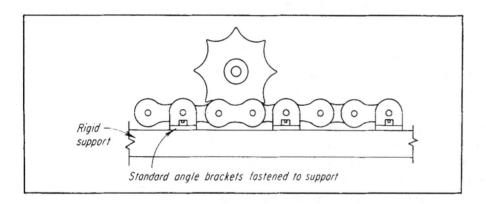

1 LOW-COST RACK-AND-PINION device is easily assembled from standard parts.

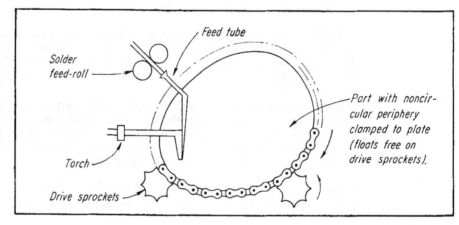

2 AN EXTENSION OF RACK-AND-PINION PRINCIPLE—soldering fixture for noncircular shells. Positive-action cams can be similarly designed. Standard angle brackets attach chain to cam or fixture plate.

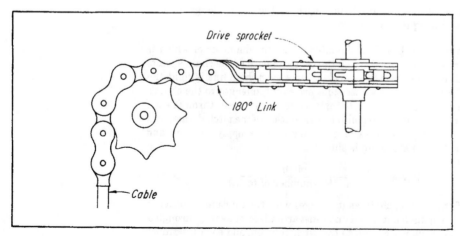

3 CONTROL-CABLE DIRECTION-CHANGER extensively used in aircraft.

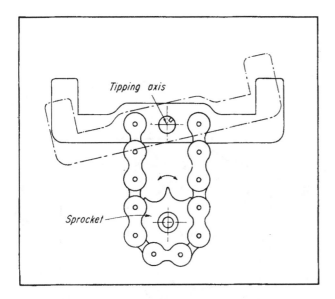

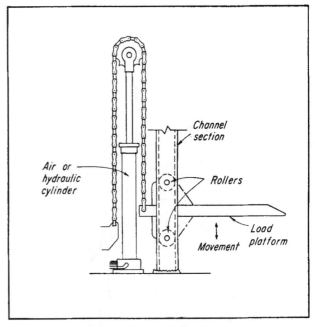

4 TRANSMISSION OF TIPPING OR ROCKING MOTION.
Can be combined with previous example (3) to transmit
this type of motion to a remote location and around
obstructions. Tipping angle should not exceed 40° approx.

5 LIFTING DEVICE is simplified by roller chain.

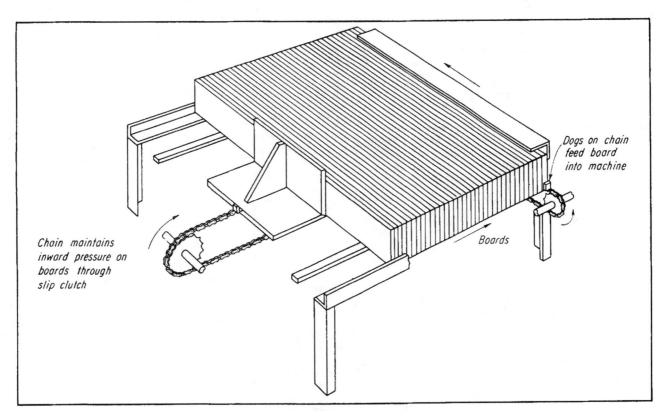

6 TWO EXAMPLES OF INDEXING AND FEEDING uses of roller chain are shown here in a setup that feeds plywood
strips into a brush-making machine. Advantages of roller chain as used here are flexibility and long feed.

Examples of how this low-cost but precision-made product can be arranged to do tasks other than transmit power.

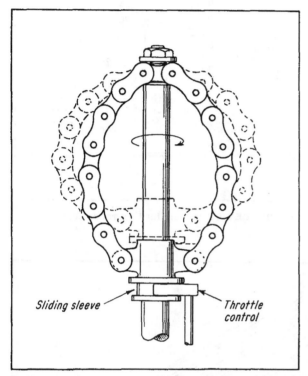

1 SIMPLE GOVERNOR—weights can be attached by means of standard brackets to increase response force when rotation speed is slow.

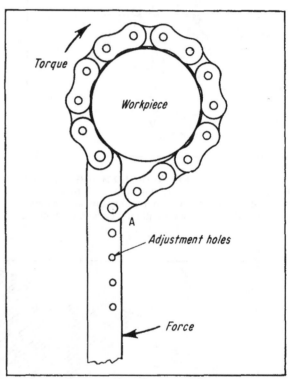

2 WRENCH—pivot A can be adjusted to grip a variety of regularly or irregularly shaped objects.

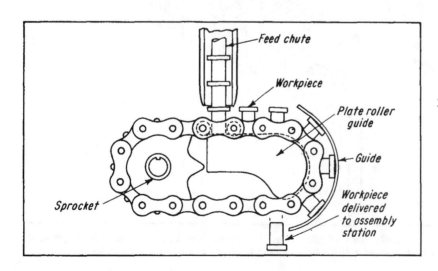

3 SMALL PARTS CAN BE CONVEYED, fed, or oriented between spaces of roller chain.

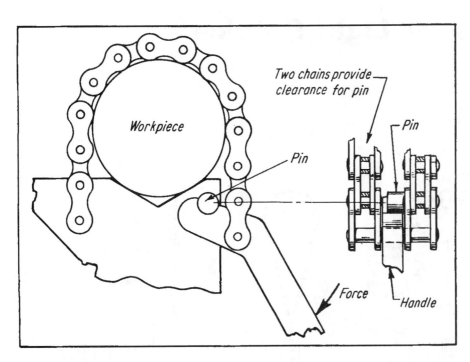

Two chains provide clearance for pin

Workpiece

Pin

Pin

Force

Handle

4 CLAMP—toggle action is supplied by two chains, thus clearing pin at fulcrum.

5 LIGHT-DUTY TROLLEY CONVEYORS can be made by combining standard roller-chain components with standard curtain-track components. Small gearmotors are used to drive the conveyor.

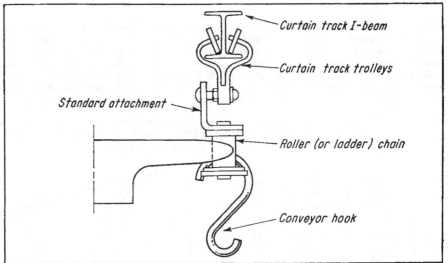

Curtain track I-beam

Curtain track trolleys

Standard attachment

Roller (or ladder) chain

Conveyor hook

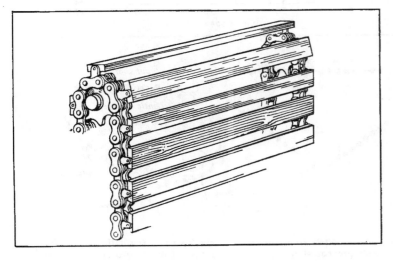

6 SLATTED BELT, made by attaching wood, plastic or metal slats, can serve as adjustable safety guard, conveyor belt, fast-acting security-wicket window.

Bead Chains for Light Service

Bernard Wasko

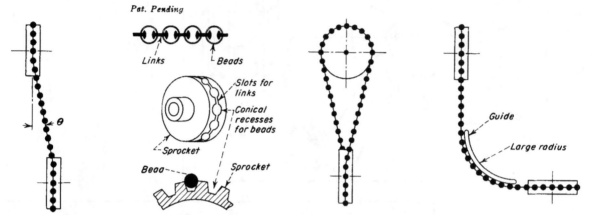

Fig. 1—Misaligned sprockets. Nonparallel planes usually occur when alignment is too expensive to maintain. Bead chain can operate at angles up to θ=20 degrees.

Fig. 2—Details of bead chain and sprocket. Beads of chain seat themselves firmly in conical recesses in the face of sprocket. Links ride freely in slots between recesses in sprocket.

Fig. 3—Skewed shafts normally acquire two sets of spiral gears to bridge space between shafts. Angle misalignment does not interfere with qualified bead chain operation on sprockets.

Fig. 4—Right angle drive does not require idler sprockets to go around corner. Suitable only for very low torque application because of friction drag of bead chain against guide.

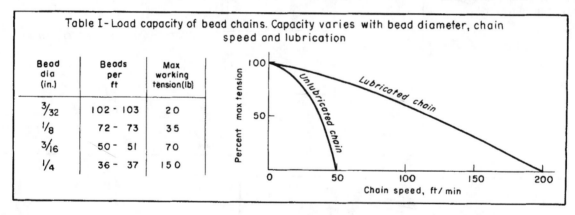

Table I-Load capacity of bead chains. Capacity varies with bead diameter, chain speed and lubrication

Bead dia (in.)	Beads per ft	Max working tension (lb)
3/32	102 - 103	20
1/8	72 - 73	35
3/16	50 - 51	70
1/4	36 - 37	150

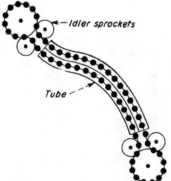

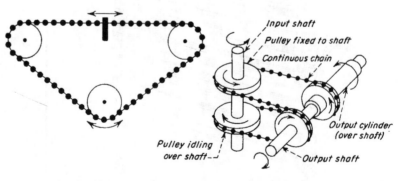

Fig. 5—Remote control through rigid or flexible tube has almost no backlash and can keep input and output shafts synchronized.

Fig. 6—Linear output from rotary input. Beads prevent slippage and maintain accurate ratio between the input and output displacements.

Fig. 7—Counter-rotating shafts. Input shaft drives two counter-rotating outputs (shaft and cylinder) through a continuous chain.

Where torque requirements and operating speeds are low, qualified bead chains offer a quick and economical way to: Couple misaligned shafts; convert from one type of motion to another; counter-rotate shafts; obtain high ratio drives and overload protection; control switches and serve as mechanical counters.

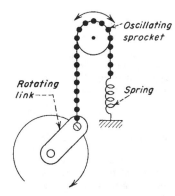

Fig. 8—Angular oscillations from rotary input. Link makes complete revolutions causing sprocket to oscillate. Spring maintains chain tension.

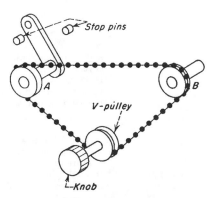

Fig. 9—Restricted angular motion. Pulley, rotated by knob, slips when limit stop is reached; shafts A and B remain stationary and synchronous.

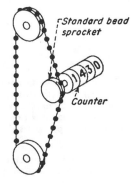

Fig. 10—Remote control of counter. For applications where counter cannot be coupled directly to shaft, bead chain and sprockets can be used.

Fig. 11—High-ratio drive less expensive than gear trains. Qualified bead chains and sprockets will transmit power without slippage.

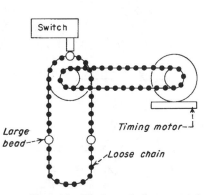

Fig. 12—Timing chain containing large beads at desired intervals operates micro-switch. Chain can be lengthened to contain thousands of intervals for complex timing.

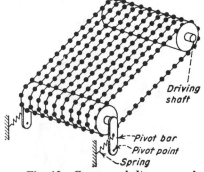

Fig. 13—Conveyor belt composed of multiple chains and sprockets. Tension maintained by pivot bar and spring. Width of belt easily changed.

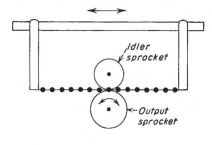

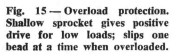

Fig. 14—Gear and rack duplicated by chain and two sprockets. Converts linear motion into rotary motion.

Fig. 15 — Overload protection. Shallow sprocket gives positive drive for low loads; slips one bead at a time when overloaded.

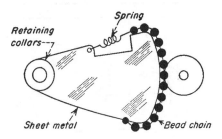

Fig. 16—Gear segment inexpensively made with bead chain and spring wrapped around edge of sheet metal. Retaining collars keep sheet metal sector from twisting on the shaft.

Methods for Reducing Pulsations in Chain Drives

Pulsations in chain motion created by the chordal action of chain and sprockets can be minimized or avoided by introducing a compensating cyclic motion in driving sprockets. Mechanisms for reducing fluctuating dynamic loads in chain and the pulsations resulting therefrom include non-circular gears, eccentric gears, and cam activated intermediate shafts.

Eugene I. Radzimovsky

Fig. 1—The large cast-tooth non-circular gear, mounted on the chain sprocket shaft, has wavy outline in which number of waves equals number of teeth on sprocket. Pinion has a corresponding noncircular shape. Although requiring special-shaped gears, drive completely equalizes chain pulsations.

Fig. 2—This drive has two eccentrically mounted spur pinions (1 and 2). Input power is through belt pulley keyed to same shaft as pinion 1. Pinion 3 (not shown), keyed to shaft of pinion 2, drives large gear and sprocket. However, mechanism does not completely equalize chain velocity unless the pitch lines of pinions 1 and 2 are non-circular instead of eccentric.

Fig. 3—Additional sprocket 2 drives noncircular sprocket 3 through fine-pitch chain 1. This imparts pulsating velocity to shaft 6 and to long-pitch conveyor sprocket 5 through pinion 7 and gear 4. Ratio of the gear pair is made same as number of teeth of sprocket 5. Spring-

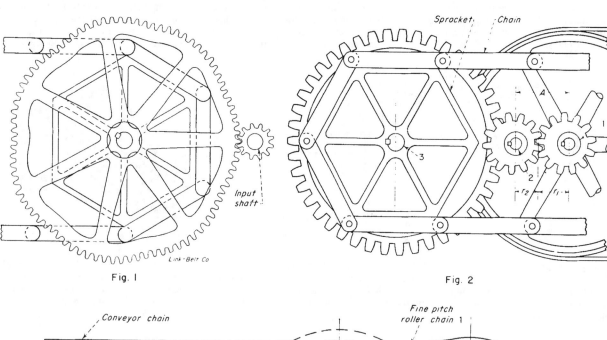

Fig. 1

Fig. 2

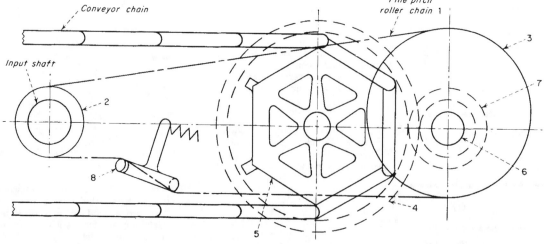

Fig. 3

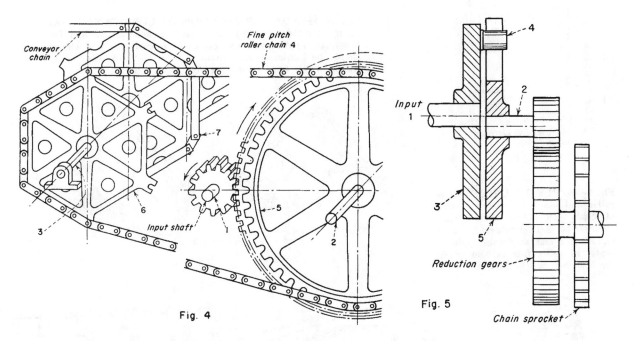

Fig. 4

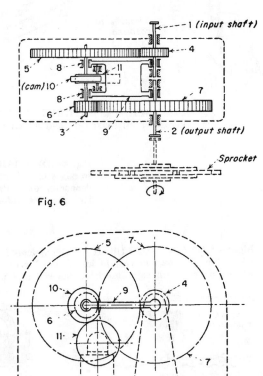

Fig. 5

Fig. 6

actuated lever and rollers 8 take up slack. Conveyor motion is equalized but mechanism has limited power capacity because pitch of chain 1 must be kept small. Capacity can be increased by using multiple strands of fine-pitch chain.

Fig. 4—Power is transmitted from shaft 2 to sprocket 6 through chain 4, thus imparting a variable velocity to shaft 3, and through it, to the conveyor sprocket 7. Since chain 4 has small pitch and sprocket 5 is relatively large, velocity of 4 is almost constant which induces an almost constant conveyor velocity. Mechanism requires rollers to tighten slack side of chain and has limited power capacity.

Fig. 5—Variable motion to sprocket is produced by disk 3 which supports pin and roller 4, and disk 5 which has a radial slot and is eccentrically mounted on shaft 2. Ratio of rpm of shaft 2 to sprocket equals number of teeth in sprocket. Chain velocity is not completely equalized.

Fig. 6—Integrated "planetary gear" system (gears 4, 5, 6 and 7) is activated by cam 10 and transmits through shaft 2 a variable velocity to sprocket synchronized with chain pulsations thus completely equalizing chain velocity. The cam 10 rides on a circular idler roller 11; because of the equilibrium of the forces the cam maintains positive contact with the roller. Unit uses standard gears, acts simultaneously as a speed reducer, and can transmit high horsepower.

Patent applied for # 425,076

Lubrication of Roller Chains

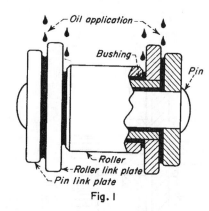

Fig. 1

Fig. 1—APPLY OIL DROPS between roller and pin links on lower strand of chain just before chain engages sprocket so that centrifugal force carries oil into clearances. Oil applied at center of roller face seldom reaches the area between bushing and roller.

Fig. 2—MANUAL APPLICATION OF LUBRICANT by (A) flared-lip oil can, or (B) hand brush, is simplest method for low-speed applications not enclosed in casings. New chains should be lubricated daily until sufficiently "broken-in," after which weekly lubrication programs should suffice.

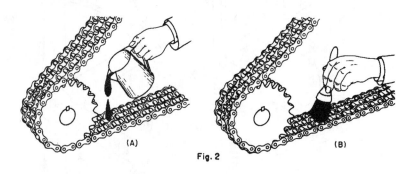

Fig. 2

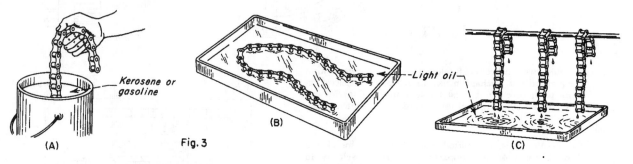

Fig. 3

Fig. 3—CHAINS WITHOUT CASING should be: (A) removed periodically and washed in kerosene, (B) soaked in light oil after cleaning, and (C) draped to permit excess oil to drain.

Fig. 4—DRIP LUBRICATION can be adjusted to feed oil to edges of link plates at rate of 4 to 20 drops per minute depending on chain speed. Pipe contains oil-soaked wick to feed multiple-width chains.

Fig. 5—CONTINUOUS LUBRICATION systems for open chains: (A) Wick lubrication is lowest in cost to install; (B) Friction wheel lubrication uses wheel covered with soft absorbent material and pressured by flat spring.

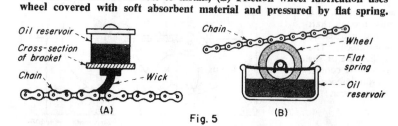

Fig. 5

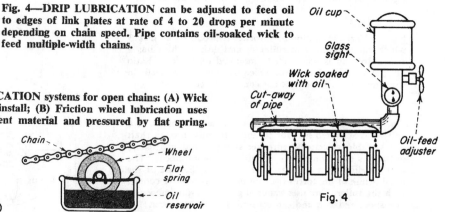

Fig. 4

Unsatisfactory chain life is usually the result of poor or ineffective lubrication. More damage is caused by faulty lubrication than by years of normal service. Illustrated below are 9 methods for lubricating roller chains. Selection should be made on basis of chain speed as shown in Table I. Recommended lubricants are listed in Table II.

Table I—Recommended Methods

Chain Speed, ft/min	Method
0–600	Manual: brush, oil can Slow Drip: 4–10 drops, min Continuous: wick, wheel
600–1500	Rapid Drip—20 drops, min Shallow Bath, Disk
over 1500	Force Feed Systems

Table II—Recommended Lubricants

Pitch of chain, in.	Viscosity at 100 F, SUS	SAE No.
¼–⅝	240–420	20
¾–1¼	420–620	30
1½–up	620–1300	40

Note: For ambient temperatures between 100 to 500 F use SAE 50.

Fig. 6—SHALLOW BATH LUBRICATION uses casing as reservoir for oil. Lower part of chain just skims through oil pool. Levels of oil must be kept tangent to chain sprocket to avoid excessive churning. Should not be used at high speeds because of tendency to generate excessive heat.

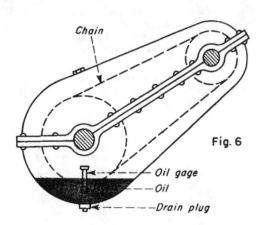

Fig. 6

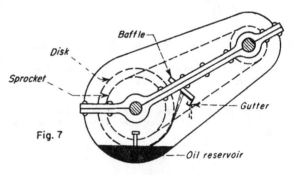

Fig. 7

Fig. 7—DISK OR SLINGER can be attached to lower sprocket to give continuous supply of oil. Disk scoops up oil from reservoir and throws it against baffle. Gutter catches oil dripping down from baffle and directs it on to chain.

Fig. 8—FORCE-FEED LUBRICATION for chains running at extremely high speeds. Pump driven by motor delivers oil under pressure to nozzles that direct spray on to chain. Excess oil collects in reservoir which has wide area to cool oil.

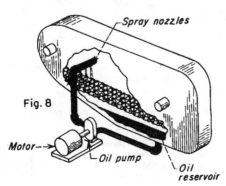

Fig. 8

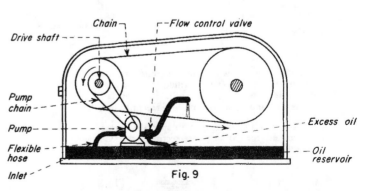

Fig. 9

Fig. 9—CHAIN-DRIVEN FORCE-FEED system has pump driven by main drive shaft. Flow control valve, regulated from outside of casing, by-passes excess oil back to reservoir. Inlet hose contains filter. Oil should be changed periodically—especially when hue is brown instead of black.

Sheet Metal Gears, Sprockets, Worms & Ratchets

Haim Murro

When a specified motion must be transmitted at intervals rather than continuously, and the loads are light, these mechanisms are ideal because of their low cost and adaptability to mass production. Although not generally considered precision parts, ratchets and gears can be stamped to tolerances of ±0.007 in. and if necessary, shaved to closer dimensions. Sketches indicate some variations used on toys, household appliances and automobile components.

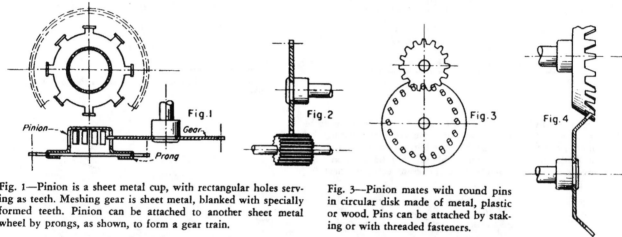

Fig. 1—Pinion is a sheet metal cup, with rectangular holes serving as teeth. Meshing gear is sheet metal, blanked with specially formed teeth. Pinion can be attached to another sheet metal wheel by prongs, as shown, to form a gear train.

Fig. 2—Sheet metal wheel gear meshes with a wide face pinion, which is either extruded or machined. Wheel is blanked with teeth of conventional form.

Fig. 3—Pinion mates with round pins in circular disk made of metal, plastic or wood. Pins can be attached by staking or with threaded fasteners.

Fig. 4—Two blanked gears, conically formed after blanking, make bevel gears meshing on parallel axis. Both have specially formed teeth.

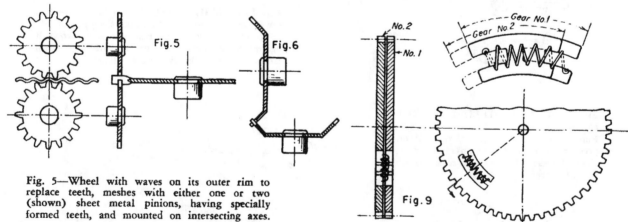

Fig. 5—Wheel with waves on its outer rim to replace teeth, meshes with either one or two (shown) sheet metal pinions, having specially formed teeth, and mounted on intersecting axes.

Fig. 6—Two bevel type gears, with specially formed teeth, mounted for 90 deg intersecting axes. Can be attached economically by staking to hubs.

Fig. 7—Blanked and formed bevel type gear meshes with solid machined or extruded pinion. Conventional form of teeth can be used on both gear and pinion.

Fig. 8—Blanked, cup-shaped wheel meshes with solid pinion for 90 deg intersecting axes.

Fig. 9—Backlash can be eliminated from stamped gears by stacking two identical gears and displacing them by one tooth. Spring then bears one projection on each gear taking up lost motion.

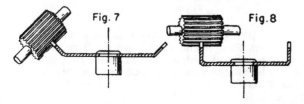

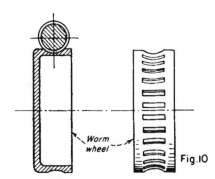

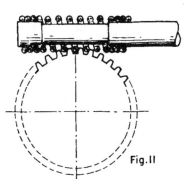

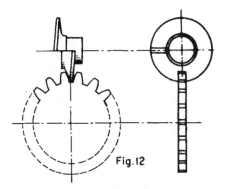

Fig. 10—Sheet metal cup which has indentations that take place of worm wheel teeth, meshes with a standard coarse thread screw.

Fig. 11—Blanked wheel, with specially formed teeth, meshes with a helical spring mounted on a shaft, which serves as the worm.

Fig. 12—Worm wheel is sheet metal blanked, with specially formed teeth. Worm is made of sheet metal disk, split and helically formed.

Fig. 13—Blanked ratchets with one sided teeth stacked to fit a wide, sheet metal finger when single thickness is not adequate. Ratchet gears can be spot welded.

Fig. 14—To avoid stacking, single ratchet is used with a U-shaped finger also made of sheet metal.

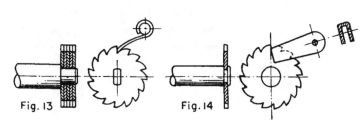

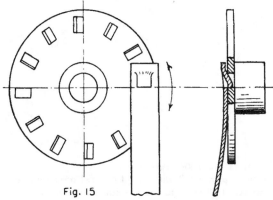

Fig. 15—Wheel is a punched disk with square punched holes to selve as teeth. Pawl is spring steel.

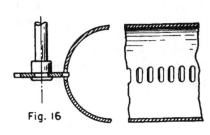

Fig. 16—Sheet metal blanked pinion, with specially formed teeth, meshes with windows blanked in a sheet metal cylinder, to form a pinion and rack assembly.

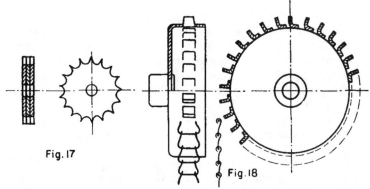

Fig. 17—Sprocket, like Fig. 13, can be fabricated from separate stampings.
Fig. 18—For a wire chain as shown, sprocket is made by bending out punched teeth on a drawn cup.

Ratchet Layout Analyzed

Here, in a brief but comprehensive rundown, are generally unavailable formulas and data for precise ratchet layout.

Emery E. Rossner

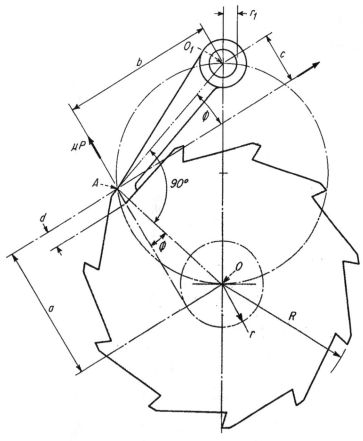

Pawl in compression . . .
has tooth pressure P and weight of pawl producing a moment that tends to engage pawl. Friction-force μP and pivot friction tend to oppose pawl engagement.

Symbols

a = moment arm of wheel torque

M = moment about O_1 caused by weight of pawl

$0, O_1$ = ratchet and pawl pivot centers respectively

P = tooth pressure = wheel torque/a

$P\sqrt{(1+\mu^2)}$ = load on pivot pin

μ, μ_1 = friction coefficients

Other symbols as defined in diagrams

The ratchet wheel is widely used in machinery, mainly to transmit intermittent motion or to allow shaft rotation in one direction only. Ratchet-wheel teeth can be either on the perimeter of a disc or on the inner edge of a ring.

The pawl, which engages the ratchet teeth, is a beam pivoted at one end; the other end is shaped to fit the ratchet-tooth flank. Usually a spring or counterweight maintains constant contact between wheel and pawl.

It is desirable in most designs to keep the spring force low. It should be just enough to overcome the separation forces—inertia, weight and pivot friction. Excess spring force should not be relied on to bring about and maintain pawl engagement against the load.

To insure that the pawl is automatically pulled in and kept in engagement independently of the spring, a properly layed out tooth flank is necessary.

The requirement for self-engagement is

$$Pc + M > \mu Pb + P \sqrt{(1+\mu^2)} \, \mu_1 r_1$$

Neglecting weight and pivot friction

$$Pc > \mu Pb$$

or

$$c/b > \mu$$

but $c/b = r/a = \tan \phi$, and since $\tan \phi$ is approximately equal to $\sin \phi$

$$c/b = r/R$$

Substituting in term (1)

$$rR > \mu$$

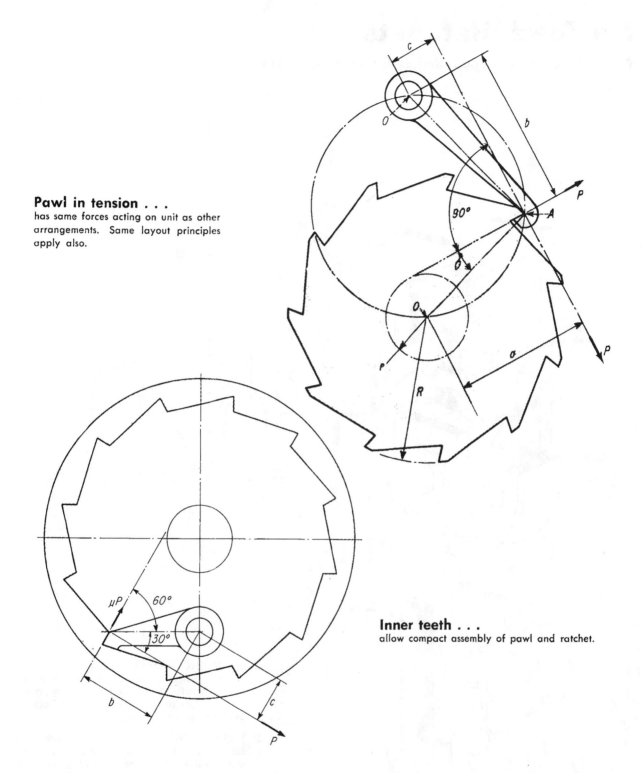

Pawl in tension . . .
has same forces acting on unit as other arrangements. Same layout principles apply also.

Inner teeth . . .
allow compact assembly of pawl and ratchet.

For steel on steel, dry, $\mu = 0.15$. Therefore, using

$$r/R = 0.20 \text{ to } 0.25$$

the margin of safety is large; the pawl will slide into engagement easily. For internal teeth with ϕ of 30°, c/b is tan 30° or 0.577 which is larger than μ, and the teeth are therefore self engaging.

When laying out the ratchet wheel and pawl, locate points O, A and O_1 on the same circle. AO and AO_1 will then be perpendicular to one another; this will

insure that the smallest forces are acting on the system.

Ratchet and pawl dimensions are governed by design sizes and stress. If the tooth, and thus pitch, must be larger than required in order to be strong enough, a multiple pawl arrangement can be used. The pawls can be arranged so that one of them will engage the ratchet after a rotation of less than the pitch.

A fine feed can be obtained by placing a number of pawls side by side, with the corresponding ratchet wheels uniformly displaced and interconnected.

No Teeth Ratchets

With springs, rollers and other devices they keep going one way.

L. Kasper

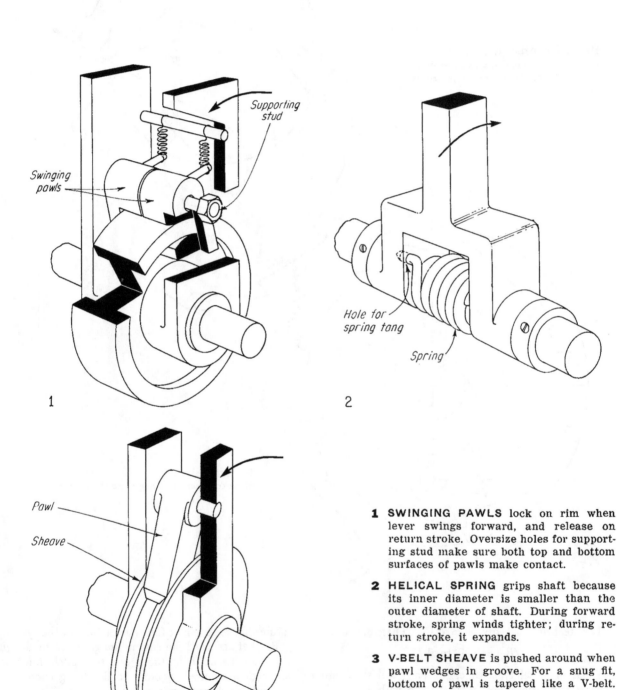

Swinging pawls — Supporting stud

1

Hole for spring tang — Spring

2

Pawl — **Sheave**

3

1 SWINGING PAWLS lock on rim when lever swings forward, and release on return stroke. Oversize holes for supporting stud make sure both top and bottom surfaces of pawls make contact.

2 HELICAL SPRING grips shaft because its inner diameter is smaller than the outer diameter of shaft. During forward stroke, spring winds tighter; during return stroke, it expands.

3 V-BELT SHEAVE is pushed around when pawl wedges in groove. For a snug fit, bottom of pawl is tapered like a V-belt.

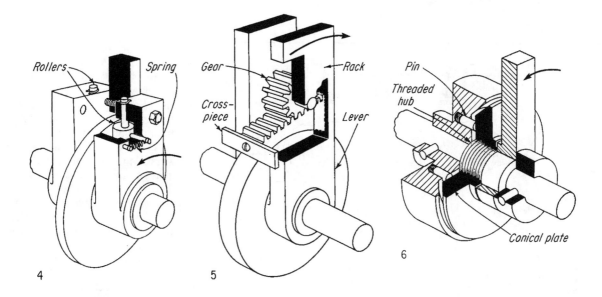

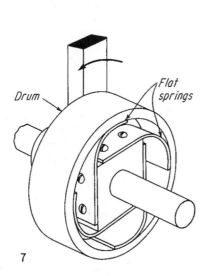

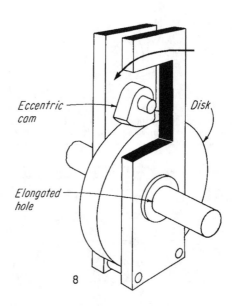

4 **ECCENTRIC ROLLERS** squeeze disk on forward stroke. On return stroke, rollers rotate backwards and release their grip. Springs keep rollers in contact with disk.

5 **RACK** is wedge-shape so that it jams between the rolling gear and the disk, pushing the shaft forward. When the driving lever makes its return stroke, it carries along the unattached rack by the cross-piece.

6 **CONICAL PLATE** moves as a nut back and forth along the threaded center hub of the lever. Light friction of spring-loaded pins keeps the plate from rotating with the hub.

7 **FLAT SPRINGS** expand against inside of drum when lever moves one way, but drag loosely when lever turns drum in opposite direction.

8 **ECCENTRIC CAM** jams against disk during motion half of cycle. Elongated holes in the levers allow cam to wedge itself more tightly in place.

One-Way Drive Chain Solves Problem of Sprocket Skip

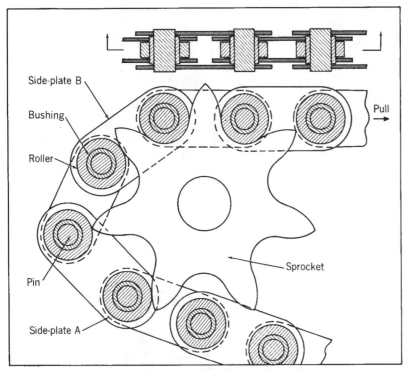

Link can skip a tooth in conventional chain (above), because tension on chain squeezes bushing, which forces next link to rotate and rise. But if all links are offset (below) and chain is properly applied, it will drape freely around sprockets.

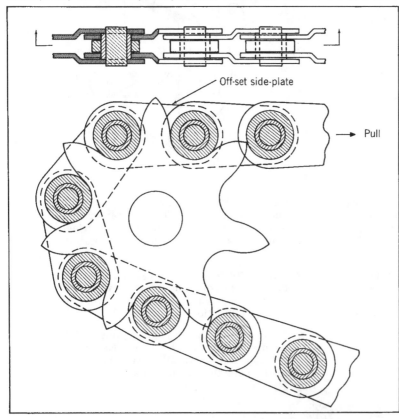

Double or nothing—that's the principle Milton Morse, president of APM Hexseal Corp., Englewood, N. J., followed when he developed his grease-free, skip-free, Kleen-chain drive-chain for bicycles. And depending upon the way it is used, it can be either the best chain drive or the worst. Morse uses it the best way and plans to make it available soon.

Alternate pairs of side plates "B" (drawing left) are attached to bushings and are carried within side-plates "A." These outer side-plates are secured in pairs to pins that are carried, but still free to move in the bushings. The bushings also carry rollers that contact, and tend to rotate with, the sprocket teeth. But it is this tendency for the bushings to rotate, plus the friction between the roller and the bushing when the chain is under tension, that causes the chain to rise and skip over sprocket teeth in conventional design.

This condition occurs only on alternate teeth where pin-supported side-plates are applying tension to the chain. The adjacent bushing-supported side plate is forced to roll on the tooth and rise as the bushing is squeezed between roller and pin.

No-skip link. On the next tooth, bushing-supported side-plates are transferring the tension to the chain and only the roller is squeezed—this time between the tooth and the bushing. The pins supporting the link following are not under pressure and are free to let the approaching links drape in the ideal manner around the sprocket. Regardless of how small the sprocket is, the chain will not attempt to skip.

Morse's chain doesn't skip, because he made it out of only one type of link, by using offset links (drawing left). One end of the side-plates is connected to bushings, and the other is connected to pins. If the chain is applied so the tension is transferred to the tooth only by bushings, the adjacent link will always be pin-supported, and the chain will drape freely around the smallest sprocket. By using smaller sprockets, it is possible to make larger reductions with fewer stages.

Find the Length of Open and Closed Belts

The following formulas give the answers (see the illustrations for notation):

Open length, $L = \pi D + (\tan \theta - \theta)(D - d)$

Closed length, $L = (D + d)[\pi + (\tan \theta - \theta)]$

You can find θ from (for open belts): $\cos \theta = (D - d)/2C$; (for closed belts) $\cos \theta = (D + d)/2C$.

When you want to find the **center distance of belt drives**, however, it is much quicker if you have a table that gives you $y = \cos \theta$ in terms of $x = (\tan \theta) - \theta$. **Sidney Kravitz**, of Picatinny Arsenal, has compiled such a table. Now, all you need do to find C is first calculate $x = [L/(D + d)] - \pi$ for open drives.

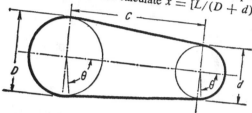

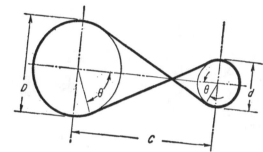

y values[1]

x	0.00	0.01	0.02	0.03	0.04	0.05	0.06	0.07	0.08	0.09
0.0	1.00000	0.95332	0.92731	0.90626	0.88804	0.87175	0.85690	0.84318	0.83039	0.81839
0.1	.80705	.79630	.78606	.77629	.76693	.75795	.74931	.74098	.73295	.72518
0.2	.71767	.71038	.70332	.69646	.68980	.68332	.67701	.67086	.66487	.65902
0.3	.65331	.64774	.64230	.63698	.63177	.62668	.62169	.61681	.61202	.60733
0.4	.60274	.59822	.59380	.58946	.58520	.58101	.57690	.57286	.56889	.56499
0.5	0.56116	0.55738	0.55367	0.55002	0.54643	0.54289	0.53941	0.53598	0.53260	0.52927
0.6	.52600	.52277	.51958	.51645	.51336	.51031	.50730	.50433	.50141	.49852
0.7	.49567	.49286	.49009	.48735	.48465	.48198	.47935	.47675	.47417	.47164
0.8	.46913	.46665	.46420	.46179	.45940	.45703	.45470	.45239	.45011	.44785
0.9	.44562	.44342	.44123	.43908	.43694	.43483	.43274	.43068	.42863	.4266?

x	.0	.1	.2	.3	.4	.5	.6	.7	.8
1	0.42461	0.40568	0.38850	0.37284	0.35848	0.34526	0.33304	0.32170	0.31115
2	.29208	.28344	.27531	.26766	.26043	.25359	.24712	.24098	.23515
3	.22431	.21926	.21445	.20984	.20544	.20121	.19717	.19328	.189??
4	.18251	.17918	.17598	.17289	.16991	.16703	.16424	.16156	.1??
5	0.15400	0.15163	0.14935	0.14712	0.14497	0.14287	0.14084	0.13886	?
6	.13326	.13149	.12977	.12810	.12646	.12487	.12332	.121??	
7	.11748	.11611	.11477	.11346	.11217	.11092	.10970	.10??	
8	.10506	.10396	.10289	.10183	.10080	.09979	.09880		
9	.09503	.09413	.09325	.09238	.09153	.09070	.08988		

x	0	1	2	3	4	5	6
10	0.08675	0.07980	0.07389	0.06879	0.06436	0.06046	0.0?
20	.04641	.04435	.04246	.04073	.03914	.03766	
30	.03169	.03072	.02980	.02894	.02812	.02735	
40	.02406	.02350	.02296	.02244	.02195	.0214?	
50	0.01939	0.01903	0.01867	0.01833	0.01800	0.01768	0.?
60	.01624	.01598	.01573	.01549	.01525	.01502	.0?
70	.01397	.01378	.01359	.01341	.01323	.01310	.012?
80	.01226	.01211	.01197	.01183	.01169	.01155	.0114?
90	.01092	.01080	.01069	.01057	.01046	.01036	.01025
100	0.00985	(see note below for x > 100)					

[1]If $x = (\tan \psi) - \psi$; then $y = \cos \psi$.

If $x > 100$, calculate C from $C = \dfrac{L}{2} - \dfrac{\pi}{4}(D + d)$ for both open and closed belts.

SECTION 9

BELTS & BELTING

Ten Types of Belt Drives

Although countless types of belt drives are possible, these ten will solve most
industrial applications. These pertain to power transmission only; the tooth
type of timing belt is not included. For each drive are given: design pitfalls;
speed and capacity ranges; and suggestions for application.

George R. Lederer

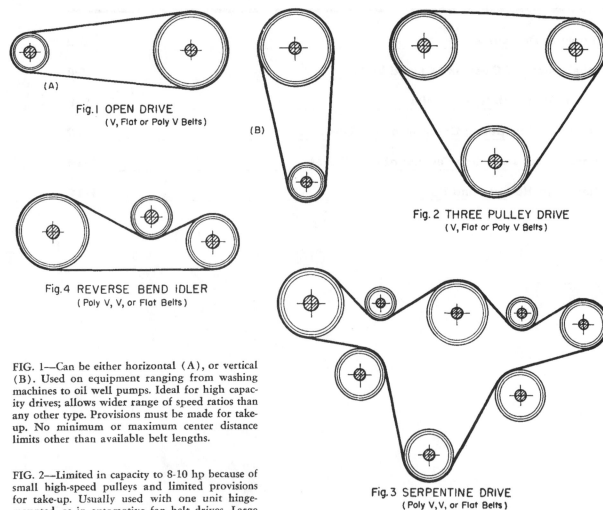

Fig. 1 OPEN DRIVE
(V, Flat or Poly V Belts)

(A)

(B)

Fig. 2 THREE PULLEY DRIVE
(V, Flat or Poly V Belts)

Fig. 4 REVERSE BEND IDLER
(Poly V, V, or Flat Belts)

Fig. 3 SERPENTINE DRIVE
(Poly V, V, or Flat Belts)

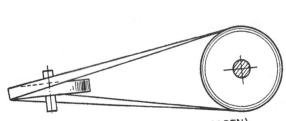

Fig. 5 QUARTER-TURN DRIVE (OPEN)
(Poly V, V, or Flat Belts)

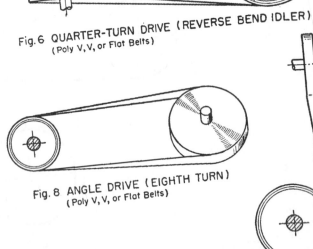

Fig. 6 QUARTER-TURN DRIVE (REVERSE BEND IDLER)
(Poly V, V, or Flat Belts)

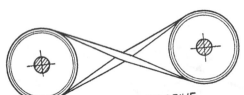

Fig. 7 CROSSED BELT DRIVE
(Flat Only)

Fig. 8 ANGLE DRIVE (EIGHTH TURN)
(Poly V, V, or Flat Belts)

Fig. 9 MULE DRIVE
(Poly V, V, or Flat Belts)

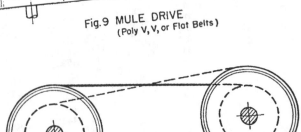

Fig. 10 VARIABLE SPEED DRIVE
(V Belts Only)

FIG. 1—Can be either horizontal (A), or vertical
(B). Used on equipment ranging from washing
machines to oil well pumps. Ideal for high capacity drives; allows wider range of speed ratios than
any other type. Provisions must be made for take-
up. No minimum or maximum center distance
limits other than available belt lengths.

FIG. 2—Limited in capacity to 8-10 hp because of
small high-speed pulleys and limited provisions
for take-up. Usually used with one unit hinge-
mounted, as in automotive fan belt drives. Large
belts would stretch beyond capacity of hinged
unit to take up slack.

FIG. 3—Useful where several units are driven
from a central shaft. The Vv belt that resembles
two v-belts joined back-to-back was developed
especially for this drive. For Vv operation all pulleys must be grooved. For regular V-, Poly-V or
flat-belt operation, only those driven by the belt
face are grooved, others are flat and are driven by
the back of the belt. Driving capacity range from
15-25 hp. Sheaves are small, speed is slow and belt
flex is extremely high, which affects belt life.

FIG. 4—Used where driver and driven sheaves are
fixed and there is no provision for take-up. Idler

is placed on the slack side of the belt near the point where the
belt leaves the driver sheave. Idler also gives increased wrap and
increased arc of contact. Applications range from agricultural
jackshaft drives to machine tools and large oil field drives. Idler
can be spring loaded to keep belt tight if drive is subject to shock.
For maximum belt life, the larger the idler, the better.

FIG. 5—Driver and driven sheaves are at right angles; belt must
travel around horizontal sheave, turn, go over vertical sheave and
return. Bend must be gradual to prevent belt from leaving sheave.
Minimum center distance for V-belts is 5.5 in. × (pitch dia of
largest sheave + width of sheave).
For Poly-V, minimum center distance = 13 × pitch dia of small
sheave or 5.5 × (pitch dia + belt width). For flat belts it is 8 ×

(pitch dia + width). V-belt sheaves must be deep
grooved and close matching is essential. Speed usually
ranges from 3,000-5,000 rpm; hp from 75-150.

FIG. 6—Similar to Open Quarter-Turn but has higher
capacity with shorter centers and increased wrap. Tracking is a problem with flat belts. With Poly-V drive,
speed ratio is unlimited. Angle of entry (angle between
belt and a line perpendicular to face of the sheave) is
limited to 3 deg or less.

FIG. 7—Limited to flat belts because either V-belt or
Poly-V would rub against itself and burn or wear
rapidly. Desirable only where the direction of rotation
must be reversed such as on planers, woodworking tools
in general and line shaft drives.

FIG. 8—Used where driver and driven sheave cannot be
in the same plane. Has same center distance and angle
of belt entry limitations as Quarter-Turn. Drive can be
even if take-up can be accomplished at either end or it
be fitted with a reverse bend idler, but not an inside
Angle between shaft can be from zero to 90 deg.

—Especially developed for drill presses and spe-
cations where driver and driven sheaves are at
les to each other and yet on the same plane.
e around a corner or from one floor to an-
center sheave is 90 deg from the driver and
and acts as an idler. Twists affect belt life.

FIG. 10—Sheaves must be grooved to change the pitch
diameter for variable or adjustable pitch operation. With
two sheaves and one belt, it is possible to have a range of
four different speeds. Widely used on propulsion drives and
cylinder drives on agricultural combines and machine tools.
Drive has same high capacity and advantage as standard
open drive, with wider speed range. They are mostly single
belt drives 1¼ to 2 in. wide. Small pulleys are not advisable. Most applications require special vari-speed cylinder
or traction belt.

$$x = \frac{L}{D + d} \qquad \text{for closed drives}$$

Then

$$C = \frac{D - d}{2y} \qquad \text{for open drives}$$

$$C = \frac{D + d}{2y} \qquad \text{for closed drives}$$

Example: $L = 60.0$, $D = 15.0$, $d = 10.0$, $x = (L - \pi D)/(D - d) = 2.575$, $y = 0.24874$ by linear interpolation in the table. $C = (D - d)/2y = 10.051$.

• • • •

Morton P. Matthew's letter on fractional derivatives (*PE*—July 22 '63, p 105) drew several interesting comments from readers. Here's what Professor Komkov of the University of Utah had to say on the subject. He pointed out that the question raised by Mr. Matthew is well known in mathematics, but very little publicized.

"The definition of fractional derivatives goes back to Abel, who developed around 1840 this fascinating little formula:

$$D^s(f) = \frac{d^s f(x)}{dx^s} = \frac{1}{\Gamma(-s)} \int_0^{\xi} (\xi - t)^{(-s-1)} f(t)\, dt$$

($\Gamma(n)$ is the Euler's Gamma Function).

An elementary proof of this formula is given for example in Courant's *Differential and Integral Calculus*, Part II, page 340. Abel claimed that the formula works for all real values of S, although there is no guarantee that the range of values obtained can be bounded. For a negative S Abel's operator D^s becomes an integral operator:

$$\iota^s(f(x)) = D^{-s}(f(x)) = \frac{1}{\Gamma(s)} \int_0^{\xi} (\xi - t)^{s-1} f(t)\, dt.$$

All results quoted by Mr Matthew may be easily obtained by application of Abel's Formula.

"There exists a generalization to partial differential equations of the fractional derivative. This is the so-called Riesz Operator. In one dimensional case it becomes Abel's derivative of fractional order.

"Details of the Riesz technique are explained, for example, in Chapter 10 of *Partial Differential Equations* by Duff. Unfortunately I know of no textbook which devotes more than a few pages to the subject of fractional derivatives. However, there exists a large number of papers on the subject in mathematical journals. I remember reading one by Professor John Barrett in the Pacific Journal of Mathematics (I think it was 1947) which discussed the equation:

$$\frac{d^s y}{dx^s} + \iota y = 0 \quad \text{where } 1 \le s \le 2$$

"There are some interesting applications in engineering and science for this theory. I was interested some years ago in formulation of elasticity equations for some plastics. I have never completed that investigation but I have established that in some cases, the behavior of plastics may be better simulated by assuming stress-strain relationship to be of the type:

$$\iota_{ij} = C_{ijkl} \frac{d^s \epsilon^{kl}}{dt^s}$$

where s is some number between 0 and 1, than by the usual assumption of linear superposition of Hooke's law and Newtonian Fluid properties. In case of some rubbers s worked to be close to 0.7."

Getting in Step with Hybrid Belts

Imaginative fusions of belts, cables, gears and chains are expanding the horizons for light-duty synchronous drives

Belts have long been used for the transfer of mechanical power. Today's familiar flat belts and V-belts are relatively light, quiet, inexpensive, and tolerant of alignment errors. They transmit power solely through frictional contacts. However, they function best at moderate speeds (4000 to 6000 fpm) under static loads. Their efficiencies drop slightly at low speeds, and centrifugal effects limit their capacities at high speeds. Moreover, they are inclined to sip under shock loads or when starting and braking. Even under constant rotation, standard belts tend to creep. Thus, these drives must be kept under tension to function properly, increasing loads on pulley shaft bearings.

Gears and chains, on the other hand, transmit power through bearing forces between positively engaged surfaces. They do not slip or creep, as measured by the relative motions of the driving and driven shafts. But the contacts themselves can slip significantly as the chain rollers and gear teeth move in and out of mesh.

Positive drives are also very sensitive to the geometries of the mating surfaces. A gear's load is borne by one or two teeth, thus magnifying small tooth-to-tooth errors. A chain's load is more widely distributed, but chordal variations in the driving wheel's effective radius produce small oscillations in the chain's velocity.

To withstand these stresses, chains and gears must be carefully made from hard materials and must then be lubri-

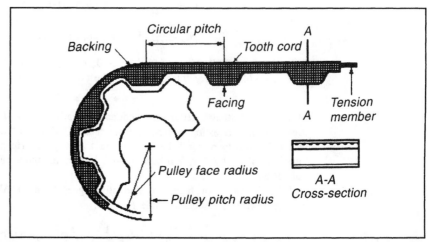

Fig. 1 **Conventional timing belts** have fiberglass or polyester tension members, bodies of neoprene or polyurethane, and trapezoidal tooth profiles.

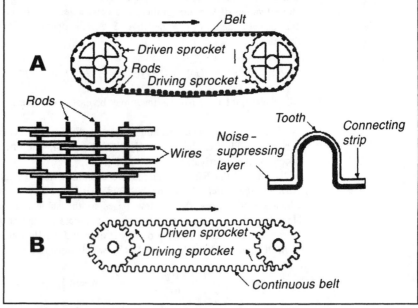

Fig. 2 **NASA metal timing belts** exploit stainless steel's strength and flexibility, and are coated with sound-and friction-reducing plastic.

Source: *Mechanisms and Mechanical Devices Sourcebook*, 3E, by Chironis & Sclater, © 2001 McGraw-Hill

cated in operations. Nevertheless, their operating noise betrays sharp impacts and friction between mating surfaces.

The cogged timing belt, with its trapezoidal teeth (Fig. 1), is the best-known fusion of belt, gear, and chain. Though these well-established timing belts can handle high powers (up to 800 hp), many of the newer ideas in synchronous belting have been incorporated into low and fractional horsepower drives for instruments and business machines.

Steel Belts for Reliability

Researchers at NASA's Goddard Space Flight Center (Greenbelt, MD) turned to steel in the construction of long-lived toothed transmission belts for spacecraft instrument drives.

The NASA engineers looked for a belt design that would retain its strength and hold together for long periods of sustained or intermittent operation in hostile environments, including extremes of heat and cold.

Two steel designs emerged. In the more chain-like version (Fig. 2A), wires running along the length of the belt are wrapped at intervals around heavier rods running across the belt. The rods do double duty, serving as link pins and as teeth that mesh with cylindrical recesses cut into the sprocket. The assembled belt is coated with plastic to reduce noise and wear.

In the second design (Fig. 2B), a strip of steel is bent into a series of U-shaped teeth. The steel is supple enough to flex as it runs around the sprocket with its protruding transverse ridges, but the material resists stretching. This belt, too, is plastic-coated to reduce wear and noise.

The V-belt is best formed from a continuous strip of stainless steel "not much thicker than a razor blade," according to the agency, but a variation can be made by welding several segments together.

NASA has patented both belts, which are now available for commercial licensing. Researchers predict that they will be particularly useful in machines that must be dismantled to uncover the belt pulleys, in permanently encased machines, and in machines installed in remote places. In addition, stainless-steel belts might find a place in high-precision instrument drives because they neither stretch nor slip.

Though plastic-and-cable belts don't have the strength or durability of the NASA steel belts, they do offer versatility and production-line economy. One of the least expensive and most adaptable is

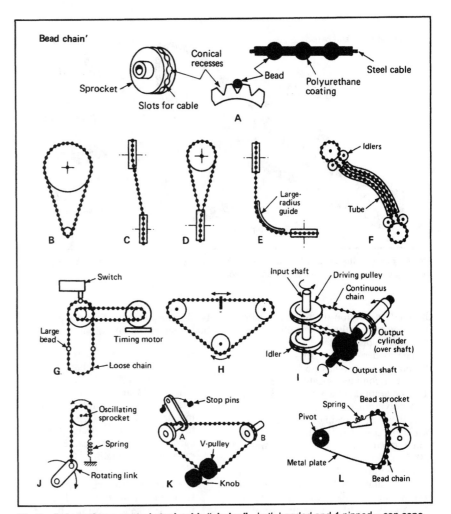

Fig. 3 Polyurethane-coated steel-cable "chains"—both beaded and 4-pinned—can cope with conditions unsuitable for most conventional belts and chains.

Table 1. Conventional Timing Belts

Type	Circular pitch, in.	Wkg. tension lb/in. width	Centr. loss const., K_c
Standard (Fig 1)			
MXL	0.080	32	10×10^{-9}
XL	0.200	41	27×10^{-9}
L	0.375	55	38×10^{-9}
H	0.500	140	53×10^{-9}
40DP	0.0816	13	—
High-torque (Fig 7)			
3 mm	0.1181	60	15×10^{-9}
5 mm	0.1968	100	21×10^{-9}
8 mm	0.3150	138	34×10^{-9}

Courtesy Stock Drive Products

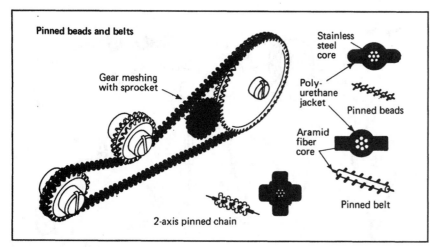

Fig. 4 Plastic pins eliminate the bead chain's tendency to cam out of pulley recesses, and permit greater precision in angular transmission.

the modern version of the bead chain, now common only in key chains and light-switch pull-cords.

The modern bead chain—if chain is the proper word—has no links. It has, instead, a continuous cable of stainless steel or aramid fiber which is covered with polyurethane. At controlled intervals, the plastic coating is molded into a bead (Fig. 3A). The length of the pitches thus formed can be controlled to within 0.001 in.

In operation, the cable runs in a grooved pulley; the beads seat in conical recesses in the pulley face. The flexibility, axial symmetry, and positive drive of

bead chain suit a number of applications, both common and uncommon:

- An inexpensive, high-ratio drive that resists slipping and requires no lubrication (Fig. 3B). As with other chains and belts, the bead chain's capacity is limited by its total tensile strength (typically 40 to 80 lb for a single-strand steel-cable chain), by the speed-change ratio, and by the radii of the sprockets or pulleys.

- Connecting misaligned sprockets. If there is play in the sprockets, or if the sprockets are parallel but lie

in different planes, the bead chain can compensate for up to 20° of misalignment (Fig. 3C).

- Skewed shafts, up to 90° out of phase (Fig. 3D).

- Right-angle and remote drives using guides or tubes (Figs. 3E and 3F). These methods are suitable only for low-speed, low-torque applications. Otherwise, frictional losses between the guide and the chain are unacceptable.

- Mechanical timing, using oversize beads at intervals to trip a microswitch (Fig. 3G). The chain can be altered or exchanged to give different timing schemes.

- Accurate rotary-to-linear motion conversion (Fig. 3H).

- Driving two counter-rotating outputs from a single input, using just a single belt (Fig. 3I).

- Rotary-to-oscillatory motion conversion (Fig. 3J).

- Clutched adjustment (Fig. 3K). A regular V-belt pulley without recesses permits the chain to slip when it reaches a pre-set limit. At the same time, bead-pulleys keep the output shafts synchronized. Similarly, a pulley or sprocket with shallow recesses permits the chain to slip one bead at a time when overloaded.

- Inexpensive "gears" or gear segments fashioned by wrapping a bead chain round the perimeter of a disk or solid arc of sheet metal (Fig. 3L). The sprocket then acts as a pinion. (Other designs are better for gear fabrication.)

A More Stable Approach

Unfortunately, bead chains tend to cam out of deep sprocket recesses under high loads. In its first evolutionary step, the simple spherical bead grew limbs—two pins projecting at right angles to the cable axis (Fig. 4). The pulley or sprocket looks like a spur gear grooved to accommodate the belt; in fact, the pulley can mesh with a conventional spur gear of proper pitch.

Versions of the belt are also available with two sets of pins, one projecting vertically and the other horizontally. This arrangement permits the device to drive a series of perpendicular shafts without twisting the cable, like a bead chain but without the bead chain's load limitations. Reducing twist increases the transmission's lifetime and reliability.

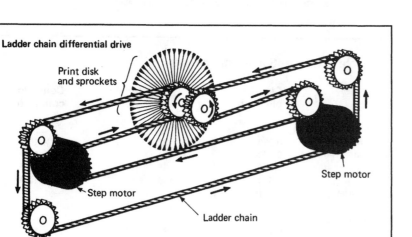

Fig. 5 A plastic-and-cable ladder chain in an impact-printer drive. In extreme conditions, such hybrids can serve many times longer than steel.

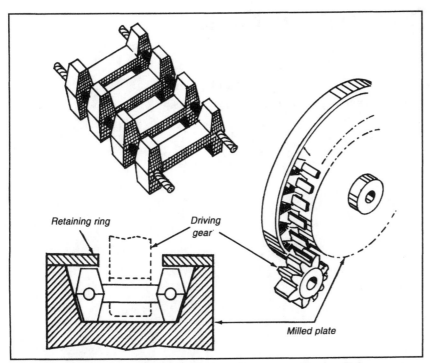

Fig. 6   A gear chain can function as a ladder chain, as a wide V-belt, or, as here, a gear surrogate meshing with a standard pinion.

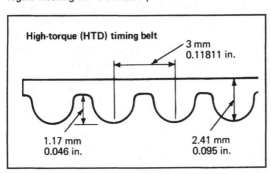

Fig. 7   Curved high-torque tooth profiles (just introduced in 3-mm and 5-mm pitches) increase load capacity of fine-pitch neoprene belts.

These belt-cable-chain hybrids can be sized and connected in the field, using metal crimp-collars. However, nonfactory splices generally reduce the cable's tensile strength by half.

Parallel-Cable Drives

Another species of positive-drive belt uses parallel cables, sacrificing some flexibility for improved stability and greater strength. Here, the cables are connected by rungs molded into the plastic coating, giving the appearance of a ladder (Fig. 6). This "ladder chain" also meshes with toothed pulleys, which need not be grooved.

A cable-and-plastic ladder chain is the basis for the differential drive system in a Hewlett-Packard impact printer (Fig. 5). When the motors rotate in the same direction at the same speed, the carriage moves to the right or left. When they rotate in opposite directions, but at the same speed, the carriage remains stationary and the print-disk rotates. A differential motion of the motors produces a combined translation and rotation of the print-disk.

The hybrid ladder chain is also well suited to laboratory of large spur gears from metal plates or pulleys (Fig. 6). Such a "gear" can run quietly in mesh with a pulley or a standard gear pinion of the proper pitch.

Another type of parallel-cable "chain," which mimics the standard chain, weighs just 1.2 oz/ft, requires no lubrication, and runs almost silently.

A Traditional Note

A new high-capacity tooth profile has been tested on conventional cogged belts. It has a standard cord and elastic body construction, but instead of the usual trapezoid, it has curved teeth (Fig. 7). Both 3-mm and 5-mm pitch versions have been introduced.

Equations for Computing Creep in Belt Drives

Don't confuse slippage caused by overloading with creep found in all belt drives. These equations give the creep rate and power loss in various pulley systems.

Peter L. Garrett

WHEN a belt is transmitting power there exists a tension T_1 on the tight side that is greater than the tension T_2 on the slack side. This is due to friction between belt and pulley. By the rules of equilibrium, the maximum possible tension ratio, T_1/T_2, that can be transmitted without slip is given by the following equation (refer to figure below):

$$\left(\frac{T_1}{T_2}\right)_{max} = e^{\mu\theta} \qquad (1)$$

where

T_1 = tension at tight side, lb
T_2 = tension at slack side, lb
e = natural logarithm base = 2.718

μ = coefficient of friction
θ = angle of wrap, radians

For velocities substantially below the speed of sound this equation has been confirmed by tests. For those sufficiently curious to note the effect of extremely high speeds, Eq 1 can be modified to include the Mach number, M.

$$\left(\frac{T_1}{T_2}\right)_{max} = e^{\left(\frac{\mu\theta}{1-M^2}\right)} \qquad (1A)$$

Any load greater than $e^{\mu\theta}$ in Eq 1 will cause belt slippage, loss of synchronism and serious wear. Coefficients of friction for belts and ropes are listed in the table of belt and rope properties on page 89.

The relationship between the driving torque, τ, and the belt tension is

$$\tau = (T_1 - T_2)R_B \qquad (2)$$

where R_B = radius of pulley B plus one half the belt thickness.

Therefore, the initial belt tension and ratio T_1/T_2 can be increased to prevent slip without affecting the torque equation (or, more specifically, the difference in tensions, $T_1 - T_2$). The limiting factors in this case are the belt stress and bearing loads.

Yet even when there is no slip the belt inexorably creeps backward over the pulleys, an effect that is particu-

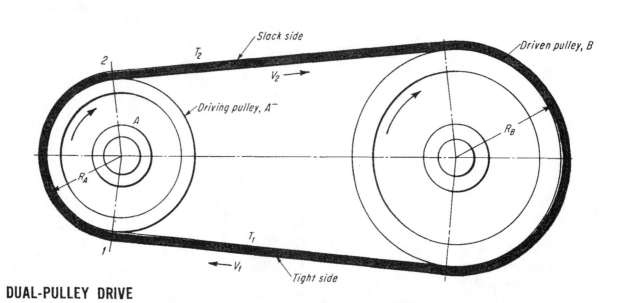

DUAL-PULLEY DRIVE

Equations for Computing Creep in Belt Drives

Don't confuse slippage caused by overloading with creep found in all belt drives. These equations give the creep rate and power loss in various pulley systems.

Peter L. Garrett

WHEN a belt is transmitting power there exists a tension T_1 on the tight side that is greater than the tension T_2 on the slack side. This is due to friction between belt and pulley. By the rules of equilibrium, the maximum possible tension ratio, T_1/T_2, that can be transmitted without slip is given by the following equation (refer to figure below):

$$\left(\frac{T_1}{T_2}\right)_{max} = e^{\mu\theta} \qquad (1)$$

where

$T_1 =$ tension at tight side, lb
$T_2 =$ tension at slack side, lb
$e =$ natural logarithm base = 2.718

$\mu =$ coefficient of friction
$\theta =$ angle of wrap, radians

For velocities substantially below the speed of sound this equation has been confirmed by tests. For those sufficiently curious to note the effect of extremely high speeds, Eq 1 can be modified to include the Mach number, M.

$$\left(\frac{T_1}{T_2}\right)_{max} = e^{\left(\frac{\mu\theta}{1-M^2}\right)} \qquad (1A)$$

Any load greater than $e^{\mu\theta}$ in Eq 1 will cause belt slippage, loss of synchronism and serious wear. Coefficients of friction for belts and ropes

are listed in the table of belt and rope properties on page 89.

The relationship between the driving torque, τ, and the belt tension is

$$\tau = (T_1 - T_2)R_B \qquad (2)$$

where $R_B =$ radius of pulley B plus one half the belt thickness.

Therefore, the initial belt tension and ratio T_1/T_2 can be increased to prevent slip without affecting the torque equation (or, more specifically, the difference in tensions, $T_1 - T_2$). The limiting factors in this case are the belt stress and bearing loads.

Yet even when there is no slip the belt inexorably creeps backward over the pulleys, an effect that is particu-

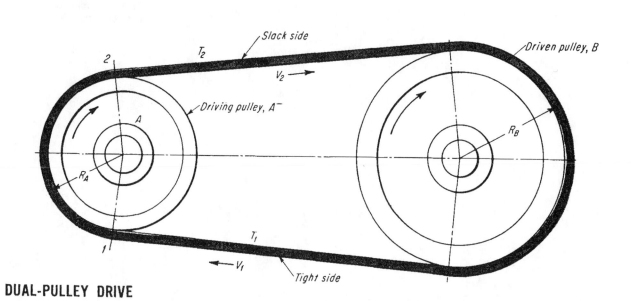

DUAL-PULLEY DRIVE

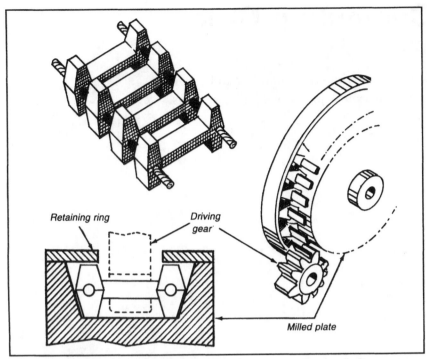

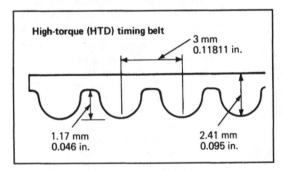

Fig. 6 A gear chain can function as a ladder chain, as a wide V-belt, or, as here, a gear surrogate meshing with a standard pinion.

Fig. 7 Curved high-torque tooth profiles (just introduced in 3-mm and 5-mm pitches) increase load capacity of fine-pitch neoprene belts.

These belt-cable-chain hybrids can be sized and connected in the field, using metal crimp-collars. However, nonfactory splices generally reduce the cable's tensile strength by half.

Parallel-Cable Drives

Another species of positive-drive belt uses parallel cables, sacrificing some flexibility for improved stability and greater strength. Here, the cables are connected by rungs molded into the plastic coating, giving the appearance of a ladder (Fig. 6). This "ladder chain" also meshes with toothed pulleys, which need not be grooved.

A cable-and-plastic ladder chain is the basis for the differential drive system in a Hewlett-Packard impact printer (Fig. 5). When the motors rotate in the same direction at the same speed, the carriage moves to the right or left. When they rotate in opposite directions, but at the same speed, the carriage remains stationary and the print-disk rotates. A differential motion of the motors produces a combined translation and rotation of the print-disk.

The hybrid ladder chain is also well suited to laboratory of large spur gears from metal plates or pulleys (Fig. 6). Such a "gear" can run quietly in mesh with a pulley or a standard gear pinion of the proper pitch.

Another type of parallel-cable "chain," which mimics the standard chain, weighs just 1.2 oz/ft, requires no lubrication, and runs almost silently.

A Traditional Note

A new high-capacity tooth profile has been tested on conventional cogged belts. It has a standard cord and elastic body construction, but instead of the usual trapezoid, it has curved teeth (Fig. 7). Both 3-mm and 5-mm pitch versions have been introduced.

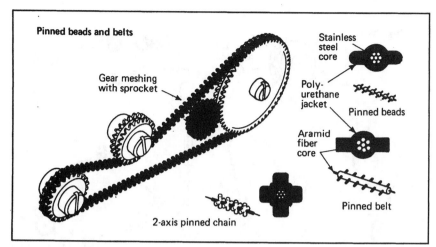

Fig. 4 Plastic pins eliminate the bead chain's tendency to cam out of pulley recesses, and permit greater precision in angular transmission.

the modern version of the bead chain, now common only in key chains and light-switch pull-cords.

The modern bead chain—if chain is the proper word—has no links. It has, instead, a continuous cable of stainless steel or aramid fiber which is covered with polyurethane. At controlled intervals, the plastic coating is molded into a bead (Fig. 3A). The length of the pitches thus formed can be controlled to within 0.001 in.

In operation, the cable runs in a grooved pulley; the beads seat in conical recesses in the pulley face. The flexibility, axial symmetry, and positive drive of

bead chain suit a number of applications, both common and uncommon:

- An inexpensive, high-ratio drive that resists slipping and requires no lubrication (Fig. 3B). As with other chains and belts, the bead chain's capacity is limited by its total tensile strength (typically 40 to 80 lb for a single-strand steel-cable chain), by the speed-change ratio, and by the radii of the sprockets or pulleys.

- Connecting misaligned sprockets. If there is play in the sprockets, or if the sprockets are parallel but lie

in different planes, the bead chain can compensate for up to 20° of misalignment (Fig. 3C).

- Skewed shafts, up to 90° out of phase (Fig. 3D).

- Right-angle and remote drives using guides or tubes (Figs. 3E and 3F). These methods are suitable only for low-speed, low-torque applications. Otherwise, frictional losses between the guide and the chain are unacceptable.

- Mechanical timing, using oversize beads at intervals to trip a microswitch (Fig. 3G). The chain can be altered or exchanged to give different timing schemes.

- Accurate rotary-to-linear motion conversion (Fig. 3H).

- Driving two counter-rotating outputs from a single input, using just a single belt (Fig. 3I).

- Rotary-to-oscillatory motion conversion (Fig. 3J).

- Clutched adjustment (Fig. 3K). A regular V-belt pulley without recesses permits the chain to slip when it reaches a pre-set limit. At the same time, bead-pulleys keep the output shafts synchronized. Similarly, a pulley or sprocket with shallow recesses permits the chain to slip one bead at a time when overloaded.

- Inexpensive "gears" or gear segments fashioned by wrapping a bead chain round the perimeter of a disk or solid arc of sheet metal (Fig. 3L). The sprocket then acts as a pinion. (Other designs are better for gear fabrication.)

A More Stable Approach

Unfortunately, bead chains tend to cam out of deep sprocket recesses under high loads. In its first evolutionary step, the simple spherical bead grew limbs—two pins projecting at right angles to the cable axis (Fig. 4). The pulley or sprocket looks like a spur gear grooved to accommodate the belt; in fact, the pulley can mesh with a conventional spur gear of proper pitch.

Versions of the belt are also available with two sets of pins, one projecting vertically and the other horizontally. This arrangement permits the device to drive a series of perpendicular shafts without twisting the cable, like a bead chain but without the bead chain's load limitations. Reducing twist increases the transmission's lifetime and reliability.

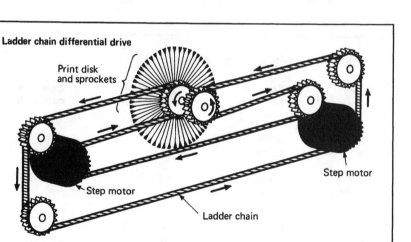

Fig. 5 A plastic-and-cable ladder chain in an impact-printer drive. In extreme conditions, such hybrids can serve many times longer than steel.

cated in operations. Nevertheless, their operating noise betrays sharp impacts and friction between mating surfaces.

The cogged timing belt, with its trapezoidal teeth (Fig. 1), is the best-known fusion of belt, gear, and chain. Though these well-established timing belts can handle high powers (up to 800 hp), many of the newer ideas in synchronous belting have been incorporated into low and fractional horsepower drives for instruments and business machines.

Steel Belts for Reliability

Researchers at NASA's Goddard Space Flight Center (Greenbelt, MD) turned to steel in the construction of long-lived toothed transmission belts for spacecraft instrument drives.

The NASA engineers looked for a belt design that would retain its strength and hold together for long periods of sustained or intermittent operation in hostile environments, including extremes of heat and cold.

Two steel designs emerged. In the more chain-like version (Fig. 2A), wires running along the length of the belt are wrapped at intervals around heavier rods running across the belt. The rods do double duty, serving as link pins and as teeth that mesh with cylindrical recesses cut into the sprocket. The assembled belt is coated with plastic to reduce noise and wear.

In the second design (Fig. 2B), a strip of steel is bent into a series of U-shaped teeth. The steel is supple enough to flex as it runs around the sprocket with its protruding transverse ridges, but the material resists stretching. This belt, too, is plastic-coated to reduce wear and noise.

The V-belt is best formed from a continuous strip of stainless steel "not much thicker than a razor blade," according to the agency, but a variation can be made by welding several segments together.

NASA has patented both belts, which are now available for commercial licensing. Researchers predict that they will be particularly useful in machines that must be dismantled to uncover the belt pulleys, in permanently encased machines, and in machines installed in remote places. In addition, stainless-steel belts might find a place in high-precision instrument drives because they neither stretch nor slip.

Though plastic-and-cable belts don't have the strength or durability of the NASA steel belts, they do offer versatility and production-line economy. One of the least expensive and most adaptable is

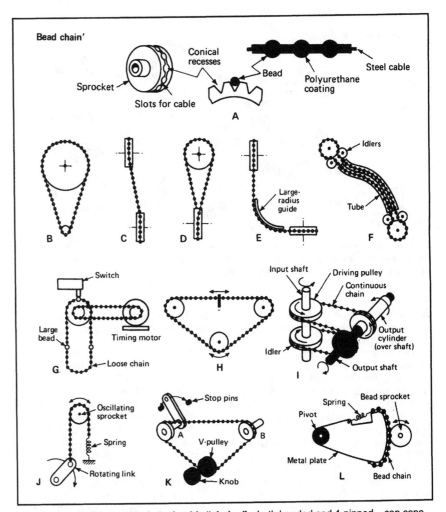

Fig. 3   Polyurethane-coated steel-cable "chains"—both beaded and 4-pinned—can cope with conditions unsuitable for most conventional belts and chains.

Table 1. Conventional Timing Belts

Type	Circular pitch, in.	Wkg. tension lb/in. width	Centr. loss const., K_c
Standard (Fig 1)			
MXL	0.080	32	10×10^{-9}
XL	0.200	41	27×10^{-9}
L	0.375	55	38×10^{-9}
H	0.500	140	53×10^{-9}
40DP	0.0816	13	—
High-torque (Fig 7)			
3 mm	0.1181	60	15×10^{-9}
5 mm	0.1968	100	21×10^{-9}
8 mm	0.3150	138	34×10^{-9}

Courtesy Stock Drive Products

Getting in Step with Hybrid Belts

Imaginative fusions of belts, cables, gears and chains are expanding the horizons for light-duty synchronous drives

Belts have long been used for the transfer of mechanical power. Today's familiar flat belts and V-belts are relatively light, quiet, inexpensive, and tolerant of alignment errors. They transmit power solely through frictional contacts. However, they function best at moderate speeds (4000 to 6000 fpm) under static loads. Their efficiencies drop slightly at low speeds, and centrifugal effects limit their capacities at high speeds. Moreover, they are inclined to sip under shock loads or when starting and braking. Even under constant rotation, standard belts tend to creep. Thus, these drives must be kept under tension to function properly, increasing loads on pulley shaft bearings.

Gears and chains, on the other hand, transmit power through bearing forces between positively engaged surfaces. They do not slip or creep, as measured by the relative motions of the driving and driven shafts. But the contacts themselves can slip significantly as the chain rollers and gear teeth move in and out of mesh.

Positive drives are also very sensitive to the geometries of the mating surfaces. A gear's load is borne by one or two teeth, thus magnifying small tooth-to-tooth errors. A chain's load is more widely distributed, but chordal variations in the driving wheel's effective radius produce small oscillations in the chain's velocity.

To withstand these stresses, chains and gears must be carefully made from hard materials and must then be lubri-

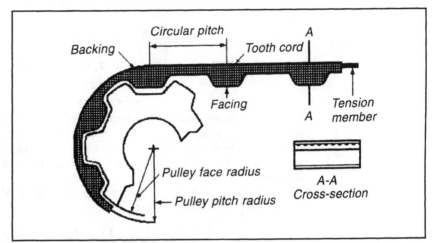

Fig. 1 Conventional timing belts have fiberglass or polyester tension members, bodies of neoprene or polyurethane, and trapezoidal tooth profiles.

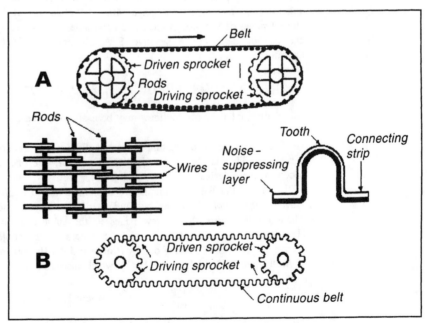

Fig. 2 NASA metal timing belts exploit stainless steel's strength and flexibility, and are coated with sound-and friction-reducing plastic.

$$x = \frac{L}{D + d} \quad \text{for closed drives}$$

Then

$$C = \frac{D - d}{2y} \quad \text{for open drives}$$

$$C = \frac{D + d}{2y} \quad \text{for closed drives}$$

Example: $L = 60.0$, $D = 15.0$, $d = 10.0$, $x = (L - \pi D)/(D - d) = 2.575$, $y = 0.24874$ by linear interpolation in the table. $C = (D - d)/2y = 10.051$.

•   •   •   •

Morton P. Matthew's letter on fractional derivatives (*PE*—July 22 '63, p 105) drew several interesting comments from readers. Here's what Professor Komkov of the University of Utah had to say on the subject. He pointed out that the question raised by Mr. Matthew is well known in mathematics, but very little publicized.

"The definition of fractional derivatives goes back to Abel, who developed around 1840 this fascinating little formula:

$$D^s (f) = \frac{d^s f(x)}{dx^s} = \frac{1}{\Gamma (-s)} \int_0^x (\xi - t)^{(-s-1)} f(t) \, dt$$

($\Gamma (n)$ is the Euler's Gamma Function).

An elementary proof of this formula is given for example in Courant's *Differential and Integral Calculus*, Part II, page 340. Abel claimed that the formula works for all real values of S, although there is no guarantee that the range of values obtained can be bounded. For a negative S Abel's operator D^s becomes an integral operator:

$$\iota^s (f(x)) = D^{-s} (f(x)) = \frac{1}{\Gamma (s)} \int_0^x (\xi - t)^{s-1} f(t) \, dt.$$

All results quoted by Mr Matthew may be easily obtained by application of Abel's Formula.

"There exists a generalization to partial differential equations of the fractional derivative. This is the so-called Riesz Operator. In one dimensional case it becomes Abel's derivative of fractional order.

"Details of the Riesz technique are explained, for example, in Chapter 10 of *Partial Differential Equations* by Duff. Unfortunately I know of no textbook which devotes more than a few pages to the subject of fractional derivatives. However, there exists a large number of papers on the subject in mathematical journals. I remember reading one by Professor John Barrett in the Pacific Journal of Mathematics (I think it was 1947) which discussed the equation:

$$\frac{d^s y}{dx^s} + \iota y = 0 \quad \text{where } 1 \leq s \leq 2$$

"There are some interesting applications in engineering and science for this theory. I was interested some years ago in formulation of elasticity equations for some plastics. I have never completed that investigation but I have established that in some cases, the behavior of plastics may be better simulated by assuming stress-strain relationship to be of the type:

$$\iota_{ij} = C_{ijkl} \frac{d^s \epsilon^{kl}}{dt^s}$$

where s is some number between 0 and 1, than by the usual assumption of linear superposition of Hooke's law and Newtonian Fluid properties. In case of some rubbers s worked to be close to 0.7."

SECTION 9

BELTS & BELTING

Ten Types of Belt Drives

Although countless types of belt drives are possible, these ten will solve most industrial applications. These pertain to power transmission only; the tooth type of timing belt is not included. For each drive are given: design pitfalls; speed and capacity ranges; and suggestions for application.

George R. Lederer

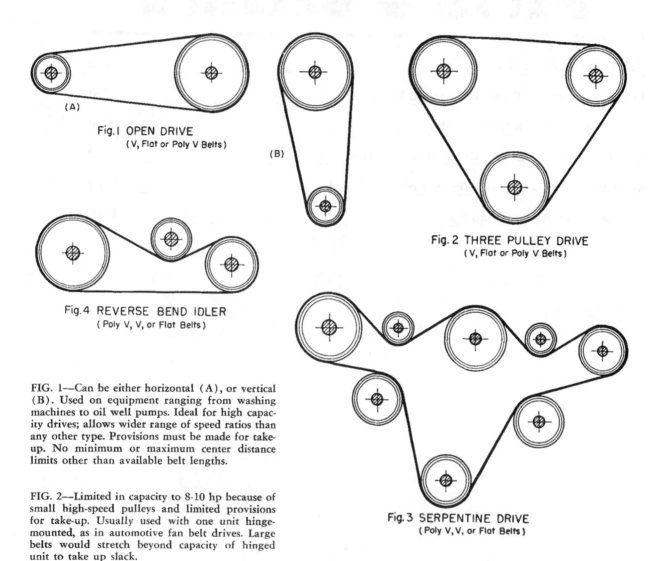

Fig.1 OPEN DRIVE
(V, Flat or Poly V Belts)

(A)

(B)

Fig. 2 THREE PULLEY DRIVE
(V, Flat or Poly V Belts)

Fig.4 REVERSE BEND IDLER
(Poly V, V, or Flat Belts)

Fig. 3 SERPENTINE DRIVE
(Poly V, V, or Flat Belts)

FIG. 1—Can be either horizontal (A), or vertical (B). Used on equipment ranging from washing machines to oil well pumps. Ideal for high capacity drives; allows wider range of speed ratios than any other type. Provisions must be made for take-up. No minimum or maximum center distance limits other than available belt lengths.

FIG. 2—Limited in capacity to 8-10 hp because of small high-speed pulleys and limited provisions for take-up. Usually used with one unit hinge-mounted, as in automotive fan belt drives. Large belts would stretch beyond capacity of hinged unit to take up slack.

FIG. 3—Useful where several units are driven from a central shaft. The Vv belt that resembles two v-belts joined back-to-back was developed especially for this drive. For Vv operation all pulleys must be grooved. For regular V-, Poly-V or flat-belt operation, only those driven by the belt face are grooved, others are flat and are driven by the back of the belt. Driving capacity range from 15-25 hp. Sheaves are small, speed is slow and belt flex is extremely high, which affects belt life.

FIG. 4—Used where driver and driven sheaves are fixed and there is no provision for take-up. Idler

is placed on the slack side of the belt near the point where the belt leaves the driver sheave. Idler also gives increased wrap and increased arc of contact. Applications range from agricultural jackshaft drives to machine tools and large oil field drives. Idler can be spring loaded to keep belt tight if drive is subject to shock. For maximum belt life, the larger the idler, the better.

FIG. 5—Driver and driven sheaves are at right angles; belt must travel around horizontal sheave, turn, go over vertical sheave and return. Bend must be gradual to prevent belt from leaving sheave. Minimum center distance for V-belts is 5.5 in. × (pitch dia of largest sheave + width of sheave).

For Poly-V, minimum center distance = 13 × pitch dia of small sheave or 5.5 × (pitch dia + belt width). For flat belts it is 8 ×

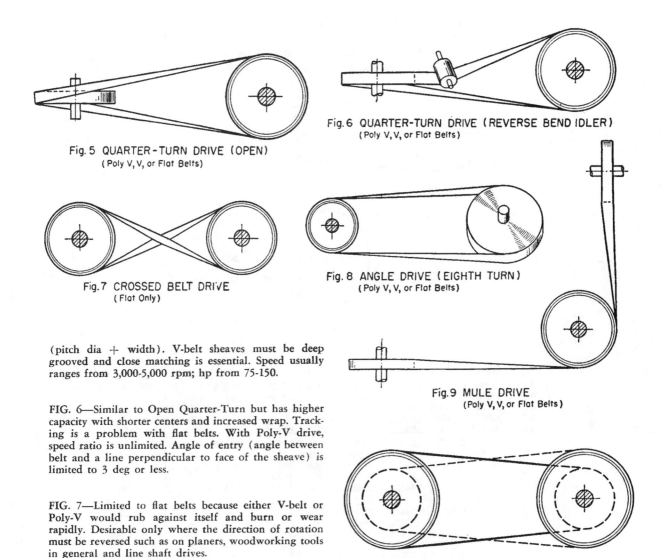

Fig. 5 QUARTER-TURN DRIVE (OPEN)
(Poly V, V, or Flat Belts)

Fig. 6 QUARTER-TURN DRIVE (REVERSE BEND IDLER)
(Poly V, V, or Flat Belts)

Fig. 7 CROSSED BELT DRIVE
(Flat Only)

Fig. 8 ANGLE DRIVE (EIGHTH TURN)
(Poly V, V, or Flat Belts)

Fig. 9 MULE DRIVE
(Poly V, V, or Flat Belts)

Fig. 10 VARIABLE SPEED DRIVE
(V Belts Only)

(pitch dia + width). V-belt sheaves must be deep grooved and close matching is essential. Speed usually ranges from 3,000-5,000 rpm; hp from 75-150.

FIG. 6—Similar to Open Quarter-Turn but has higher capacity with shorter centers and increased wrap. Tracking is a problem with flat belts. With Poly-V drive, speed ratio is unlimited. Angle of entry (angle between belt and a line perpendicular to face of the sheave) is limited to 3 deg or less.

FIG. 7—Limited to flat belts because either V-belt or Poly-V would rub against itself and burn or wear rapidly. Desirable only where the direction of rotation must be reversed such as on planers, woodworking tools in general and line shaft drives.

FIG. 8—Used where driver and driven sheave cannot be on the same plane. Has same center distance and angle of belt entry limitations as Quarter-Turn. Drive can be open if take-up can be accomplished at either end or it can be fitted with a reverse bend idler, but not an inside idler. Angle between shaft can be from zero to 90 deg.

FIG. 9—Especially developed for drill presses and special applications where driver and driven sheaves are at right angles to each other and yet on the same plane. Can operate around a corner or from one floor to another. The center sheave is 90 deg from the driver and driven sheave and acts as an idler. Twists affect belt life.

FIG. 10—Sheaves must be grooved to change the pitch diameter for variable or adjustable pitch operation. With two sheaves and one belt, it is possible to have a range of four different speeds. Widely used on propulsion drives and cylinder drives on agricultural combines and machine tools. Drive has same high capacity and advantage as standard open drive, with wider speed range. They are mostly single belt drives 1¼ to 2 in. wide. Small pulleys are not advisable. Most applications require special vari-speed cylinder or traction belt.

Find the Length of Open and Closed Belts

The following formulas give
the answers (see the illustrations for notation):
Open length, $L = \pi D + (\tan\theta - \theta)(D - d)$
Closed length, $L = (D + d)[\pi + (\tan\theta - \theta)]$
You can find θ from (for open belts): $\cos\theta = (D - d)/2C$; (for closed belts) $\cos\theta = (D + d)/2C$.

When you want to find the **center distance of belt drives,** however, it is much quicker if you have a table that gives you $y = \cos\theta$ in terms of $x = (\tan\theta) - \theta$. **Sidney Kravitz,** of Picatinny Arsenal, has compiled such a table. Now, all you need do to find C is first calculate $x = [L/(D + d)] - \pi$ for open drives.

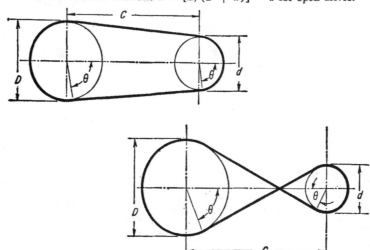

y values[1]

x	0.00	0.01	0.02	0.03	0.04	0.05	0.06	0.07	0.08	0.09
0.0	1.00000	0.95332	0.92731	0.90626	0.88804	0.87175	0.85690	0.84318	0.83039	0.81839
0.1	.80705	.79630	.78606	.77629	.76693	.75795	.74931	.74098	.73295	.72518
0.2	.71767	.71038	.70332	.69646	.68980	.68332	.67701	.67086	.66487	.65902
0.3	.65331	.64774	.64230	.63698	.63177	.62668	.62169	.61681	.61202	.60733
0.4	.60274	.59822	.59380	.58946	.58520	.58101	.57690	.57286	.56889	.56499
0.5	0.56116	0.55738	0.55367	0.55002	0.54643	0.54289	0.53941	0.53598	0.53260	0.52927
0.6	.52600	.52277	.51958	.51645	.51336	.51031	.50730	.50433	.50141	.49852
0.7	.49567	.49286	.49009	.48735	.48465	.48198	.47935	.47675	.47417	.47164
0.8	.46913	.46665	.46420	.46179	.45940	.45703	.45470	.45239	.45011	.44785
0.9	.44562	.44342	.44123	.43908	.43694	.43483	.43274	.43068	.42863	.42661

	.0	.1	.2	.3	.4	.5	.6	.7	.8	.9
1	0.42461	0.40568	0.38850	0.37284	0.35848	0.34526	0.33304	0.32170	0.31115	0.30130
2	.29208	.28344	.27531	.26766	.26043	.25359	.24712	.24098	.23515	.22960
3	.22431	.21926	.21445	.20984	.20544	.20121	.19717	.19328	.18955	.18596
4	.18251	.17918	.17598	.17289	.16991	.16703	.16424	.16156	.15895	.15644
5	0.15400	0.15163	0.14935	0.14712	0.14497	0.14287	0.14084	0.13886	0.13694	0.13508
6	.13326	.13149	.12977	.12810	.12646	.12487	.12332	.12181	.12033	.11889
7	.11748	.11611	.11477	.11346	.11217	.11092	.10970	.10850	.10733	.10618
8	.10506	.10396	.10289	.10183	.10080	.09979	.09880	.09783	.09688	.09594
9	.09503	.09413	.09325	.09238	.09153	.09070	.08988	.08908	.08829	.08751

	0	1	2	3	4	5	6	7	8	9
10	0.08675	0.07980	0.07389	0.06879	0.06436	0.06046	0.05701	0.05393	0.05116	0.04867
20	.04641	.04435	.04246	.04073	.03914	.03766	.03629	.03502	.03384	.03273
30	.03169	.03072	.02980	.02894	.02812	.02735	.02663	.02594	.02528	.02466
40	.02406	.02350	.02296	.02244	.02195	.02148	.02103	.02059	.02018	.01978
50	0.01939	0.01903	0.01867	0.01833	0.01800	0.01768	0.01737	0.01708	0.01679	0.01651
60	.01624	.01598	.01573	.01549	.01525	.01502	.01480	.01459	.01438	.01417
70	.01397	.01378	.01359	.01341	.01323	.01310	.01289	.01273	.01257	.01241
80	.01226	.01211	.01197	.01183	.01169	.01155	.01142	.01129	.01117	.01104
90	.01092	.01080	.01069	.01057	.01046	.01036	.01025	.01015	.01004	.00994
100	0.00985 (see note below for x > 100)									

[1]If $x = (\tan\psi) - \psi$; then $y = \cos\psi$.

If $x > 100$, calculate C from $C = \dfrac{L}{2} - \dfrac{\pi}{4}(D + d)$ for both open and closed belts.

larly obvious if the drive never reverses. For zero slip, one revolution of pulley A will pull $2\pi R_A$ inches of belt at tension T_1. However, as the belt approaches point 1, the tension drops to a lower value, T_2, and the belt contracts by an amount d equal to

$$d = (T_1 - T_2)\frac{2\pi R_A}{K} \quad (3)$$

where K = spring constant of belt. Typical K values are included in the table of properties on page 89.

This change in length causes localized slippage between belt and pulley with consequent power dissipation. Thus, pulley A unwinds only $(2\pi R_A - d)$ inches of belt per revolution, and the belt creeps backward an amount equal to d inches each revolution.

The K factors

Factor K in Eq 3 is the spring constant of the belt in units of force per unit strain. Thus

$$K = \frac{\text{lb}}{\text{in./in.}}$$

These units for K simplify calculations with belts of different length. The K factor is constant for all belts of a particular material and cross section. It can be calculated easily for belts of uniform material and measurable section. For example, for a steel band with a ¼ x 0.020-in. cross section

$$K = \frac{F}{FL/AE}$$

where

A = area of belt cross section, in.²
L = segment of belt length, in.
E = modulus of elasticity, psi
F = tensile loading

If $L = 1$ in., then

$$K = AE$$

or

$$\begin{aligned} K &= (\tfrac{1}{4})(0.020)(30 \times 10^6) \\ &= 150{,}000 \text{ lb/in./in.} \end{aligned}$$

For laminated belts, the calculations are more complex. Belt manufacturers usually can give you the K factor for a belt, but it is a simple matter to obtain the value by test if they do not.

For example, 30 in. of belting elongates ⅛ in. under a 20-lb load. What is the K factor?

$$\begin{aligned} \text{Strain} &= \frac{\text{deflection}}{\text{length}} \\ &= \frac{1/8}{30} = \frac{1}{240} \text{ in./in.} \end{aligned}$$

$$\begin{aligned} K &= \frac{20}{1/240} \\ &= 4800 \text{ lb/in./in.} \end{aligned}$$

Velocity relationships

Returning to Eq 3, this relationship can be put into terms of velocity. During a specific time, Δt, pulley A will have rotated through an angular displacement of θ_A. Hence the velocity of the tight side is

$$V_1 = \frac{R_A \theta_A}{\Delta t} \quad (4)$$

and of the slack side

$$V_2 = \frac{R_A \theta_A - R_A \theta_A (T_1 - T_2)\left(\dfrac{1}{K}\right)}{\Delta t} \quad (5)$$

Hence

$$\frac{V_1}{V_2} =$$

$$\frac{R_A \theta_A / \Delta t}{R_A \theta_A \left[1 - (T_1 - T_2)\left(\dfrac{1}{K}\right)\right]\Big/ \Delta t}$$

$$\frac{V_1}{V_2} = \frac{1}{1 - \dfrac{T_1 - T_2}{K}} \quad (6)$$

Velocity V_1 can be considered as the nominal belt velocity, and V_2 as being equal to V_1 minus the creep rate, or

$$V_2 = V_1 - V_1 C$$

$$\frac{V_2}{V_1} = 1 - C \quad (7)$$

where creep factor, C, is the fractional loss in speed due to elastic effects. Combining Eq 6 and 7 gives

$$1 - \frac{T_1 + T_2}{K} = 1 - C$$

$$C_A = \frac{T_1 - T_2}{K} \quad (8)$$

Since the torque load is equal to $\tau = (T_1 - T_2)R_B$, then

$$C_A = \frac{\tau}{R_A K} \quad (9)$$

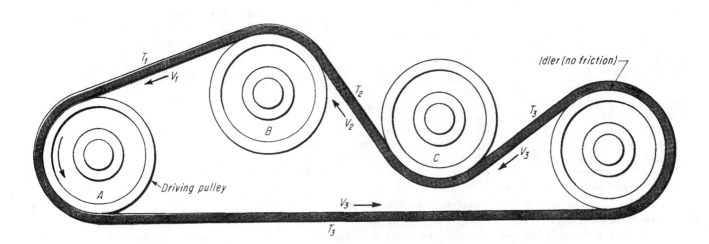

MULTIPLE-PULLEY DRIVE

SYMBOLS

C = CREEP FACTOR, DIMENSIONLESS

d = CHANGE IN BELT LENGTH, IN.

K = SPRING CONSTANT OF BELT, LB/(IN./IN.)

L = SEGMENT OF BELT LENGTH, IN.

M = MACH NUMBER — THE RATIO OF THE VELOCITY OF A MOVING BODY TO THAT OF SOUND

P = POWER, LB-IN./SEC.

R = PULLEY RADIUS + ½ BELT THICKNESS, IN.

t = TIME, SEC

T = TENSION, LB

V = LINEAR VELOCITY OF BELT, IN./SEC

VC = CREEP RATE, IN./SEC

θ = ANGLE OF WRAP, RADIANS

θ' = FINITE ANGULAR DISPLACEMENT, RADIANS

τ = TORQUE OF LOAD, LB-IN.

μ = FRICTION COEFFICIENT

SUBSCRIPTS A, B, AND C REFER TO PULLEYS A, B, AND C, RESPECTIVELY.

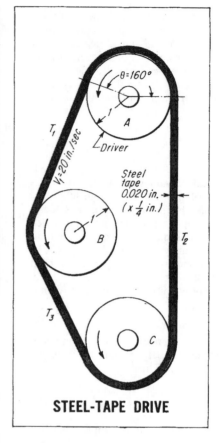

STEEL-TAPE DRIVE

Factor $1/K$ is frequently referred to as the "compliance" of the belt, in units of strain per pound. Thus from Eq 9 it can be noted that *creep is directly proportional to the product of load and compliance.*

Power losses

Ignoring windage and hysteresis losses (the energy converted into heat by the belt during the stretch and relaxation cycles), the power relationships are

Driven pulley

$$P_{in} = T_1 V_1 - T_2 V_2 \quad (10)$$

$$P_{out} = V_2(T_1 - T_2) \quad (11)$$

The power loss is

$$P_{loss} = P_{in} - P_{out}$$

Thus

$$P_{loss} = T_1(V_1 - V_2) \quad (12)$$

Driving pulley

$$P_{out} = T_1 V_1 - T_2 V_2 \quad (13)$$

An ideal driving pulley will recover the power $T_2 V_1$ and its power output will be

$$P_{out} = T_1 V_1 - T_2 V_1 \quad (14)$$

Hence, subtracting Eq 14 from Eq 13, the power loss is equal to

$$P_{loss} = T_2(V_1 - V_2) \quad (15)$$

Because T_1 is greater than T_2, comparing Eq 15 with Eq 12 shows that the *power loss at the driven pulley is slightly greater than the loss at the driver.*

Belt creep—multiple pulleys

Employing the same analytical approach to arrangements with three or more pulleys results in the following equations (refer also to the illustration of the multiple-pulley drive on previous page):

$$\frac{V_1}{V_2} = \frac{1}{1 - C_B} \quad (16)$$

$$\frac{V_2}{V_3} = \frac{1}{1 - C_C} \quad (17)$$

where

$$C_B = \frac{T_1 - T_2}{K} \quad (18)$$

$$C_C = \frac{T_2 - T_3}{K} \quad (19)$$

Hence

$$V_3 = V_2(1 - C_C) \quad (20)$$

$$V_2 = V_1(1 - C_B) \quad (21)$$

Therefore

$$V_3 = V_1(1 - C_B)(1 - C_C) \quad (22)$$

and

$$V_n = V_1(1 - C_1)(1 - C_2) \cdots (1 - C_{n-1}) \quad (23)$$

From this it can be concluded that the creep with respect to the driving pulley speed will increase as we go from pulley to pulley through the driven series of belts. Therefore it is not possible to have a reversible drive with a constant ratio between any two pulleys if there are more than two pulleys in the chain (not counting idlers).

Power loss—multiple pulleys

The power loss is due to friction in the areas, where the belt must slip to transmit the difference in tension. For all drivers and loads, this area of slippage begins at the unreeling line of tangency and ends *short* of the other line of tangency.

Pulley A (driver)

Similar to Eq 12:

$$P_{loss} = T_1(V_1 - V_3)$$

Pulley B (1st driven)

$$P_{in} = T_1 V_1 - T_2 V_2$$
$$P_{out} = V_2(T_1 - T_2)$$
$$P_{loss} = T_1(V_1 - V_2)$$

Pulley C (2nd driven)

$$P_{in} = T_2 V_2 - T_3 V_3$$
$$P_{out} = V_3(T_2 - T_3)$$
$$P_{loss} = T_2(V_2 - V_3)$$

To add the power losses of the two driven pulleys, note that $T_1 > T_2 > T_3$, and $V_1 > V_2 > V_3$; hence

$$T_1(V_1 - V_2) + T_2(V_2 - V_3)$$
$$< T_1(V_1 - V_2) + T_1(V_2 - V_3)$$

or

$$T_1(V_1 - V_2) + T_2(V_2 - V_3)$$
$$< T_1(V_1 - V_3)$$

Thus, the power losses of all the driven pulleys together is slightly greater than the loss at the driver.

Example—steel-tape drive

To show the use of the design equa-

BELT AND ROPE PROPERTIES

TYPE	MATERIAL	COEFFICIENT OF FRICTION, μ (ON CAST IRON PULLEYS)	DENSITY LB/IN.^{3}	MODULUS OF ELASTICITY, PSI	WORKING STRENGTH (MAXIMUM TENSION), PSI
FLAT BELT	LEATHER	0.25 TO 0.35	0.035	20,000	300 TO 500
	WOVEN COTTON	0.2 TO 0.3	0.035	40,000	300 TO 400
	WOVEN HAIR	0.2 TO 0.3	0.035	30,000	300 TO 400
	BALATA	0.3 TO 0.4	0.04	50,000	400 TO 500
	RUBBER	0.2 TO 0.3	0.045	30,000	400 TO 500
	STEEL	0.15 TO 0.25	0.28	30 X 10^6	10,000
V-BELT	FABRIC SET IN RUBBER	0.25 TO 0.35	0.04	35,000	400
COTTON ROPES	STRANDED COTTON	0.2 TO 0.3	0.28D^2 LB/FT D IN.=DIA	VARIABLE	200
WIRE ROPES	STRANDED STEEL WIRE	0.15 TO 0.25	1.5D^2 LB/FT D IN.=DIAMETER	VARIABLE	4000

From G. H. Ryder

tions, assume that the three-pulley drive, at left, employs a 0.020 x ¼-in. steel tape. To avoid overstressing the tape, the drive uses 2-in.-dia pulleys.

Pulley A drives the belt system.

Pulley B is coupled to an indicator with a 0.5-in.-lb load.

Pulley C is coupled to a mechanism with a 5-in.-lb load.

Other design specifications are:

Coefficient of friction for steel tape (see table), $\mu = 0.2$.

Angle θ for pulley A = 160 deg.

Belt speed, $V_1 = 20$ in./sec.

Tension, T_1, by preloading the belt = 10 lb. Thus

$$T_1 = \frac{10}{(0.020)\,(0.25)} = 2000 \text{ psi}$$

This amount is sufficient for proper belt seating, yet within the 10,000-psi limit (from the table) for the steel belt.

Calculations for the tape tensions are as follows:

$$T_2 = T_1 e^{\mu \theta_A}$$
$$T_2 = 10\, e^{(0.2)\,(2.79)} = 17.5 \text{ lb}$$

This value is the maximum permissible tension at the high-tension side of the drive—the belt will slip with higher tensions. (The actual T_2 value will depend on the driven loads and should be less than 17.5 lb.) Thus the tension drop available for work is

$$(T_2 - T_1)_{\max} = 17.5 - 10 = 7.5 \text{ lb}$$

Because $R = 1$ in., the maximum torque that can be transmitted is (1) (7.5) = 7.5 in.-lb, which is greater than the 5.5 in.-lb total load requirement.

It has been shown that the design is adequate at the driver pulley. The two driven pulleys will now be examined.

Pulley C—major load

To determine the minimum angle of wrap at pulley C, it is convenient and conservative to ignore the tension drop across the ½-in.-lb load. In other words, assume that

$$T_3 = T_1 = 10 \text{ lb}$$

To support this contention, note that for the equation

$$\frac{T_3}{T_1} = e^{\mu \theta_B}$$

when $\theta_B = 0$, $T_3/T_1 = 1$. But at any positive value of θ, $T_3/T_1 > 1$, or $T_3 > T_1$.

Returning to the analysis of pulley C, the maximum possible torque is 5 in.-lb, or 5 lb at the 1-in. radius. Hence the actual T_2 tension will be

$$T_2 = T_3 + 5 \text{ lb} = 10 + 5 = 15 \text{ lb}$$

$$e^{\mu \theta_C} = \frac{T_2}{T_3} = \frac{15}{10} = 1.5$$

$$\mu \theta_C = \ln 1.5 = 0.405$$
$$\theta_C = 0.405/0.2 = 2.02 \text{ rad} = 115 \text{ deg}$$

A wrap angle of less than 115 deg may cause slippage at pulley C.

Pulley B—minor load

In a similar manner

$$e^{\mu \theta_B} = \frac{0.5 + 10}{10}$$

$$\theta_B = 0.25 = 15 \text{ deg}$$

The power loss at the driver is

$$C_A = \frac{\tau}{R_A K}$$

$$= \frac{5.5 \text{ in.-lb}}{(1 \text{ in.} \times 150{,}000 \text{ lb/in./in.})}$$
$$C_A = 3.7 \times 10^{-5}$$

where the value for K was previously computer on page 87.

$$\frac{V_2}{V_1} = \frac{1}{1 - 3.7 \times 10^{-5}}$$
$$\approx 1.000{,}037$$

$$P_{loss} = T_1(V_2 - V_1)$$
$$= (15.5)(20)(1.000{,}037 - 1)$$
$$= 310(0.37)(10^{-4})$$
$$= 0.011 \text{ in.-lb/sec}$$

It is safe to assume that the power losses for both driven pulleys are very nearly equal to the loss at the driver. Hence, the total power loss is approximately 0.022 in.-lb/sec.

Input power = 20 in./sec x 5.5 lb = 110 in.-lb/sec; therefore the efficiency of the tape drive is

$$\frac{110 - 0.022}{110} \approx$$

$$1 - 0.00022 \text{ or } 99.98\%$$

This example shows why X-Y curve plotters and other instruments often use steel tapes for drives. The low creep rate resulting from the high modulus of steel permits use of a band that is not indexed to any pulley in the drive and can still indicate position repeatably.

Mechanisms for Adjusting Tension of Belt Drives

Sketches show devices for both manual and automatic take-up as required by wear or stretch. Some are for installations having fixed center distances; others are of the expanding center take-up types. Many units provide for adjustment of speed as well as tension.

Joseph H. Gepfect

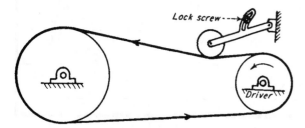

Fig. 1—Manually adjusted idler run on slack side of chain or flat belt. Useful where speed is constant, load is uniform and the tension adjustment is not critical. Can be adjusted while drive is running. Horsepower capacity depends upon belt tension.

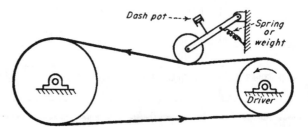

Fig. 2—Spring or weight loaded idler run on slack side of flat belt or chain provides automatic adjustment. For constant speed but either uniform or pulsating loads. Adjustments should be made while drive is running. Capacity limited by spring or weight value.

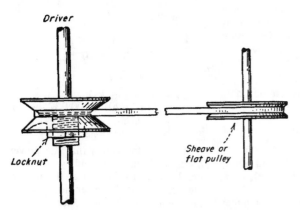

Fig. 5—Screw type split sheave for V-belts when tension adjustment is not critical. Best suited for installations with uniform loads. Running speed increases with take up. Drive must be stopped to make adjustments. Capacity depends directly upon value of belt tension.

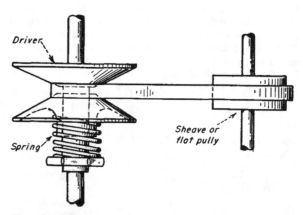

Fig. 6—Split sheave unit for automatic adjustment of V-belts. Tension on belt remains constant; speed increases with belt take up. Spring establishes maximum torque capacity of the drive. Hence, this can be used as a torque limiting or overload device.

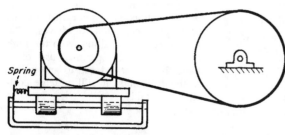

Fig. 9—Spring actuated base for automatic adjustment of uniformly loaded chain drive. With belts, it provides slipping for starting and suddenly applied torque. Can also be used to establish a safety limit for the horsepower capacity of belts.

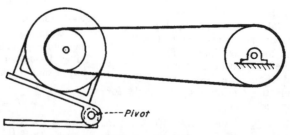

Fig. 10—Gravity actuated pivoting motor base for uniformly loaded belts or chains, only. Same safety and slipping characteristics as that of Fig. 9. Position of motor from pivot controls the proportion of motor weight effective in producing belt tension.

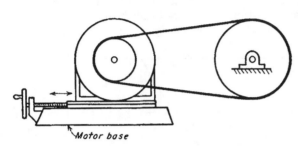

Fig. 3—Screw-base type unit provides normal tension control of belt or chain drive for motors. Wide range of adjustments can be made either while unit is running or stopped. With split sheaves, this device can be used to control speed as well as tension.

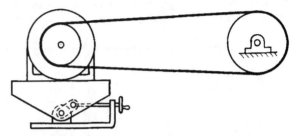

Fig. 4—Pivoting screw base for normal adjustment of motor drive tension. Like that of Fig. 3, this design can be adjusted either while running or stopped and will provide speed adjustment when used with split sheaves. Easier to adjust than previous design.

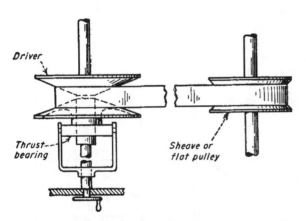

Fig. 7—Another manually adjusted screw type split sheave for V-belts. However, this unit can be adjusted while the drive is running. Other characteristics similar to those of Fig. 6. Like Fig. 6, sheave spacing can be changed to maintain speed or to vary speed.

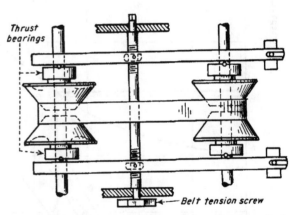

Fig. 8—Special split sheaves for accurate tension and speed control of V-belts or chains. Applicable to parallel shafts on short center distances. Manually adjusted with belt tension screw. No change in speed with changes in tension.

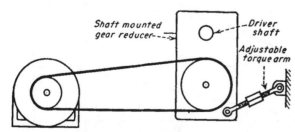

Fig. 11—Torque arm adjustment for use with shaft mounted speed reducer. Can be used as belt or chain take up for normal wear and stretch within the swing radius of reducer; or for changing speed while running when spring type split sheave is used on motor.

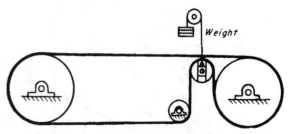

Fig. 12—Wrapping type automatic take-up for flat and wire belts of any width. Used for maximum driving capacity. Size of weight determines tension put on belt. Maximum value should be established to protect the belt from being overloaded.

Leather Belts–Hp Loss and Speeds

From 0-10,000 ft/min and 435-3450 rpm, for pulley diameters up to 30 in.

Douglas C. Greenwood

Horsepower ratings and correction factors for various leather belt sizes, tensions, and operating conditions are given by most engineering handbooks or manufacturers' catalogs. Such data, however, are usually not corrected for centrifugal force. This chart may be entered at any axis or pulley-speed curve. As shown, secants parallel to the axes connect any four values in correct relationship. In the sample construction, a 12 in. dia pulley at 1150 rpm gives a belt velocity of about 3620 fps at which speed there is a 12% hp reduction. Consult belt manufacturer regarding suitability, efficiency and other factors in high-speed applications.

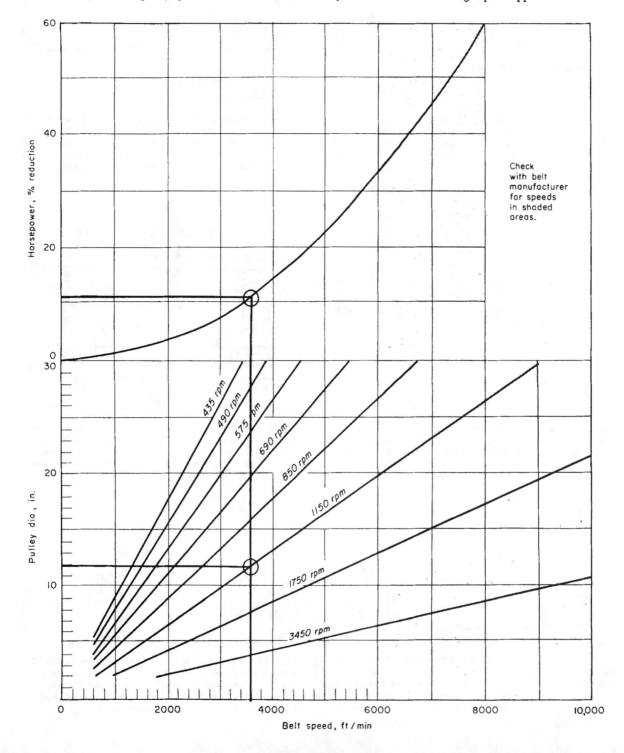

SECTION 10

SHAFTS & COUPLINGS

Overview of Shafts & Couplings

Robert O. Parmley (from *Mechanical Components Handbook*, © 1985)

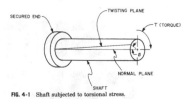

FIG. 4-1 Shaft subjected to torsional stress.

SHAFTS

A rotating bar, usually cylindrical in shape, which transmits power is called a shaft. Power is delivered to the shaft through the action of an outside tangential force, resulting in a torsional action set up in the shaft. The resultant torque allows the power to be distributed to other machines or to various components connected to the shaft.

4-1 Usage and Classification

Shafts and shafting may be classified according to their general usage. The following categories are presented here for discussion only and are basic in nature.

Engine Shafts An engine shaft may be described as a shaft directly connected to the power delivery of a motor.

Generator Shafts Generator shafts, along with engine shafts and turbine shafts, are called prime movers. There is a wide range of shaft diameters, depending on power transmission required.

Turbine Shafts Also prime movers, turbine shafts have a tremendous range of diameter size.

Machine Shafts General category of shafts. Variation in sizes of stock diameters ranges from ½ to 2½ in (increments of 1/16 in), 2½ to 4 in (increments of ⅛ in), 4 to 6 in (increments of ¼ in).

Line Shafts Line shafting is a term employed to describe long and continuous "lines of shafting," generally seen in factories, paper or steel mills, and shops where power distribution over an extended distance is required. Stock lengths of line shafting generally are 12 ft, 20 ft, and 24 ft.

Jackshafts Jackshafts are used where a shaft is connected directly to a source of power from which other shafts are driven.

Countershafts Countershafts are placed between a line shaft and a machine. The countershaft receives power from a line shaft and transmits it to the drive shaft.

4-2 Torsional Stress

A shaft is said to be under torsional stress when one end is securely held and a twisting force acts at the opposite end. Figure 4-1 illustrates this action. Note that the only deformation in the shaft is the rotation of the cross sections with respect to each other, as shown by angle ϕ.

Shafts which are subjected to torsional force only, or those with a minimal bending moment that can be disregarded, may use the following formula to obtain torque in inch-pounds, where horsepower P and rotational speed N in revolutions per minute are known.

$$T = \frac{12 \times 33,000\, P}{2\pi N} \tag{4-1}$$

4-3 Twisting Moment

Twisting moment T is equal to the product of the resultant P_r of the twisting forces multiplied by its distance from the axis R. See Fig. 4-2.

$$T = P_r \times R \tag{4-2}$$

4-4 Resisting Moment

Resisting moment T_r equals the sum of the moments of the unit shearing stresses acting along the cross section of the shaft. This moment is the force which "resists" the twisting force exerted to rotate the shaft.

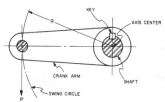

FIG. 4-2 Typical crank arm forces.

4-5 Torsion Formula for Round Shafts

Torsion formulas apply to solid or hollow circular shafts, and only when the applied force is perpendicular to the shaft's axis, if the shearing proportional limit (of the material) is not exceeded.

Conditions of equilibrium, therefore, require the "twisting" moment to be opposed by an equal "resisting" moment. The following formulas may be used to solve the allowable unit shearing stress τ if twisting moment T, diameter of solid shaft D, outside diameter of hollow shaft d, and inside diameter of hollow shaft d_1 are known.

Solid round shafts:

$$\tau = \frac{16T}{\pi D^3} \tag{4-3}$$

Hollow round shafts:

$$\tau = \frac{16Td}{\pi(d^4 - d_1^4)} \tag{4-4}$$

4-6 Shear Stress

In terms of horsepower, for shafts used in the transmission of power, shearing stress may be calculated as follows, where P = horsepower to be transmitted, N = rotational speed in revolutions per minute, and the shaft diameters are those described previously. Maximum unit shearing stress τ is in pounds per square inch.

Solid round shafts:

$$\tau = \frac{321,000P}{ND^3} \tag{4-5}$$

Hollow round shafts:

$$\tau = \frac{321,000Pd}{N(d^4 - d_1^4)} \tag{4-6}$$

The foregoing formulas do not consider any loads other than torsion. Weight of shaft and pulleys or belt tensions are not included.

4-7 Critical Speeds of Shafts

Shafts in rotation become very unstable at certain speeds, and damaging vibrations are likely to occur. The revolution at which this mechanical phenomenon takes place is called the "critical speed."

Vibration problems may occur at a "fundamental" critical speed. The following formula is used for finding this speed for a shaft on two supports, where W_1, W_2, etc. = weights of rotating components; y_1, y_2, etc. = respective static deflection of the weights; g = gravitational constant, 386 in/s².

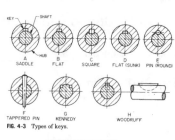

FIG. 4-3 Types of keys.

$$f = \frac{1}{2\pi}\sqrt{\frac{g(W_1 y_1 + W_2 y_2 + \cdots)}{W_1 y_1^2 + W_2 y_2^2 + \cdots}} \qquad \text{cycles/s} \tag{4-7}$$

A thorough discussion of this phenomenon is beyond the scope of this book. Readers should consult the many volumes devoted to vibration theory for an in-depth technical presentation.

4-8 Fasteners for Torque Transmission

Keys Basically keys are wedge-like steel fasteners that are positioned in a gear, sprocket, pulley, or coupling and then secured to a shaft for the transmission of power. The key is the most effective and therefore the most common fastener used for this purpose.

Figure 4-3 illustrates several standard key designs, including round and tapered pins. The saddle key (a) is hollowed to fit the shaft, without a keyway cut into the shaft. The flat key (b) is positioned on a planed surface of the shaft to give more frictional resistance. Both of these keys can transmit light loads. Square (c) and flat-sunk (d) keys fit in mating keyways, half in the shaft and half into the hub. This positive holding power provides maximum torque transfer. Round (e) and tapered (f) pins are also an excellent method of keying hubs to shafts. Kennedy (g) and Woodruff (h) keys are widely used. Figure 4-4 pictures feather keys, which are used to prevent hubs from rotating on a shaft, but will permit the component part to move along the shaft's axis. Figure 4-4a shows a key which is relatively long for axial movement and is secured in position on the shaft with two flat fillister-head matching screws. Figure 4-4b is held to the hub and moves freely with the hub along the shaft's keyseat.

A more in-depth presentation of keys will be found in Sec. 12, "Locking Components."

Set Screws Set screws may be used for light applications. A headless screw with a hexagon socket head and a conical tip should be used. Figure 4-5 illustrates both a "good" design and a "bad" design. The set screw must be threaded into the hub and tightened on the shaft to provide a positive anchor.

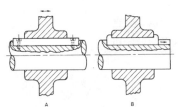

FIG. 4-4 Feather keys.

Pins Round and taper pins were briefly discussed previously, but mention should be made of the groove, spring, spiral, and shear pins. The groove pin has one or more longitudinal grooves, known as flutes, over a portion of its length. The farther you insert this pin, the tighter it becomes. The spring or slotted tubular pin is a hollow tube with a full-length slot and tapered ends. This slot allows the pin's diameter to be reduced somewhat when the pin is inserted, thus providing easy adaptation to irregular holes. Spirally coiled pins are very similar in application to spring pins. They are fabricated from a sheet of metal wrapped twice around itself, forming a spiral effect. Shear pins, of course, are used as a weak link. They are designed to fail when a predetermined force is encountered.

4-9 Splines

Spline shafts are often used instead of keys to transmit power from hub to shaft or from shaft to hub. Splines may be either square or involute.

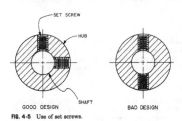

FIG. 4-5 Use of set screws.

One may think of splines as a series of teeth, cut longitudinally into the external circumference of a shaft, that match or mate with a similar series of keyways cut into the hub of a mounted component. Splines are extremely effective when a "sliding" connection is necessary, such as for a PTO (power take-off) on agricultural equipment.

Square or parallel-side splines are employed as multispline shaft fittings in series of 4, 6, 10, or 16.

Splines are especially successful when heavy torque loads and/or reversing loads are transmitted. Torque capacity (in inch-pounds) of spline fittings may be calculated by the following formula:

$$T = 1000NrhL \quad \text{in·lb} \qquad (4\text{-}8)$$

where N = number of splines
  r = mean radial distance from center of shaft/hub to center of spline
  h = depth of spline
  L = length of spline bearing surface

This gives torque based on spline side pressure of 1000 lb/in². Involute splines are similar in design to gear teeth, but modified from the standard profile. This involute contour provides greater strength and is easier to fabricate. Figure 4-6 shows five typical involute spline shapes.

SHAFT COUPLINGS

In machine design, it often becomes necessary to fasten or join the ends of two shafts axially so that they will act as a single unit to transmit power. When this parameter is required, shaft couplings are called into use. Shaft couplings

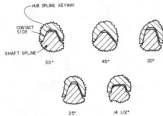

FIG. 4-6 Involute spline shapes.

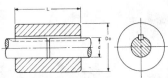

FIG. 4-7 Sleeve coupling.

are grouped into two general classifications: rigid (or solid) and flexible. A rigid coupling will not provide for shaft misalignment or reduce vibration or shock from one shaft to the other. However, flexible shaft couplings provide connection of misaligned shafts and can reduce shock and/or vibration to a degree.

4-10 Sleeve Coupling

Sleeve coupling, as illustrated in Fig. 4-7, consists of a simple hollow cylinder which is slipped over the ends of two shafts fastened into place with a key positioned into mating keyways. This is the simplest rigid coupling in use today. Note that there are no projecting parts, so that it is very safe. Additionally, this coupling is inexpensive to fabricate.

Figure 4-8 pictures two styles of sleeve couplings using standard set screws to anchor the coupling to each shaft end. One design is used for shafts of equal diameters. The other design connects two shafts of unequal diameters.

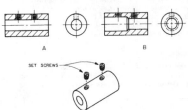

FIG. 4-8 Sleeve shaft coupling.

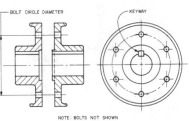

NOTE: BOLTS NOT SHOWN

FIG. 4-9 Solid coupling.

4-11 Solid Coupling

The solid coupling shown in Fig. 4-9 is a tough, inexpensive, and positive shaft connector. When heavy torque transmission is required, a rigid coupling of this design is an excellent selection.

4-12 Clamp or Compression Coupling

The rigid coupling shown in Fig. 4-10 has evolved from the basic sleeve coupling. This clamp or compression coupling simply splits into halves, which have recesses for through bolts that secure or clamp the mating parts together, producing a compression effect on the two connecting shafts. This coupling may be used for transmission of large torques because of its positive grip from frictional contact.

4-13 Flange Coupling

Flange couplings are rigid shaft connectors, also known as solid couplings. Figure 4-11 illustrates a typical design. This rigid coupling consists of two components, which are connected to the two shafts with keys. The hub halves

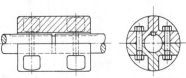

FIG. 4-10 Clamp or compression coupling.

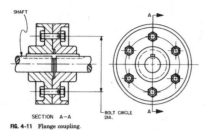

FIG. 4-11 Flange coupling.

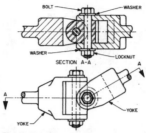

FIG. 4-13 Universal coupling.

are fastened together with a series of bolts arranged in an even pattern concentrically about the center of the shaft. A flange on the outside circumference of the hub provides a safety guard for the bolt heads and nuts, while adding strength to the total assembly.

4-14 Flexible Coupling

Flexible couplings connect two shafts which have some nonalignment between them. The couplings also absorb some shock and vibration which may be transmitted from one shaft to the other.

There are a wide variety of flexible-coupling designs. Figure 4-12 pictures a two-part cast-iron coupling which is fastened onto the shafts by keys and set

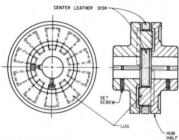

FIG. 4-12 Flexible coupling.

screws. The halves have lugs, which are cast an an integral part of each hub half. The lugs fit into entry pockets in a disk made of leather plies which are stitched and cemented together. The center leather laminated disk provides flexibility in all directions. Rotation speed, either slow or fast, will not affect the efficiency of the coupling.

4-15 Universal Coupling

If two shafts are not lined up but have intersecting centerlines or axes, a positive connection can be made with a universal coupling. Figure 4-13 details a typical universal coupling.

Note that the bolts are at right angles to each other. This makes possible the peculiar action of the universal coupling. Either yoke can be rotated about the axis of each bolt so that adjustment to the angle between connected shafts can be made. A good rule of thumb is not to exceed 15° of adjustment per coupling.

4-16 Multijawed Coupling

This rigid-type shaft coupling is a special design. The coupling consists of two halves, each of which has a series of mating teeth which lock together, forming a positive jawlike connection. Set screws secure the hubs onto the respective shafts. This style of coupling is strong and yet easily dismantled. See Fig. 4-14.

4-17 Spider-Type Coupling

The spider-type or Oldham coupling is a form of flexible coupling that was designed for connection of two shafts which are parallel but not in line. The two end hubs, which are connected to the two respective shafts, have grooved

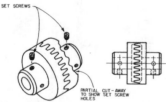

FIG. 4-14 Multijawed coupling.

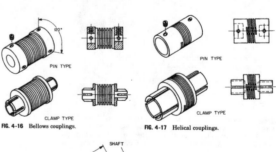

FIG. 4-16 Bellows couplings.   **FIG. 4-17** Helical couplings.

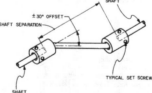

FIG. 4-18 Offset extension shaft coupling.

faces which mate with the two tongues of the center disk. This configuration and slot adjustment allow for misalignment of shafts. Figure 4-15 shows an assembled spider-type coupling.

4-18 Bellows Coupling

Two styles of bellows couplings are illustrated in Fig. 4-16. These couplings are used in applications involving large amounts of shaft misalignment, usually combined with low radial loads. Maximum permissible angular misalignment varies between 5° and 10°, depending on manufacturer's recommendation. Follow manufacturer's guidelines for maximum allowable torque. Generally, these couplings are used in small, light-duty equipment.

4-19 Helical Coupling

These couplings, also, are employed to minimize the forces acting on shafts and bearings as a result of angular and/or parallel misalignment.

These couplings are used when motion must be transmitted from shaft to shaft with constant velocity and zero backlash.

The helical coupling achieves these parameters by virtue of its patented design, which consists of a one-piece construction with a machined helical groove circling its exterior diameter. Removal of this coil or helical strip results in a flexible unit with considerable torsional strength. See Fig. 4-17, which pictures both the pin- and clamp-type designs.

4-20 Offset Extension Coupling

Figure 4-18 depicts an offset extension shaft coupling. This coupling is used to connect or join parallel drive shafts that are offset ±30° in any direction, with separations generally greater than 3 in. Shafts are secured to the coupling with set screws.

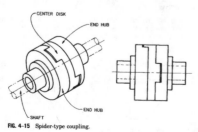

FIG. 4-15 Spider-type coupling.

REFERENCES

Master Catalog 82, Sterling Instrument Division of Designatronics, Inc., New Hyde Park, N.Y.

Levinson, Irving J.: *Machine Design*, Reston Publishing Co., Reston, Va., 1978.

Parmley, R. O.: *Standard Handbook of Fastening and Joining*, McGraw-Hill, New York, 1977.

Spotts, M. F.: *Design of Machine Elements*, 5th ed., Prentice-Hall, Englewood Cliffs, N.J., 1978.

Winston, Stanton E.: *Machine Design*, American Technical Society, Chicago, 1966.

Carmichael, Colin, ed.: *Kent's Mechanical Engineer's Handbook*, 12th ed., Wiley, New York, 1958.

Critical Speeds of End Supported Bare Shafts

L. Morgan Porter

THIS NOMOGRAM solves the equation for the critical speed of a bare steel shaft that is hinged at the bearings. For one bearing fixed and the other hinged multiply the critical speed by 1.56. For both bearings fixed, multiply the critical speed by 2.27. The scales for critical speed and length of shaft are folded; the right hand sides, or the left hand sides, of each are used together. The chart is valid for both hollow and solid shafts. For solid shafts, $D_2 = 0$.

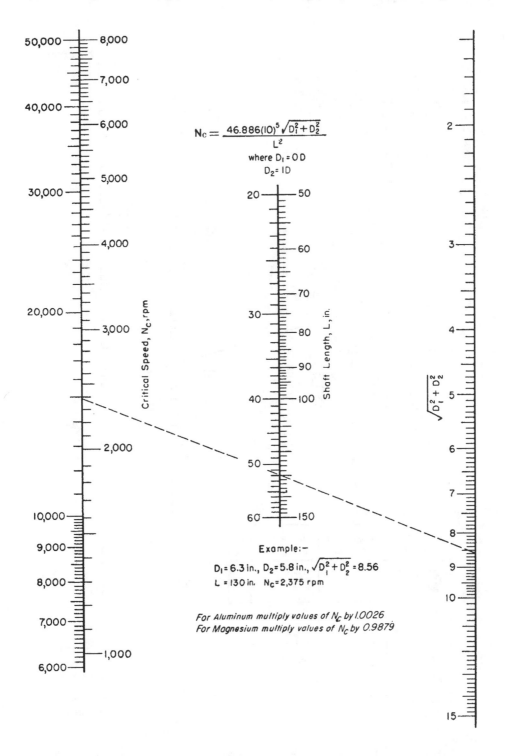

$$N_c = \frac{46.886(10)^5 \sqrt{D_1^2 + D_2^2}}{L^2}$$

where $D_1 = OD$
$D_2 = ID$

Critical Speed, N_c, rpm

Shaft Length, L, in.

$\sqrt{D_1^2 + D_2^2}$

Example:—

$D_1 = 6.3$ in., $D_2 = 5.8$ in., $\sqrt{D_1^2 + D_2^2} = 8.56$
$L = 130$ in. $N_c = 2,375$ rpm

For Aluminum multiply values of N_c by 1.0026
For Magnesium multiply values of N_c by 0.9879

Shaft Torque:
Charts Find Equivalent Sections

An easy way to convert solid circular shafts to equivalent-strength shafts of hollow circular, elliptical, square, and rectangular sections.

Dr. Biswa Nath Ghosh

1—ROUND and ELLIPTICAL SHAFTS

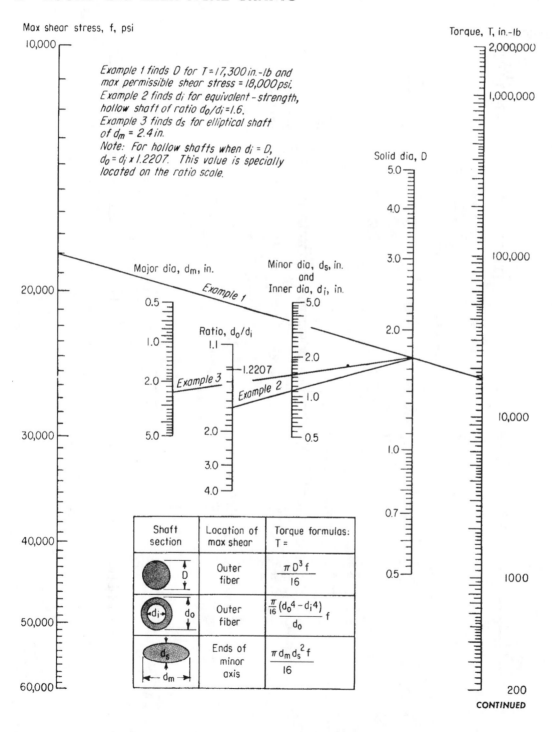

Max shear stress, f, psi

Example 1 finds D for T = 17,300 in.-lb and max permissible shear stress = 18,000 psi.
Example 2 finds d_i for equivalent-strength, hollow shaft of ratio d_o/d_i = 1.6.
Example 3 finds d_s for elliptical shaft of d_m = 2.4 in.
Note: For hollow shafts when d_i = D, d_o = d_i x 1.2207. This value is specially located on the ratio scale.

Torque, T, in.-lb

Shaft section	Location of max shear	Torque formulas: T =
D	Outer fiber	$\dfrac{\pi D^3 f}{16}$
d_i d_o	Outer fiber	$\dfrac{\frac{\pi}{16}(d_o^4 - d_i^4)}{d_o} f$
d_s d_m	Ends of minor axis	$\dfrac{\pi d_m d_s^2 f}{16}$

CONTINUED

2—SQUARE and RECTANGULAR SHAFTS

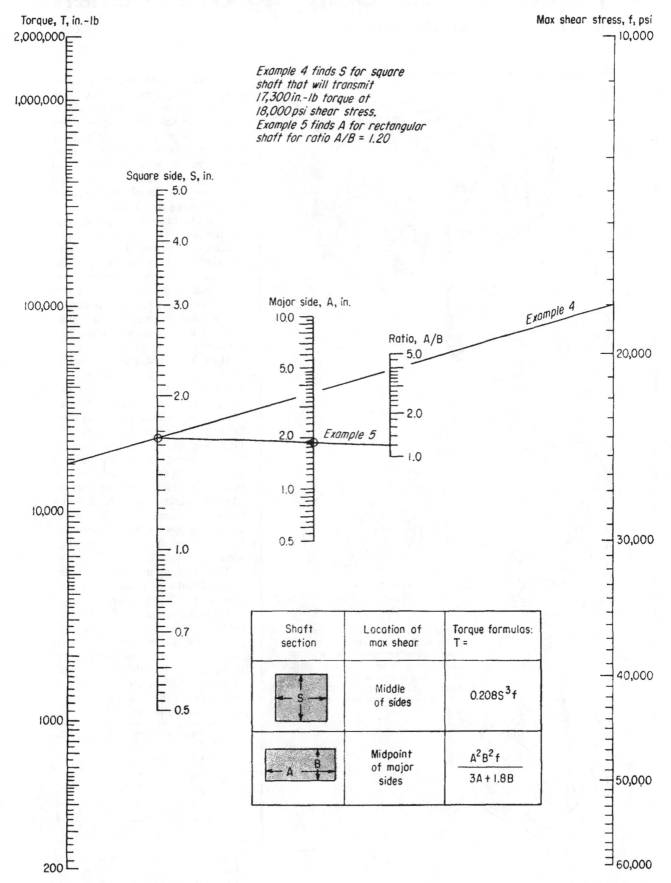

Torque, T, in.-lb

Max shear stress, f, psi

Example 4 finds S for square
shaft that will transmit
17,300 in.-lb torque at
18,000 psi shear stress.
Example 5 finds A for rectangular
shaft for ratio A/B = 1.20

Square side, S, in.

Major side, A, in.

Ratio, A/B

Example 4

Example 5

Shaft section	Location of max shear	Torque formulas: T =
S	Middle of sides	$0.208S^3 f$
A B	Midpoint of major sides	$\dfrac{A^2 B^2 f}{3A + 1.8B}$

Novel Linkage for Coupling Offset Shafts

...simplifies the design of a variety of products.

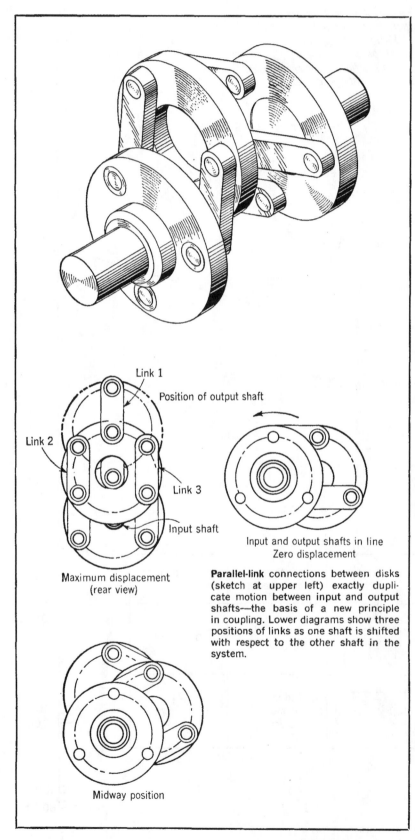

Parallel-link connections between disks (sketch at upper left) exactly duplicate motion between input and output shafts—the basis of a new principle in coupling. Lower diagrams show three positions of links as one shaft is shifted with respect to the other shaft in the system.

An unorthodox yet remarkably simple arrangement of links and disks forms the basis of a versatile type of parallel-shaft coupling. This type of coupling—essentially three disks rotating in unison and interconnected in series by six links (drawing, left)—can adapt to wide variations in axial displacement while running under load.

Changes in radial displacement do not affect the constant-velocity relationship between input and output shafts, nor do they initial radial reaction forces that might cause imbalance in the system. These features open up unusual applications in automotive, marine, machine-tool, and rolling-mill machinery (drawings, facing page).

How it works. The inventor of the coupling, Richard Schmidt of Schmidt Couplings, Inc., Madison, Ala., notes that a similar link arrangement has been known to some German engineers for years. But these engineers were discouraged from applying the theory because they erroneously assumed that the center disk had to be retained by its own bearing. Actually, Schmidt found, the center disk is free to assume its own center of rotation. In operation, all three disks rotate with equal velocity.

The bearing-mounted connections of links to disks are equally spaced at 120 deg. on pitch circles of the same diameter. The distance between shafts can be varied steplessly between zero (when the shafts are in line) and a maximum that is twice the length of the links (drawings, left). There is no phase shift between shafts while the coupling is undulating. □

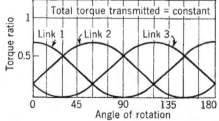

Torque transmitted by three links in group adds up to a constant value regardless of the angle of rotation.

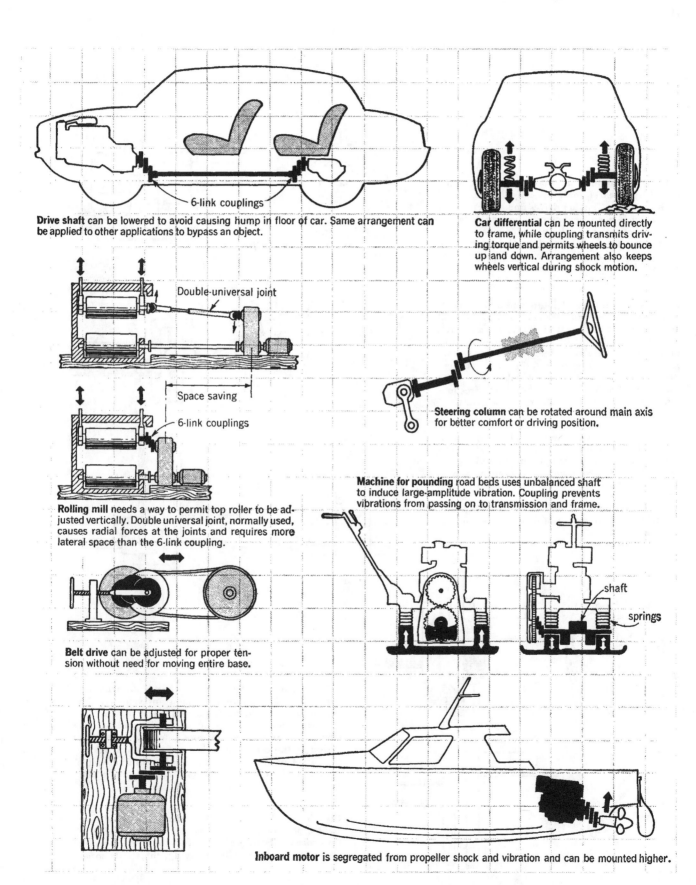

Drive shaft can be lowered to avoid causing hump in floor of car. Same arrangement can be applied to other applications to bypass an object.

6-link couplings

Car differential can be mounted directly to frame, while coupling transmits driving torque and permits wheels to bounce up and down. Arrangement also keeps wheels vertical during shock motion.

Double-universal joint

Space saving

6-link couplings

Rolling mill needs a way to permit top roller to be adjusted vertically. Double universal joint, normally used, causes radial forces at the joints and requires more lateral space than the 6-link coupling.

Steering column can be rotated around main axis for better comfort or driving position.

Machine for pounding road beds uses unbalanced shaft to induce large-amplitude vibration. Coupling prevents vibrations from passing on to transmission and frame.

shaft

springs

Belt drive can be adjusted for proper tension without need for moving entire base.

Inboard motor is segregated from propeller shock and vibration and can be mounted higher.

Coupling of Parallel Shafts

H. G. Conway

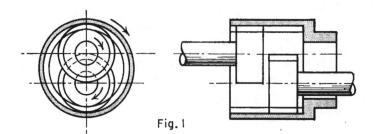

Fig. 1

FIG. 1—A common method of coupling shafts is with two gears; for gears may be substituted chains, pulleys, friction drives and others. Major limitation is need for adequate center distance; however, an idler can be used for close centers as shown. This can be a plain pinion or an internal gear. Transmission is at constant velocity and axial freedom is present.

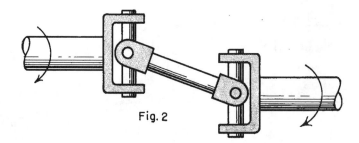

Fig. 2

FIG. 2—Two universal joints and a short shaft can be used. Velocity transmission is constant between input and output shafts if the shafts remain parallel and if the end yokes are disposed symmetrically. Velocity of the central shaft fluctuates during rotation and at high speed and angles may cause vibration. The shaft offset may be varied but axial freedom requires a splined mounting of one shaft.

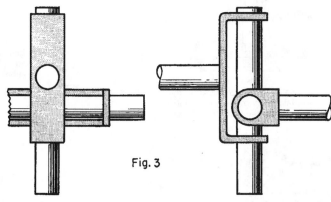

Fig. 3

FIG. 3—Crossed axis yoke coupling is a variation of the mechanism in Fig. 2. Each shaft has a yoke connected so that it can slide along the arms of a rigid cross member. Transmission is at a constant velocity but the shafts must remain parallel, although the offset may vary. There is no axial freedom. The central cross member describes a circle and is thus subjected to centrifugal loads.

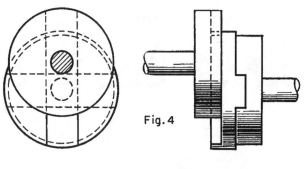

Fig. 4

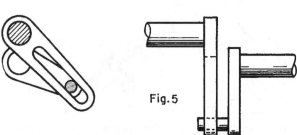

Fig. 5

FIG. 4—Another often used method is the Oldham coupling. The motion is at constant velocity, the central member describing a circle. The shaft offset may vary but the shafts must remain parallel. A small amount of axial freedom is possible. A tilting action of the central member can occur caused by the offset of the slots. This can be eliminated by enlarging the diameter and milling the slots in the same transverse plane.

FIG. 5—If the velocity does not have to be constant a pin and slot coupling can be used. Velocity transmission is irregular as the effective radius of operation is continually changing, the shafts must

The coupling of parallel shafts so that they rotate together is a common machine design problem. Illustrated are several methods where a constant 1:1 velocity ratio is possible and others where the velocity ratio may fluctuate during rotation. Some of the couplings have particular value for joining two shafts that may deflect or move relative to each other.

remain parallel unless a ball joint is used between the slot and pin. Axial freedom is possible but any change in the shaft offset will further affect the fluctuation of velocity transmission.

FIG. 6—The parallel-crank mechanism is sometimes used to drive the overhead camshaft on engines. Each shaft has at least two cranks connected by links and with full symmetry for constant velocity action and to avoid dead points. By using ball joints at the ends of the links, displacement between the crank assemblies is possible.

FIG. 7—A mechanism kinematically equivalent to Fig. 6, can be made by substituting two circular and contacting pins for each link. Each shaft has a disk carrying three or more projecting pins, the sum of the radii of the pins being equal to the eccentricity of offset of the shafts. The lines of center between each pair of pins remain parallel as the coupling rotates. Pins do not need to be of equal diameter. Transmission is at constant velocity and axial freedom is possible.

FIG. 8—Similar to the mechanism in Fig. 7, but with one set of pins being holes. The difference of radii is equal to the eccentricity or offset. Velocity transmission is constant; axial freedom is possible, but as in Fig. 7, the shaft axes must remain fixed. This type of mechanism is sometimes used in epicyclic reduction gear boxes.

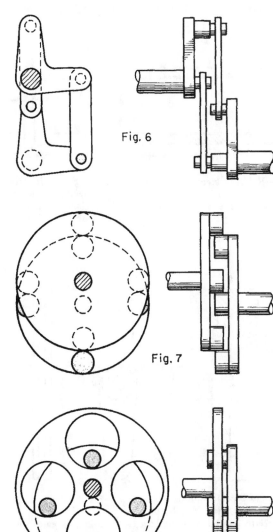

Fig. 6

Fig. 7

Fig. 8

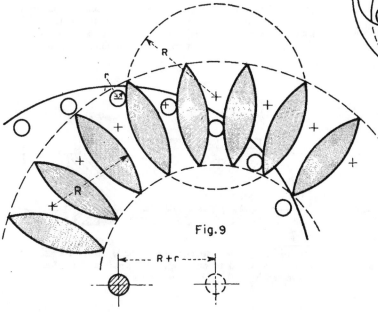

Fig. 9

FIG. 9—An unusual development of the pin coupling is shown left. A large number of pins engage lenticular or shield shaped sections formed from segments of theoretical large pins. The axes forming the lenticular sections are struck from the pitch points of the coupling and the distance $R + r$ is equal to the eccentricity between shaft centers. Velocity transmission is constant; axial freedom is possible but the shafts must remain parallel.

Low-Cost Methods of Coupling Small Diameter Shafts

Sixteen types of low-cost couplings, including flexible and non-flexible types. Most are for small diameter, lightly loaded shafts, but a few of them can also be adapted to heavy duty shafts. Some of them are currently available as standard commercial parts.

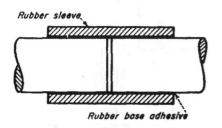

Fig 1—Rubber sleeve has inside diameter smaller than shaft diameters. Using rubber-base adhesive will increase the torque capacity.

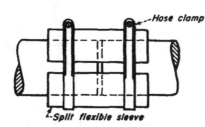

Fig 2—Slit sleeve of rubber or other flexible material is held by hose clamps. Easy to install and remove. Absorbs vibration and shock loads.

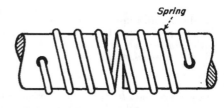

Fig 3—Ends of spring extend through holes in shafts to form coupling. Dia of spring determined by shaft dia, wire dia determined by loads.

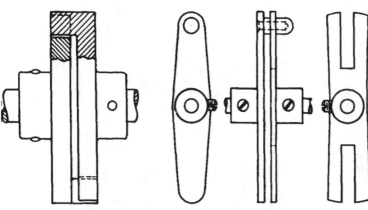

Fig 7—Jaw-type coupling is secured to shafts with straight pins. Commercially available; some have flexible insulators between jaws.

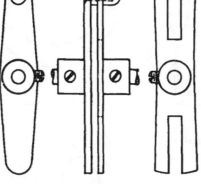

Fig 8—Removable type coupling with insulated coupling pin. A set screw in the collar of each stamped member is used to fasten it to the shaft.

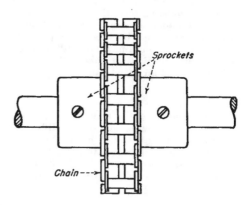

Fig 9—Sprockets mounted on each shaft are linked together with roller chain. Wide range of torque capacity. Commercially available.

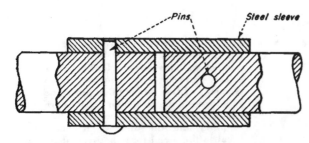

Fig 13—Steel sleeve coupling fastened to shafts with two straight pins. Pins are staggered at 90 degree intervals to reduce the stress concentration.

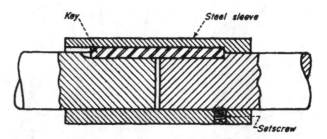

Fig 14—Single key engages both shafts and metal sleeve which is attached to one shaft with setscrew. Shoulder on sleeve can be omitted to reduce costs.

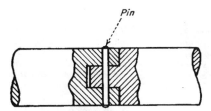

Fig 4—Tongue - and - groove coupling made from shaft ends is used to transmit torque. Pin or set screw keeps shafts in proper alignment.

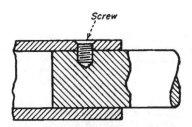

Fig 5—Screw fastens hollow shaft to inner shaft. Set screw can be used for small shafts and low torque by milling a flat on the inner shaft.

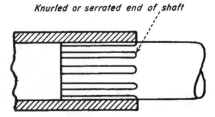

Fig 6—Knurled or serrated shaft is pressed into hollow shaft. Effects of misalignment must be checked to prevent overloading the bearings.

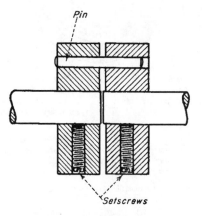

Fig 10—Coupling made of two collars fastened to shafts with set screws. Pin in one collar engages hole in other. Soft spacer can be used as cushion.

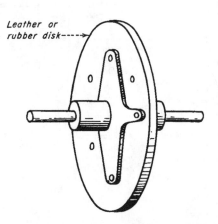

Fig 11—Coupling is made from two flanges rivited to leather or rubber center disk. Flanges are fastened to the shafts by means of setscrews.

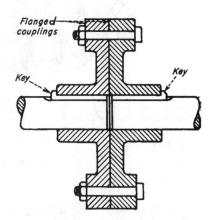

Fig 12—Bolted flange couplings are used on shafts from one to twelve inches in diameter. Flanges are joined by four bolts and are keyed to shafts.

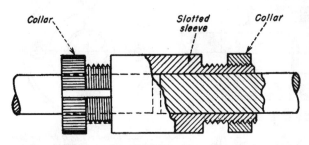

Fig 15—Screwing split collars on tapered threads of slotted sleeve tightens coupling. For light loads and small shafts, sleeve can be made of plastic material.

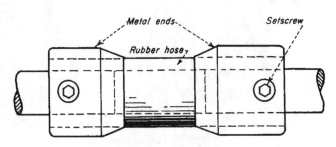

Fig 16—One-piece flexible coupling has rubber hose with metallic ends that are fastened to shafts with set screws. Commercially available in several sizes and lengths.

Typical Methods of Coupling Rotating Shafts I

Methods of coupling rotating shafts vary from simple bolted flange constructions to complex spring and synthetic rubber mechanisms. Some types incorporating chain, belts, splines, bands, and rollers are described and illustrated below.

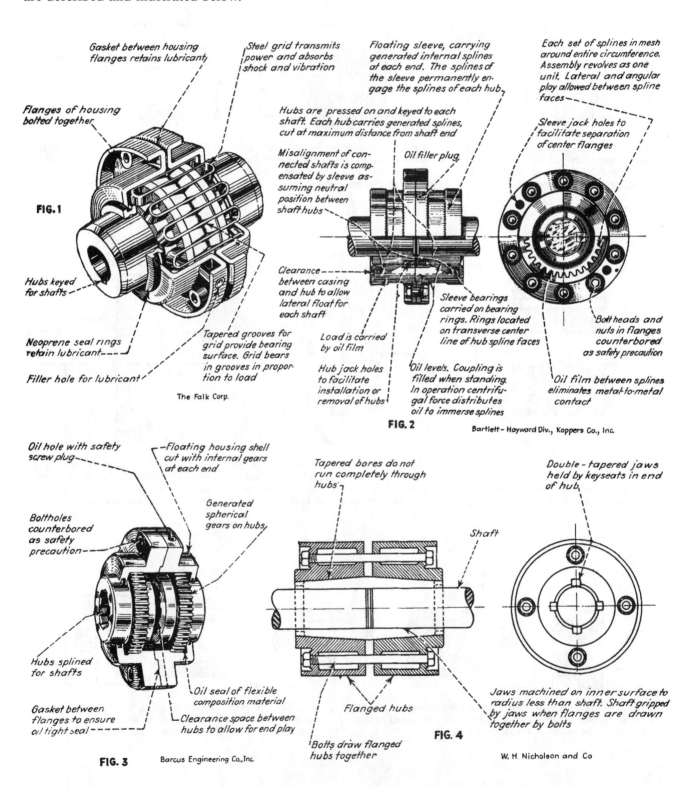

FIG. 1

Gasket between housing flanges retains lubricant

Flanges of housing bolted together

Hubs keyed for shafts

Neoprene seal rings retain lubricant

Filler hole for lubricant

Tapered grooves for grid provide bearing surface. Grid bears in grooves in proportion to load

Steel grid transmits power and absorbs shock and vibration

The Falk Corp.

FIG. 2

Floating sleeve, carrying generated internal splines at each end. The splines of the sleeve permanently engage the splines of each hub

Hubs are pressed on and keyed to each shaft. Each hub carries generated splines, cut at maximum distance from shaft end

Misalignment of connected shafts is compensated by sleeve assuming neutral position between shaft hubs

Oil filler plug

Clearance between casing and hub to allow lateral float for each shaft

Load is carried by oil film

Hub jack holes to facilitate installation or removal of hubs

Oil levels. Coupling is filled when standing. In operation centrifugal force distributes oil to immerse splines

Sleeve bearings carried on bearing rings. Rings located on transverse center line of hub spline faces

Each set of splines in mesh around entire circumference. Assembly revolves as one unit. Lateral and angular play allowed between spline faces

Sleeve jack holes to facilitate separation of center flanges

Bolt heads and nuts in flanges counterbored as safety precaution

Oil film between splines eliminates metal-to-metal contact

Bartlett - Hayward Div., Koppers Co., Inc.

FIG. 3

Oil hole with safety screw plug

Boltholes counterbored as safety precaution

Hubs splined for shafts

Gasket between flanges to ensure oil tight seal

Floating housing shell cut with internal gears at each end

Generated spherical gears on hubs

Oil seal of flexible composition material

Clearance space between hubs to allow for end play

Barcus Engineering Co., Inc.

FIG. 4

Tapered bores do not run completely through hubs

Shaft

Flanged hubs

Bolts draw flanged hubs together

Double - tapered jaws held by keyseats in end of hub

Jaws machined on inner surface to radius less than shaft. Shaft gripped by jaws when flanges are drawn together by bolts

W. H. Nicholson and Co

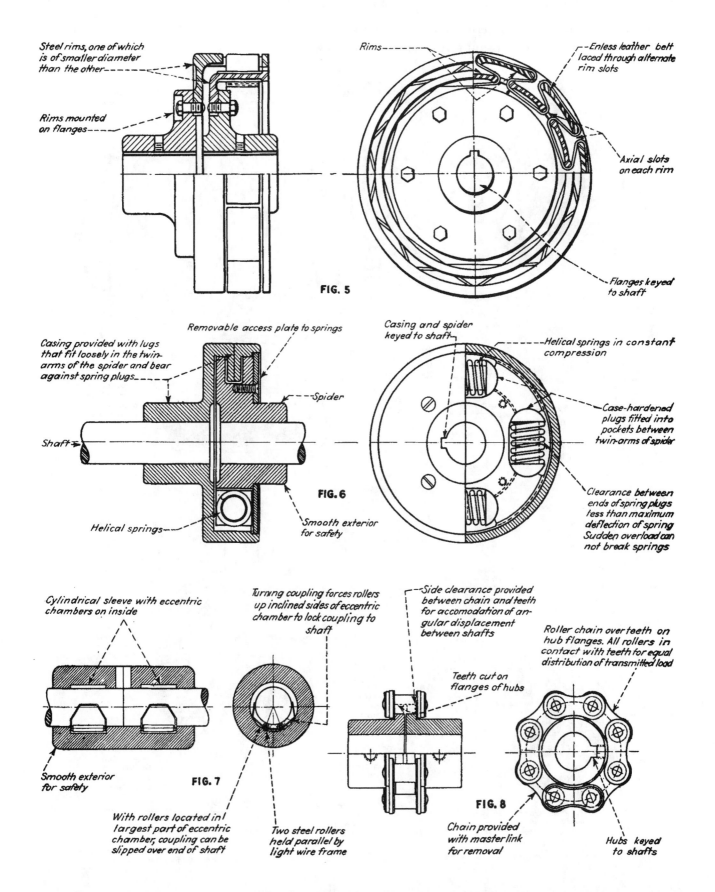

Steel rims, one of which is of smaller diameter than the other

Rims mounted on flanges

Rims

Enless leather belt laced through alternate rim slots

Axial slots on each rim

Flanges keyed to shaft

FIG. 5

Removable access plate to springs

Casing provided with lugs that fit loosely in the twin-arms of the spider and bear against spring plugs

Spider

Shaft

Helical springs

Smooth exterior for safety

Casing and spider keyed to shaft

Helical springs in constant compression

Case-hardened plugs fitted into pockets between twin-arms of spider

Clearance between ends of spring plugs less than maximum deflection of spring Sudden overload can not break springs

FIG. 6

Cylindrical sleeve with eccentric chambers on inside

Smooth exterior for safety

FIG. 7

Turning coupling forces rollers up inclined sides of eccentric chamber to lock coupling to shaft

With rollers located in largest part of eccentric chamber, coupling can be slipped over end of shaft

Two steel rollers held parallel by light wire frame

Side clearance provided between chain and teeth for accomodation of angular displacement between shafts

Teeth cut on flanges of hubs

Roller chain over teeth on hub flanges. All rollers in contact with teeth for equal distribution of transmitted load

Chain provided with master link for removal

Hubs keyed to shafts

FIG. 8

Typical Methods of Coupling Rotating Shafts II

Shafts couplings that utilize internal and external gears, balls, pins and non-metallic parts to transmit torque are shown herewith.

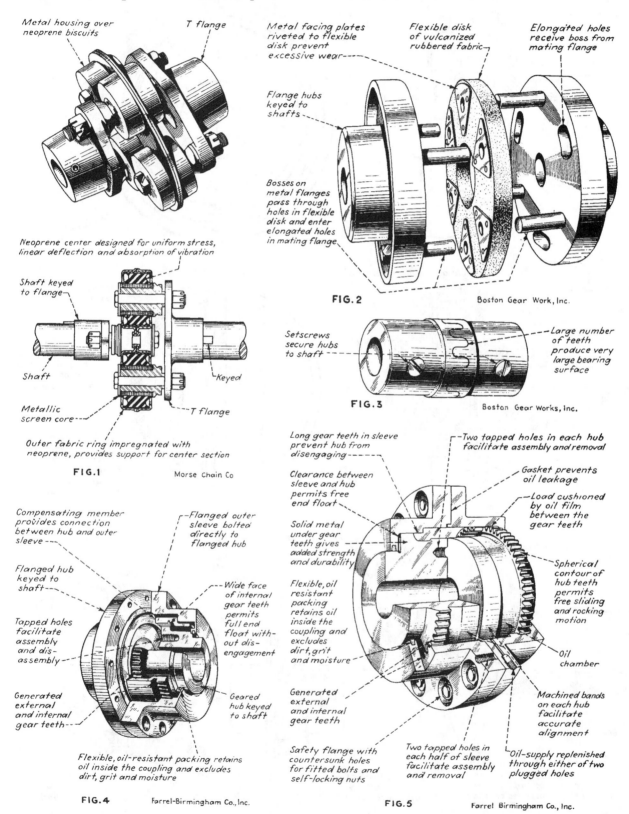

Metal housing over neoprene biscuits

T flange

Metal facing plates riveted to flexible disk prevent excessive wear---

Flexible disk of vulcanized rubbered fabric

Elongated holes receive boss from mating flange

Flange hubs keyed to shafts

Bosses on metal flanges pass through holes in flexible disk and enter elongated holes in mating flange

FIG. 2

Boston Gear Work, Inc.

Neoprene center designed for uniform stress, linear deflection and absorption of vibration

Shaft keyed to flange

Shaft

Keyed

Metallic screen core

T flange

Outer fabric ring impregnated with neoprene, provides support for center section

FIG. 1

Morse Chain Co

Setscrews secure hubs to shaft

Large number of teeth produce very large bearing surface

FIG. 3

Boston Gear Works, Inc.

Compensating member provides connection between hub and outer sleeve---

Flanged outer sleeve bolted directly to flanged hub

Flanged hub keyed to shaft---

Wide face of internal gear teeth permits full end float without disengagement

Tapped holes facilitate assembly and disassembly

Generated external and internal gear teeth---

Geared hub keyed to shaft

Flexible, oil-resistant packing retains oil inside the coupling and excludes dirt, grit and moisture

FIG. 4

Farrel-Birmingham Co., Inc.

Long gear teeth in sleeve prevent hub from disengaging

Clearance between sleeve and hub permits free end float

Solid metal under gear teeth gives added strength and durability

Flexible, oil resistant packing retains oil inside the coupling and excludes dirt, grit and moisture

Generated external and internal gear teeth

Two tapped holes in each hub facilitate assembly and removal

Gasket prevents oil leakage

Load cushioned by oil film between the gear teeth

Spherical contour of hub teeth permits free sliding and rocking motion

Oil chamber

Machined bands on each hub facilitate accurate alignment

Oil-supply replenished through either of two plugged holes

Safety flange with countersunk holes for fitted bolts and self-locking nuts

Two tapped holes in each half of sleeve facilitate assembly and removal

FIG. 5

Farrel Birmingham Co., Inc.

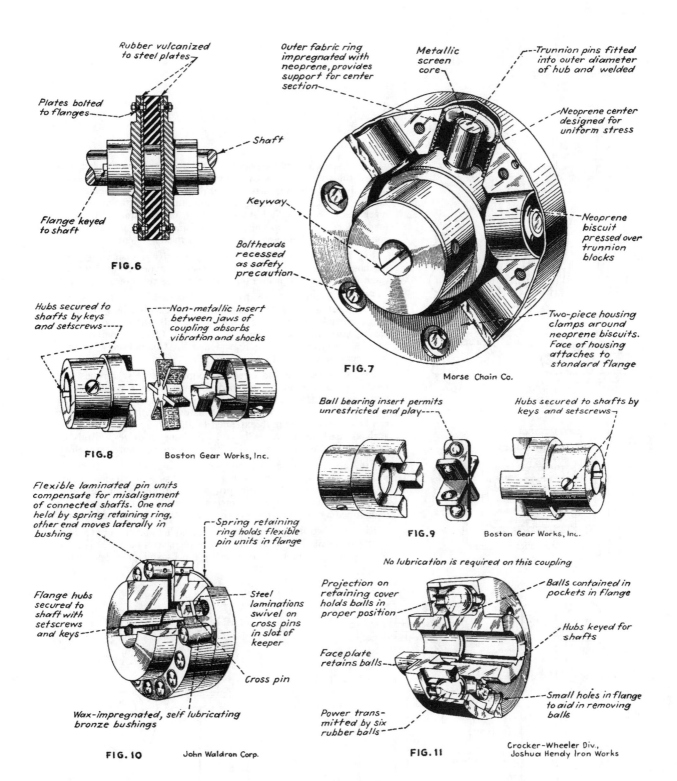

Rubber vulcanized to steel plates

Plates bolted to flanges

Shaft

Flange keyed to shaft

FIG. 6

Outer fabric ring impregnated with neoprene, provides support for center section

Metallic screen core

Trunnion pins fitted into outer diameter of hub and welded

Neoprene center designed for uniform stress

Keyway

Boltheads recessed as safety precaution

Neoprene biscuit pressed over trunnion blocks

Two-piece housing clamps around neoprene biscuits. Face of housing attaches to standard flange

FIG. 7   Morse Chain Co.

Hubs secured to shafts by keys and setscrews

Non-metallic insert between jaws of coupling absorbs vibration and shocks

FIG. 8   Boston Gear Works, Inc.

Ball bearing insert permits unrestricted end play

Hubs secured to shafts by keys and setscrews

FIG. 9   Boston Gear Works, Inc.

Flexible laminated pin units compensate for misalignment of connected shafts. One end held by spring retaining ring, other end moves laterally in bushing

Spring retaining ring holds flexible pin units in flange

Flange hubs secured to shaft with setscrews and keys

Steel laminations swivel on cross pins in slot of keeper

Cross pin

Wax-impregnated, self lubricating bronze bushings

FIG. 10   John Waldron Corp.

No lubrication is required on this coupling

Projection on retaining cover holds balls in proper position

Balls contained in pockets in flange

Hubs keyed for shafts

Face plate retains balls

Small holes in flange to aid in removing balls

Power transmitted by six rubber balls

FIG. 11   Crocker-Wheeler Div., Joshua Hendy Iron Works

Typical Designs of Flexible Couplings I

Cyril Donaldson

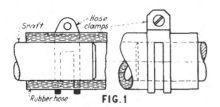

FIG. 1

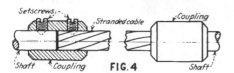

FIG. 2

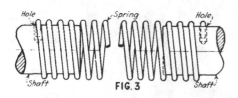

FIG. 3

FIG. 4

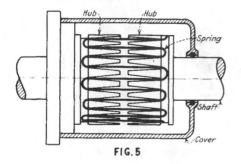

FIG. 5

Fig. 1—A rubber hose clamped to two shafts. For applications where the torque is low and slippage unimportant. It is easily assembled and disconnected without disturbing either machine element. Adaptable to changes in longitudinal distance between machines. This coupling absorbs shocks, is not damaged by overloads, does not set up end thrusts, requires no lubrication and compensates for both angular and offset misalignment.

Fig. 2—Similar to Fig. 1, but positive drive is assured by bolting hose to shafts. Has same advantages as type in Fig. 1. except there is no overload protection other than the rupture of the hose.

Fig. 3—The use of a coiled spring fastened to shafts gives the same action as a hose. Has excellent shock absorbing qualities, but torsional vibrations are possible. Will allow end play in shafts, but sets up end thrust in so doing. Other advantages are same as in types shown in Figs. 1 and 2. Compensates for misalignment in any direction.

Fig. 4—A simple and effective coupling for low torques and unidirectional rotation. Stranded cable provides a positive drive with desirable elasticity. Inertia of rotating parts is low. Easily assembled and disconnected without disturbing either shaft. Cable can be encased and length extended to allow for right angle bends such as used on dental drills and speedometer drives. Ends of cable are soldered or bound with wire to prevent unraveling.

Fig. 5—A type of Falk coupling that operates on the same principle as design shown in Fig. 6, but has a single flat spring in place of a series of coiled springs. High degree of flexibility obtained by use of tapered slots in hubs. Smooth operation is maintained by inclosing the working parts and packing with grease.

Fig. 6—Two flanges and a series of coiled springs give a high degree of flexibility. Used only where the shafts have no free end play. Needs no lubrication, absorbs shocks and provides protec-

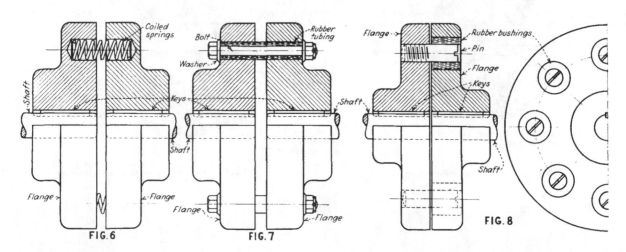

FIG. 6   FIG. 7   FIG. 8

tion against overloads, but will set up torsional vibrations. Springs can be of round or square wire with varying sizes and pitches to allow for any degree of flexibility.

Fig. 7—Is similar to Fig. 6, except that rubber tubing, reinforced by bolts, is used instead of coiled springs. Is of sturdier construction but more limited in flexibility. Has no overload protection other than shearing of the bolts. Good anti-vibration properties if thick rubber tubing is used. Can absorb minor shocks. Connection can be quickly disassembled.

Fig. 8—A series of pins engage rubber bushings cemented into flange. Coupling is easy to install. Flanges being accurately machined and of identical size makes accurate lining-up with spirit level possible. Will allow minor end play in shafts, and provides a positive drive with good flexibility in all direction.

Fig. 9—A Foote Gear Works flexible coupling which has shear pins in a separate set of bushings to provide overload protection. Construction of studs, rubber bushings and self-lubricating bronze bearings is in principle similar to that shown in Fig. 10. Replaceable shear pins are made of softer material than the shear pin bushings.

Fig. 10—A design made by the Ajax Flexible Coupling Company. Studs are firmly anchored with nuts and lock washers and bear in self-lubricating bronze bushings spaced alternately in both flanges. Thick rubber bushings cemented in flanges are forced over the bronze bushings. Life of coupling said to be considerably increased because of self-lubricated bushings.

Fig. 11—Another Foote Gear Works coupling. Flexibility is obtained by solid conically-shaped pins of metal or fiber. This type of pin is said to provide a positive drive of sturdy construction with flexibility in all directions.

Fig. 12—In this Smith & Serrell coupling a high degree of flexibility is obtained by laminated pins built-up of tempered spring steel leaves. Spring leaves secured to holder by keeper pin. Phosphor bronze bearing strips are welded to outer spring leaves and bear in rectangular holes of hardened steel bushings fastened in flange. Pins are free to slide endwise in one flange, but are locked in the other flange by a spring retaining ring. This type is used for severe duty in both marine and land service.

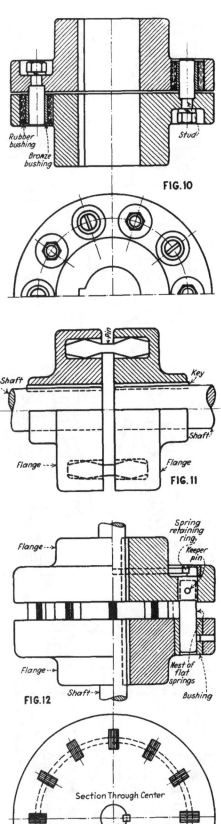

FIG. 10

FIG. 11

FIG. 12

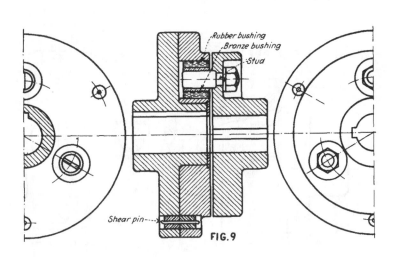

FIG. 9

Typical Designs of Flexible Couplings II

Cyril Donaldson

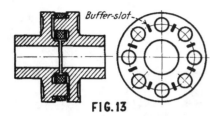

FIG. 13

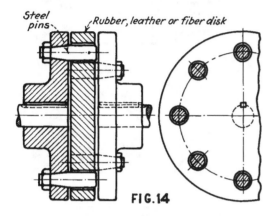

Steel pins

Rubber, leather or fiber disk

FIG. 14

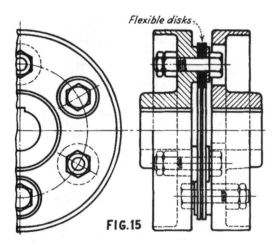

Flexible disks

FIG. 15

Fig. 13—In this Brown Engineering Company coupling flexibility is increased by addition of buffer-slots in the laminated leather. These slots also aid in the absorption of shock loads and torsional vibration. Under parallel misalignment or shock loads, buffer slots will close over their entire width, but under angular misalignment buffer slots will close only on one side.

Fig. 14—Flexibility is provided by resilience of a rubber, leather, or fiber disk in this W. A. Jones Foundry & Machine Company coupling. Degree of flexibility is limited to clearance between pins and holes in the disk plus the resilience of the disk. Has good shock absorbing properties, allows for end play and needs no lubrication.

Fig. 15—A coupling made by Aldrich Pump Company, similar to Fig. 14, except bolts are used instead of pins. This coupling permits only slight endwise movement of the shaft and allows machines to be temporarily disconnected without disturbing the flanges. Driving and driven members are flanged for protection against projecting bolts.

Fig. 16—Laminated metal disks are used in this coupling made by Thomas Flexible Coupling Company. The disks are bolted to each flange and connected to each other by means of pins supported by a steel center disk. The spring action of the center ring allows torsional flexibility and the two side rings compensate for angular and offset misalignment. This type of coupling provides a positive drive in either direction without setting up backlash. No lubrication is required.

Fig. 17—A design made by Palmer-Bee Company for heavy torques. Each flange carries two studs upon which are mounted square metal blocks. The blocks slide in the slots of the center metal disk.

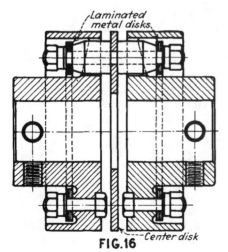

Laminated metal disks

FIG. 16

Center disk

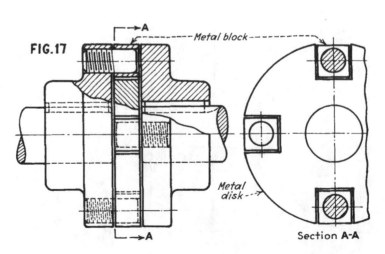

FIG. 17

Metal block

Metal disk

Section A-A

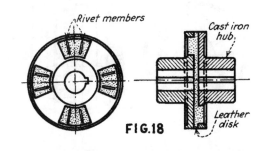

FIG.18

Fig. 18—In this Charles Bond Company coupling a leather disk floats between two identical flanges. Drive is through four laminated leather lugs cemented and riveted to the leather disk. Compensates for misalignment in all directions and sets up no end thrusts. The flanges are made of cast iron and the driving lug slots are cored.

Fig. 19—The principle of the T. B. Wood & Sons Company coupling is the same as Fig. 18, but the driving lugs are cast integrally with the metal flanges. The laminated leather disk is punched out to accommodate the metal driving lugs of each flange. This coupling has flexibility in all directions and does not require lubrication.

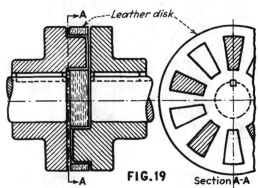

FIG.19 Section A-A

Fig. 20—Another design made by Charles Bond Company. The flanges have square recesses into which a built-up leather cube fits. Endwise movement is prevented by through bolts set at right angles. The coupling operates quietly and is used where low torque loads are to be transmitted. Die-castings can be used for the flanges.

Fig. 21—Similar to Fig. 20, being quiet in operation and used for low torques. This is also a design of Charles Bond Company. The floating member is made of laminated leather and is shaped like a cross. The ends of the intermediate member engage the two cored slots of each flange. The coupling will withstand a limited amount of end play.

Fig. 22—Pins mounted in flanges are connected by leather, canvas, or rubber bands. Coupling is used for temporary connections where large torques are transmitted, such as the driving of dynamometers by test engines. Allows for a large amount of flexibility in all directions, absorbs shocks but requires frequent inspection. Machines can be quickly disconnected, especially when belt fasteners are used on the bands. Driven member lags behind driver when under load.

Fig. 23—This Bruce-Macbeth Engine Company coupling is similar to that of Fig. 22, except that six endless wire cable links are used, made of plow-steel wire rope. The links engage small metal spools mounted on eccentric bushings. By turning these bushings the links are adjusted to the proper tension. The load is transmitted from one flange to the other by direct pull on the cable links. This type of coupling is used for severe service.

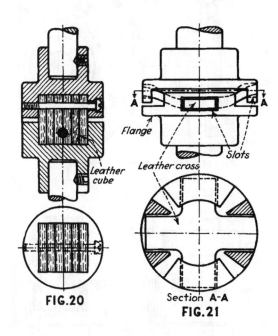

FIG.20 Section A-A **FIG.21**

FIG.23

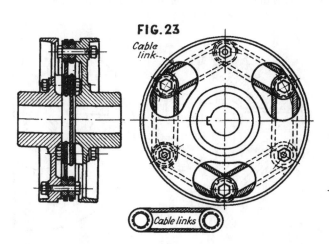

Cable link

Cable links

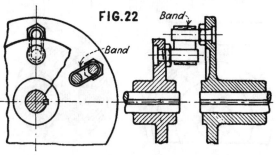

FIG.22

Typical Designs of Flexible Couplings III

Cyril Donaldson

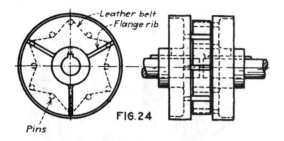

FIG. 24

Leather belt
Flange rib
Pins

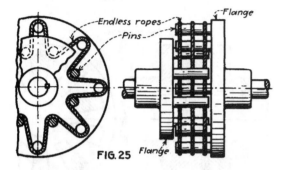

Endless ropes
Pins
Flange
FIG. 25 Flange

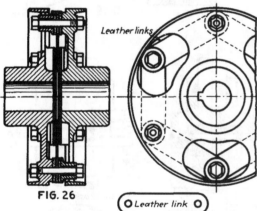

Leather links
FIG. 26
Leather link

Fig. 24—This Webster Manufacturing Company coupling uses a single endless leather belt instead of a series of bands, as in Fig. 22. The belt is looped over alternate pins in both flanges. Has good shock resisting properties because of belt stretch and the tendency of the pins to settle back into the loops of the belt.

Fig. 25—This coupling made by the Weller Manufacturing Company is similar to the design in Fig. 24, but instead of a leather belt uses hemp rope, made endless by splicing. The action under load is the same as in the endless belt type.

Fig. 26—This Bruce-Macbeth design uses leather links instead of endless wire cables, as shown in Fig. 23. The load is transmitted from one flange to the other by direct pull of the links, which at the same time allows for the proper flexibility. Intended for permanent installations requiring a minimum of supervision.

Fig. 27—The Oldham form of coupling made by W. A. Jones Foundry and Machine Company is of the two-jaw type with a metal disk. Is used for transmitting heavy loads at low speed.

Fig. 28—The Charles Bond Company star coupling is similar to the cross type shown in Fig. 21. The star-shaped floating member is made of laminated leather. Has three jaws in each flange. Torque capacity is thus increased over the two-jaw or cross type. Coupling takes care of limited end play.

Fig. 29—Combination rubber and canvas disk is bolted to two metal spiders. Extensively used for low torques where compensation for only slight angular misalignment is required. Is quiet in operation and needs no lubrication or other attention. Offset misalignment shortens disk life.

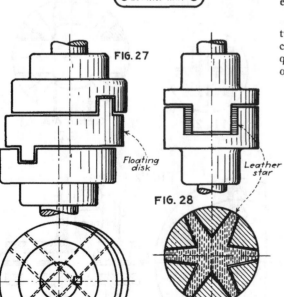

FIG. 27
Floating disk
FIG. 28
Leather star

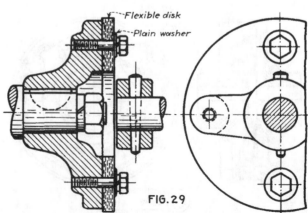

Flexible disk
Plain washer
FIG. 29

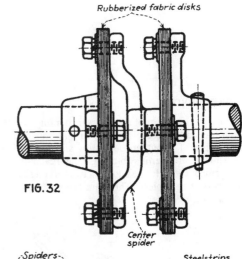

FIG. 32

Fig. 30—A metal block as a floating center is used in this American Flexible Coupling Company design. Quiet operation is secured by facing the block with removable fiber strips and packing the center with grease. The coupling sets up no end thrusts, is easy to assemble and does not depend on flexible material for the driving action. Can be built in small sizes by using hardwood block without facings, for the floating member.

Fig. 31—This Westinghouse Nuttall Company coupling is an all-metal type having excellent torsional flexibility. The eight compression springs compensate for angular and offset misalignment. Allows for some free endwise float of the shafts. Will transmit high torques in either direction. No lubrication is needed.

Fig. 32—Similar to Fig. 29, but will withstand offset misalignment by addition of the extra disk. In this instance the center spider is free to float. By use of two rubber-canvas disks, as shown, coupling will withstand a considerable angular misalignment.

Fig. 33—In this Smith & Serrell coupling a flexible cross made of laminated steel strips floats between two spiders. The laminated spokes, retained by four segmental shoes, engage lugs integral with the flanges. Coupling is intended for the transmission of light loads only.

Fig. 34—This coupling made by Brown Engineering Company is useful for improvising connections between apparatus in laboratories and similar temporary installations. Compensates for misalignment in all directions. Will absorb varying degrees of torsional shocks by changing the size of the springs. Springs are retained by threaded pins engaging the coils. Overload protection is possible by the slippage or breakage of replacable springs.

Fig. 35—In another design by Brown Engineering Company, a series of laminated spokes transmit power between the two flanges without setting up end thrusts. This type allows free end play. Among other advantages are absorption of torsional shocks, has no exposed moving parts, and is well balanced at all speeds. Wearing parts are replacable and working parts are protected from dust.

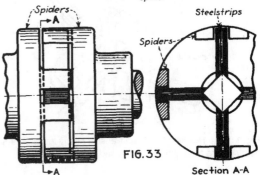

FIG. 33

Section A-A

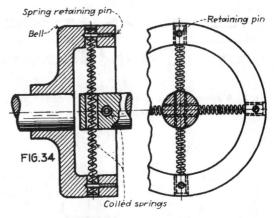

FIG. 34

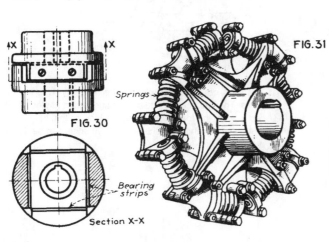

FIG. 31

FIG. 30

Section X-X

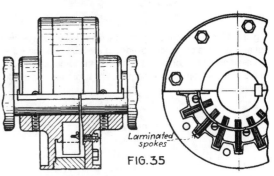

FIG. 35

Ten Universal Shaft Couplings

Hooke's Joints

The commonest form of a universal coupling is a *Hooke's joint*. It can transmit torque efficiently up to a maximum shaft alignment angle of about 36°. At slow speeds, on hand-operated mechanisms, the permissible angle can reach 45°. The simplest arrangement for a Hooke's joint is two forked shaft-ends coupled by a cross-shaped piece. There are many variations and a few of them are included here.

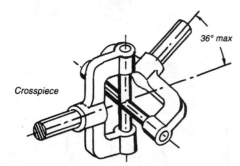

Fig. 1   The Hooke's joint can transmit heavy loads. Anti-friction bearings are a refinement often used.

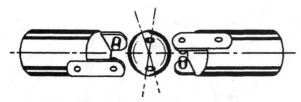

Fig. 2   A pinned sphere shaft coupling replaces a cross-piece. The result is a more compact joint.

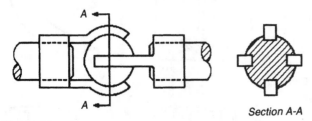

Fig. 3   A grooved-sphere joint is a modification of a pinned sphere. Torques on fastening sleeves are bent over the sphere on the assembly. Greater sliding contact of the torques in grooves makes simple lubrication essential at high torques and alignment angles.

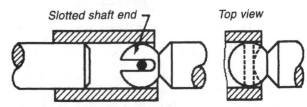

Fig. 4   A pinned-sleeve shaft-coupling is fastened to one shaft that engages the forked, spherical end on the other shaft to provide a joint which also allows for axial shaft movement. In this example, however, the angle between shafts must be small. Also, the joint is only suitable for low torques.

Constant-Velocity Couplings

The disadvantages of a single Hooke's joint is that the velocity of the driven shaft varies. Its maximum velocity can be found by multiplying driving-shaft speed by the secant of the shaft angle; for minimum speed, multiply by the cosine. An example of speed variation: a driving shaft rotates at 100 rpm; the angle between the shafts is 20°. The minimum output is 100×0.9397, which equals 93.9 rpm; the maximum output is 1.0642×100, or 106.4 rpm. Thus, the difference is 12.43 rpm. When output speed is high, output torque is low, and vice versa. This is an objectionable feature in some mechanisms. However, two universal joints connected by an intermediate shaft solve this speed-torque objection.

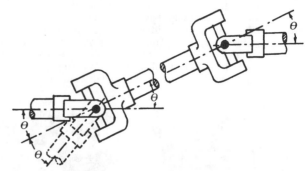

Fig. 5   A constant-velocity joint is made by coupling two Hooke's joints. They must have equal input and output angles to work correctly. Also, the forks must be assembled so that they will always be in the same plane. The shaft-alignment angle can be double that for a single joint.

Source: Mechanisms and Mechanical Devices Sourcebook, 3E, by Chironis & Sclater, © 2001 McGraw-Hill

This single constant-velocity coupling is based on the principle (Fig. 6) that the contact point of the two members must always lie on the homokinetic plane. Their rotation speed will then always be equal because the radius to the contact point of each member will always be equal. Such simple couplings are ideal for toys, instruments, and other light-duty mechanisms. For heavy duty, such as the front-wheel drives of military vehicles, a more complex coupling is shown dia-grammatically in Fig. 7A. It has two joints close-coupled with a sliding member between them. The exploded view (Fig. 7B) shows these members. There are other designs for heavy-duty universal couplings; one, known as the *Rzeppa,* consists of a cage that keeps six balls in the homokinetic plane at all times. Another constant-velocity joint, the *Bendix-Weiss,* also incorporates balls.

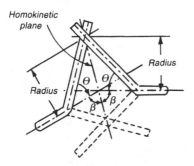

Fig. 6

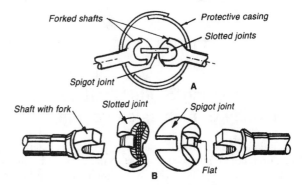

Fig. 7

Triple-strand spring

Fig. 8   This flexible shaft permits any shaft angle. These shafts, if long, should be supported to prevent backlash and coiling.

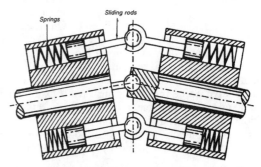

Fig. 9   This pump-type coupling has the reciprocating action of sliding rods that can drive pistons in cylinders.

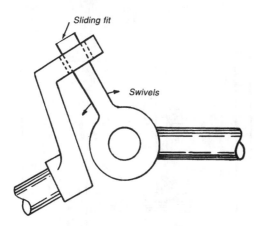

Fig. 10   This light-duty coupling is ideal for many simple, low-cost mechanisms. The sliding swivel-rod must be kept well lubricated at all times.

Novel Coupling Shifts Shafts

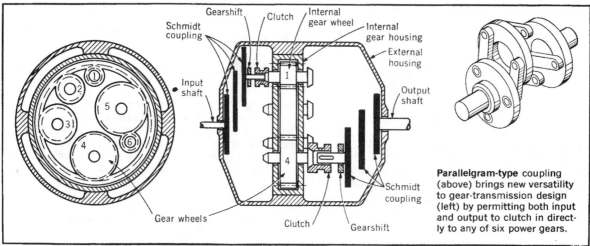

Parallelgram-type coupling (above) brings new versatility to gear-transmission design (left) by permitting both input and output to clutch in directly to any of six power gears.

Schmidt Coupling, Inc., Madison, AL

Novel Coupling Shifts Shafts: to simplify transmission design

A unique disk-and-link coupling that can handle large axial displacement between shafts, while the shafts are running under load, is opening up new approaches to transmission design.

The coupling (drawing, upper right) maintains a constant transmission ratio between input and output shafts while the shafts undergo axial shifts in their relative positions. This permits gear-and-belt transmissions to be designed that need fewer gears and pulleys.

Half as many gears. In the internal-gear transmission above, a Schmidt coupling on the input side permits the input to be "plugged-in" directly to any one of six gears, all of which are in mesh with the internal gear wheel.

On the output side, after the power flows through the gear wheel, a second Schmidt coupling permits a direct power takeoff from any of the same six gears. Thus, any one of 6 x 6 minus 5 or 31 different speed ratios can be selected while the unit is running. A more orthodox design would require almost twice as many gears.

Powerful pump. In the worm-type pump (bottom left), as the input shaft rotates clockwise, the worm rotor is forced to roll around the inside of the gear housing, which has a helical groove running from end to end. Thus, the rotor centerline will rotate counterclockwise to produce a powerful pumping action for moving heavy media.

In the belt drive (bottom right), the Schmidt coupling permits the belt to be shifted to a different bottom pulley while remaining on the same top pulley. Normally, because of the constant belt length, the top pulley would have to be shifted, too, to provide a choice of only three output speeds. With the new arrangement, nine different output speeds can be obtained. □

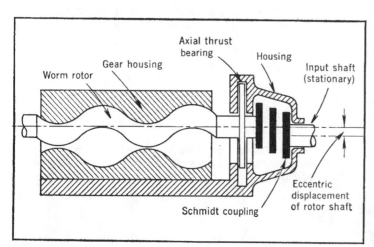

Coupling allows helical-shape rotor to wiggle for pumping purposes.

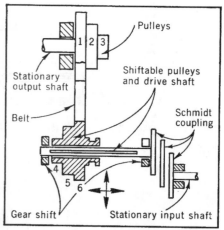

Coupling takes up slack when bottom shifts.

SECTION 11

THREADED COMPONENTS

Getting the Most from Screws

Special jobs often call for special screw arrangements; here are some examples of how this busy fastener can perform.

Federico Strasser

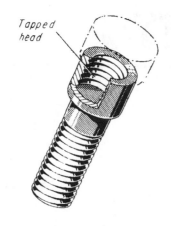

1 TAPPED HEAD LETS EXTENSIONS BE ADDED.

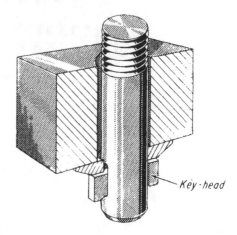

2 KEY-TYPE HEAD PROVIDES QUICK-RELEASE FEATURE.

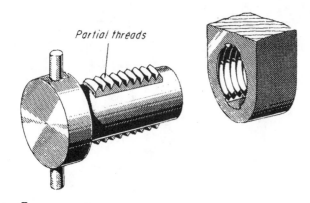

4 PARTIAL THREADS ASSEMBLE FAST, DON'T WORK LOOSE.

5 BUTTRESS THREADS PREVENT (a) radial forces from opening slotted ends; otherwise (b) a reinforcing sleeve is needed.

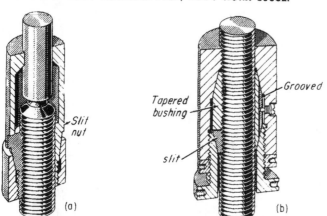

7 SLIT NUT (a) and tapered bushing (also slit) (b) allows backlash-free adjustment.

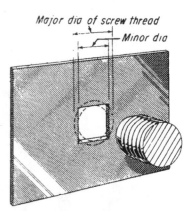

8 SQUARE HOLE for light metal or plastic substitutes well for threaded holes.

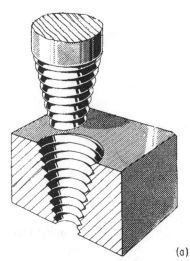

(a)

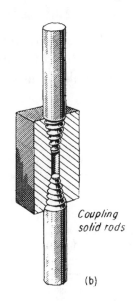

Coupling
solid rods

(b)

3 TAPERED SCREWS ASSEMBLE AND RELEASE FAST, BUT WORK LOOSE EASILY.

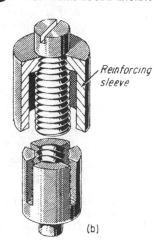

Reinforcing
sleeve

(b)

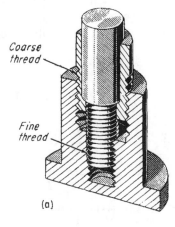

Coarse
thread

Fine
thread

(a)

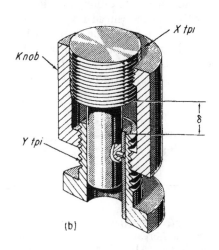

X tpi

Knob

Y tpi

δ

(b)

6 DIFFERENTIAL THREADS PROVIDE (a) extra tight fastening or (b) extra small relative movement, δ, per revolution of knob.

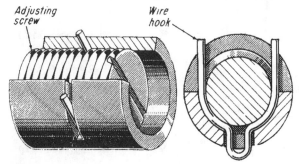

Adjusting
screw

Wire
hook

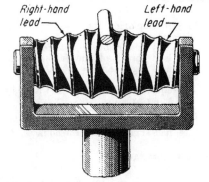

Right-hand
lead

Left-hand
lead

9 WIRE HOOK provides single-thread grip for low-cost device.

10 DOUBLE SCREW for wire guide or follower always leads wire to center.

20 Dynamic Applications for Screw Threads

Have you forgotten how simply, and economically, screw threads can be made into dynamic members of a linkage? Here are some memory-joggers, plus suggestions for simplified nuts, threads and nut guides.

Kurt Rabe

You need a threaded shaft, a nut . . . plus some way for one of these members to rotate without translating and the other to translate without rotating. That's all. Yet these simple components can do practically all of the adjusting, setting, or locking used in design.

Most such applications have low-precision requirements. That's why the thread may be a coiled wire or a twisted strip; the nut may be a notched ear on a shaft or a slotted disk. Standard screws and nuts right off your supply shelves can often serve at very low cost.

Here are the basic motion transformations possible with screw threads (Fig 1):
- transform rotation into linear motion or reverse (A),
- transform helical motion into linear motion or reverse (B),
- transform rotation into helical motion or reverse (C).

Of course the screw thread may be combined with other components: in a 4-bar linkage (Fig 2), or with multiple screw elements for force or motion amplification.

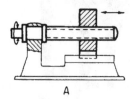
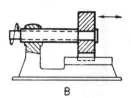
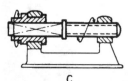
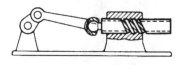

A   B   C

1 MOTION TRANSFORMATIONS of a screw thread include: rotation to translation (A), helical to translation (B), rotation to helical (C). Any of these is reversible if the thread is not self-locking (see screw-thread mathematics on following page—thread is reversible when efficiency is over 50%).

2 STANDARD 4-BAR LINKAGE has screw thread substituted for slider. Output is helical rather than linear.

A Review of Screw-Thread Mathematics

a — friction angle, $\tan a = f$
r — mean radius of thread
    $= \frac{1}{2}$ (root radius + outside radius), in inches
l — lead, thread advance in one revolution, in.
b — lead angle, $\tan b = 1/2\pi r$, deg
f — friction coefficient
P — equivalent driving force at radius r from screw axis, lb
L — axial load, lb
e — efficiency
c — half angle between thread faces, deg

SQUARE THREADS:

$$P = L \tan (b \pm a) = L \frac{(1 \pm 2\pi rf)}{(2\pi r \mp fl)}$$

Where upper signs are for motion opposed in direction to L. Screw is self-locking when $b \leqq a$.

$$e = \frac{\tan b}{\tan (b + a)} \qquad \text{(motion opposed to } L)$$

$$e = \frac{\tan (b - a)}{\tan b} \qquad \text{(motion assisted by } L)$$

V THREADS:

$$P = L \frac{(l \pm 2\pi rf \sec c)}{(2\pi r \mp lf \sec c)}$$

$$e = \frac{\tan b (1 - f \tan b \sec c)}{(\tan b + f \sec c)} \qquad \text{(motion opposed to } L)$$

$$e = \frac{\tan b - f \sec c}{\tan b (1 + f \tan b \sec c)} \qquad \text{(motion assisted by } L)$$

For more detailed analysis of screw-thread friction forces, see Marks *Mechanical Engineers' Handbook*, McGraw-Hill Book Co.

Rotation to Translation

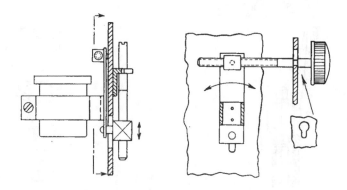

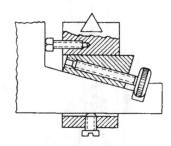

3 TWO-DIRECTIONAL LAMP ADJUSTMENT with screwdriver to move lamp up and down. Knob adjust (right) rotates lamp about pivot.

4 KNIFE-EDGE BEARING is raised or lowered by screw-driven wedge. Two additional screws locate the knife edge laterally and lock it.

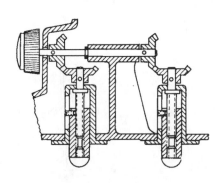

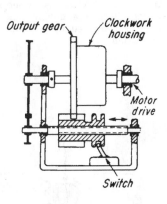

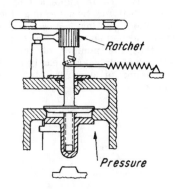

5 SIDE-BY-SIDE ARRANGEMENT of tandem screw threads gives parallel rise in this height adjustment for projector.

6 AUTOMATIC CLOCKWORK is kept wound tight by electric motor turned on and off by screw thread and nut. Note motor drive must be self-locking or it will permit clock to unwind as soon as switch turns off.

7 VALVE STEM has two oppositely moving valve cones. When opening, the upper cone moves up first, until it contacts its stop. Further turning of the valve wheel forces the lower cone out of its seat. The spring is wound up at the same time. When the ratchet is released, spring pulls both cones into their seats.

A Review of Screw-Thread Mathematics *(continued)*

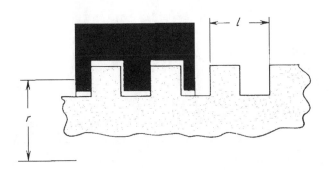

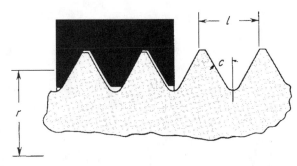

Translation to Rotation

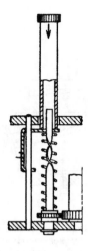

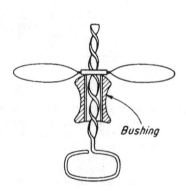

8 A METAL STRIP or square rod may be twisted to make a long-lead thread, ideal for transforming linear into rotary motion. Here a push-button mechanism winds a camera. Note that the number of turns or dwell of output gear is easily altered by changing (or even reversing) twist of the strip.

9 FEELER GAGE has its motion amplified through a double linkage and then transformed to rotation for dial indication.

10 THE FAMILIAR flying propeller-toy is operated by pushing the bushing straight up and off the thread.

Self-Locking

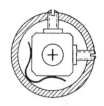

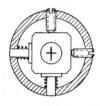

11 HAIRLINE ADJUSTMENT for a telescope, with two alternative methods of drive and spring return.

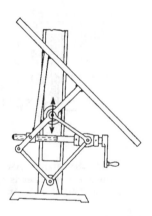

12 SCREW AND NUT provide self-locking drive for a complex linkage.

Double Threading

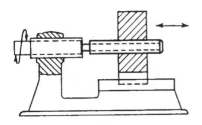

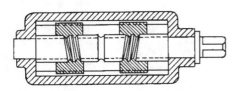

13 DOUBLE-THREADED SCREWS, when used as differentials, provide very fine adjustment for precision equipment at relatively low cost.

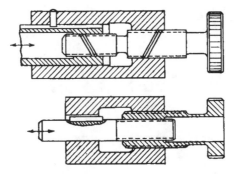

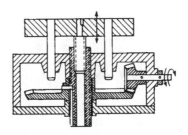

14 DIFFERENTIAL SCREWS can be made in dozens of forms. Here are two methods: above, two opposite-hand threads on a single shaft; below, same hand threads on independent shafts.

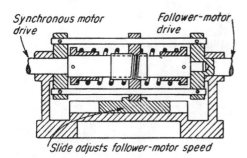

15 OPPOSITE-HAND THREADS make a high-speed centering clamp out of two moving nuts.

16 MEASURING TABLE rises very slowly for many turns of the input bevel gear. If the two threads are 1½—12 and ¾—16, in the fine-thread series, table will rise approximately 0.004 in. per input-gear revolution.

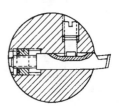

17 LATHE TURNING TOOL in drill rod is adjusted by differential screw. A special double-pin wrench turns the intermediate nut, advancing the nut and retracting the threaded tool simultaneously. Tool is then clamped by setscrew.

Synchronous motor drive   *Follower-motor drive*

Slide adjusts follower-motor speed

18 ANY VARIABLE-SPEED MOTOR can be made to follow a small synchronous motor by connecting them to the two shafts of this differential screw. Differences in number of revolutions between the two motors appear as motion of the traveling nut and slide so an electrical speed compensation is made.

19 (left) A WIRE FORK is the nut in this simple tube-and screw design.

20 (below) A MECHANICAL PENCIL includes a spring as the screw thread and a notched ear or a bent wire as the nut.

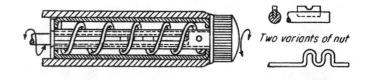

Two variants of nut

EDITOR'S NOTE: For other solutions to adjusting, setting, and locking problems in translating motion, see:

10 Ways to Employ Screw Mechanisms, May 26 '58, p 80. Shows applications in terms of three basic components—actuating member, threaded device, and sliding device.

5 Cardan-gear Mechanisms. Sep 28 '59, p 66. Gearing arrangements that convert rotation into straight-line motion.

5 Linkages for Straight-line Motion, Oct 12 '59, p 86. Linkages that convert rotation into straight-line motion.

16 Ways to Align Sheets and Plates with One Screw

Federico Strasser

Two Flat Parts

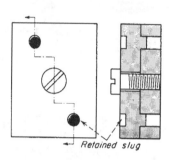

1 Dowels . . .
accurately align two plates, prevent shear stress in fastening screw. Two pins are necessary because screw can not act as aligning-pin.

2 Retained slugs . . .
act as pins, perform same function as dowels; are cheaper but not as accurate.

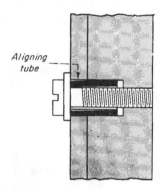

3 Aligning tube . . .
fits into counterbored hole through both parts. Screw clearance must be provided in tube.

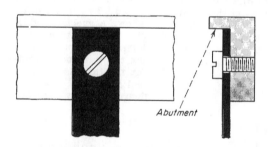

4 Abutment . . .
provides positive, cheap alignment of rectangular part.

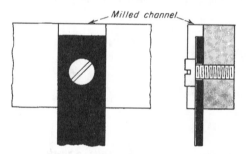

5 Matching channel . . .
milled in one part gives more efficient alignment than abutment in preceding method.

Formed Stampings Assembled with Flat Parts

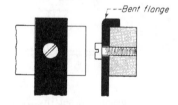

6 Bent flange . . .
performs similar function as abutment, but may be more suitable where machining or casting of abutment in large part is not desirable or practical.

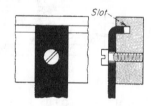

7 Narrow slot . . .
receives flange or leg on sheet metal part, allows it to be mounted remote from edge of other part.

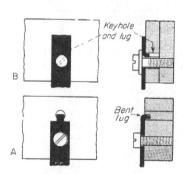

8 Bent lug . . .
(A) fits into hole, aligns parts simply and cheaply; or (B) lug formed by slitting clearance hole in sheet metal keys parts together at keyhole.

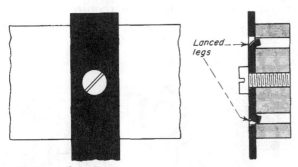

9 Two legs . . .

formed by lancing, align parts in manner similar to retained slugs in method 2, but formed legs are only an alternative for sheet too thin to partially extrude slug.

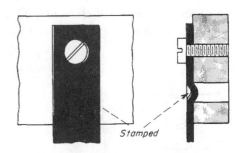

10 Aligning projection . . . .

formed by slitting and embossing is good locating method, but allows a relatively large amount of play in the assembly.

Flat Parts and Bars

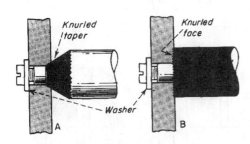

11 Knurled end . . .

of round bar (A) has taper which digs into edge of hole when screw is tightened; this gives accurate angular location of bar or sheet. (B) Radial knurling on shoulder is even more positive.

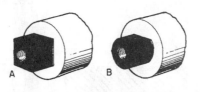

12 Noncircular end . . .

on bar may be square (A) or D-shaped (B) and introduced into a similarly-shaped hole. Screw and washer hold parts together as before.

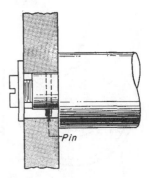

13 Transverse pin . . .

in rod end fits into slot, lets rod end be round but nonrotatable.

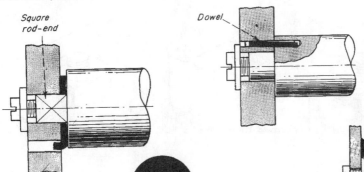

15 Dowel . . .

is simple, efficient method of preventing rotation if rod dia is big enough.

14 Washer over square rod end . . .

has leg bent to fit in small hole. Washer hole is square, preventing angular movement when all three parts are assembled and fastened with screw and washer.

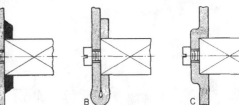

16 Double sheet thickness . . .

allows square or hexagonal locating-hole for shaft end to be provided in thin sheet. Extra thickness can be (A) welded (B) folded or (C) embossed.

Various Methods of Locking Threaded Members

Locking devices can generally be classified as either form or jam locking. Form locking units utilize mechanical interference of parts whereas the jam type depends on friction developed between the threaded elements. Thus their performance is a function of the torque required to tighten them. Both types are illustrated below.

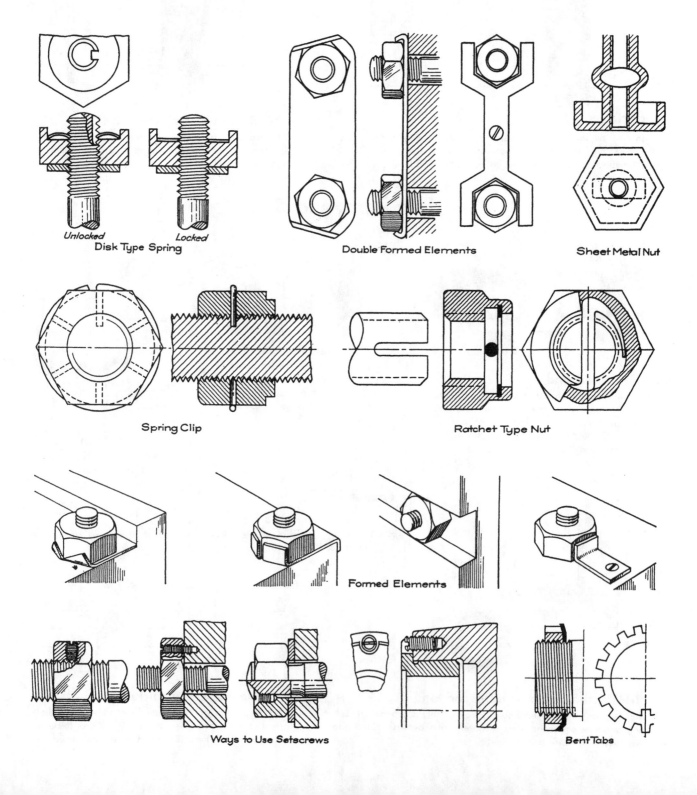

Unlocked　　　*Locked*
Disk Type Spring

Double Formed Elements

Sheet Metal Nut

Spring Clip

Ratchet Type Nut

Formed Elements

Ways to Use Setscrews

Bent Tabs

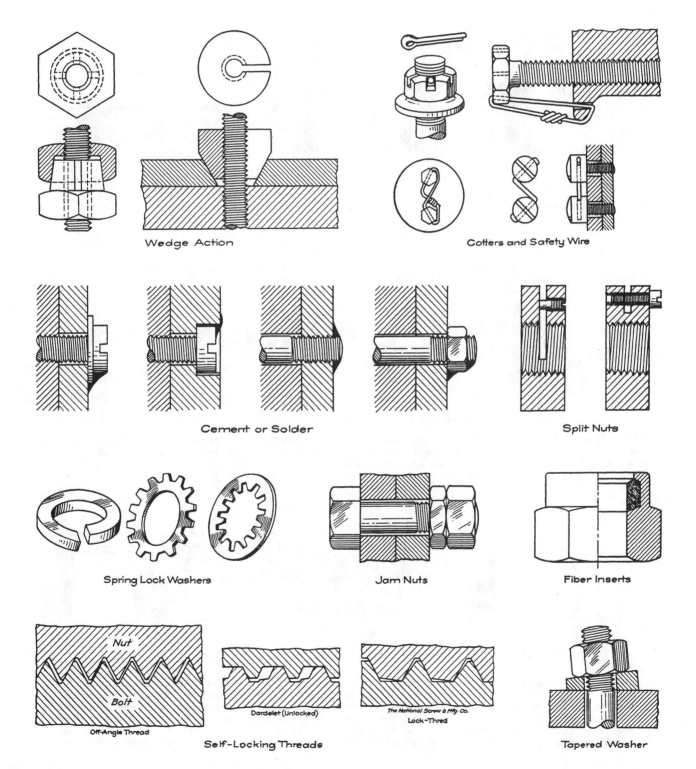

Wedge Action

Cotters and Safety Wire

Cement or Solder

Split Nuts

Spring Lock Washers

Jam Nuts

Fiber Inserts

Off-Angle Thread

Dardelet (Unlocked)

The National Screw & Mfg. Co.
Lock-Thred

Self-Locking Threads

Tapered Washer

How to Provide for Backlash in Threaded Parts

These illustrations are based on two general methods of providing for lost motion or backlash. One allows for relative movement of the nut and screw in the plane parallel to the thread axis; the other method involves a radial adjustment to compensate for clearance between sloping faces of the threads on each element.

Clifford T. Bower

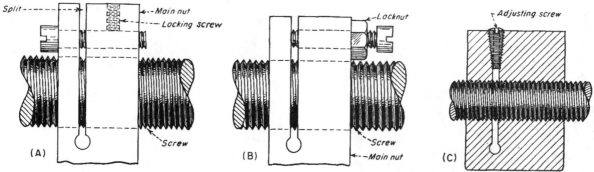

THREE METHODS of using slotted nuts. In *(A)*, nut sections are brought closer together to force left-hand nut flanks to bear on right-hand flanks of screw thread and vice versa. In *(B)*, and *(C)* nut sections are forced apart for same purpose.

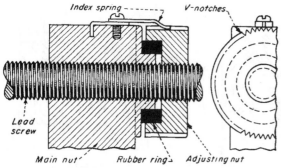

AROUND THE PERIPHERY of the backlash-adjusting nut are "v" notches of small pitch which engage the index spring. To eliminate play in the lead screw, adjusting nut is turned clockwise. Spring and adjusting nut can be calibrated for precise use.

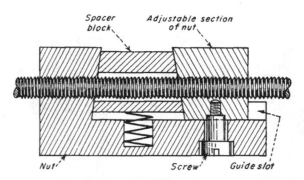

SELF-COMPENSATING MEANS of removing backlash. Slot is milled in nut for an adjustable section which is locked by a screw. Spring presses the tapered spacer block upwards, forcing the nut elements apart, thereby taking up backlash.

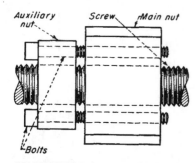

MAIN NUT is integral with base attached to part moved by screw. Auxiliary nut is positioned one or two pitches from main nut. The two are brought closer together by bolts which pass freely through the auxiliary nut.

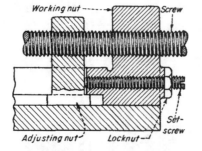

ANOTHER WAY to use an auxiliary or adjusting nut for axial adjustment of backlash. Relative movement between the working and adjusting nuts is obtained manually by the set screw which can be locked in place as shown.

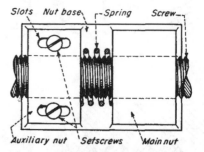

COMPRESSION SPRING placed between main and auxiliary nuts exerts force tending to separate them and thus take up slack. Set screws engage nut base and prevent rotation of auxiliary nut after adjustment is made.

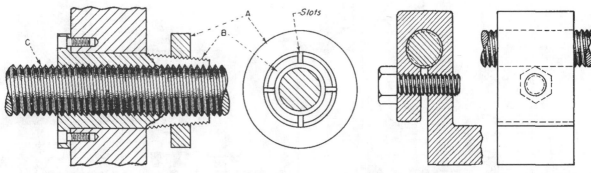

NUT *A* IS SCREWED along the tapered round nut, *B*, to eliminate backlash or wear between *B* and *C*, the main screw, by means of the four slots shown.

ANOTHER METHOD of clamping a nut around a screw to reduce radial clearance.

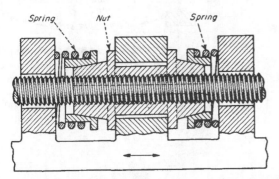

AUTOMATIC ADJUSTMENT for backlash. Nut is flanged on each end, has a square outer section between flanges and slots cut in the tapered sections. Spring forces have components which push slotted sections radially inward.

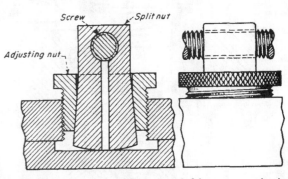

SPLIT NUT is tapered and has a rounded bottom to maintain as near as possible a fixed distance between its seat and the center line of the screw. When the adjusting nut is tightened, the split nut springs inward slightly.

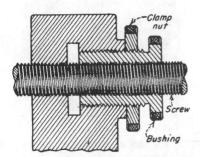

CLAMP NUT holds adjusting bushing rigidly. Bushing must have different pitch on outside thread than on inside thread. If outer thread is the coarser one, a relatively small amount of rotation will take up backlash.

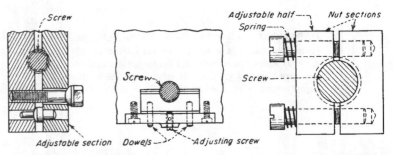

TYPICAL CONSTRUCTIONS based on the half nut principle. In each case, the nut bearing width is equal to the width of the adjustable or inserted slide piece. In the sketch at the extreme left, the cap screw with the spherical seat provides for adjustments. In the center sketch, the adjusting screw bears on the movable nut section. Two dowels insure proper alignment. The third illustration is similar to the first except that two adjusting screws are used instead of only one.

7 Special Screw Arrangements

How differential, duplex, and other types of screws can provide slow and fast feeds, minute adjustments, and strong clamping action.

Louis Dodge

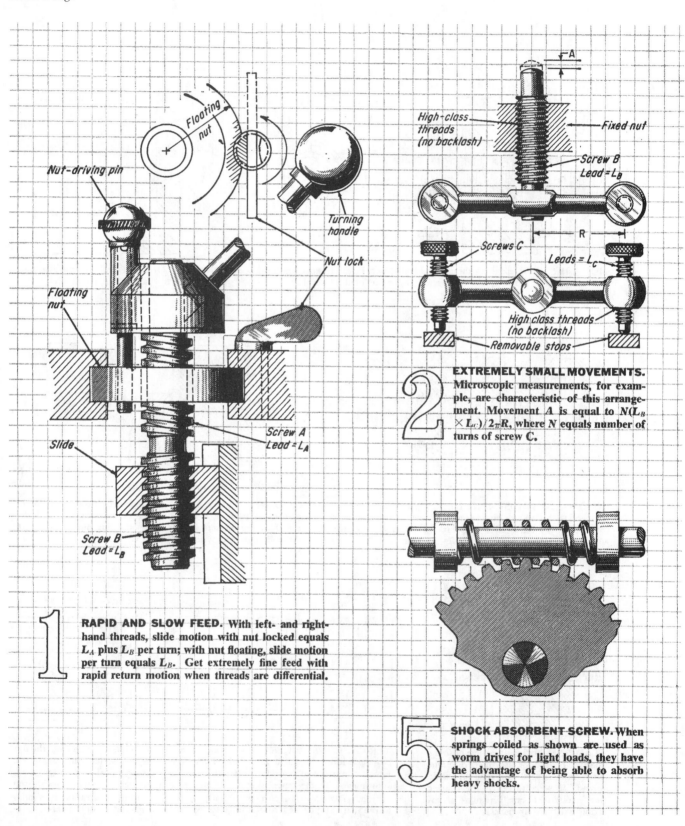

1 RAPID AND SLOW FEED. With left- and right-hand threads, slide motion with nut locked equals L_A plus L_B per turn; with nut floating, slide motion per turn equals L_B. Get extremely fine feed with rapid return motion when threads are differential.

2 EXTREMELY SMALL MOVEMENTS. Microscopic measurements, for example, are characteristic of this arrangement. Movement A is equal to $N(L_B \times L_A)/2\pi R$, where N equals number of turns of screw C.

5 SHOCK ABSORBENT SCREW. When springs coiled as shown are used as worm drives for light loads, they have the advantage of being able to absorb heavy shocks.

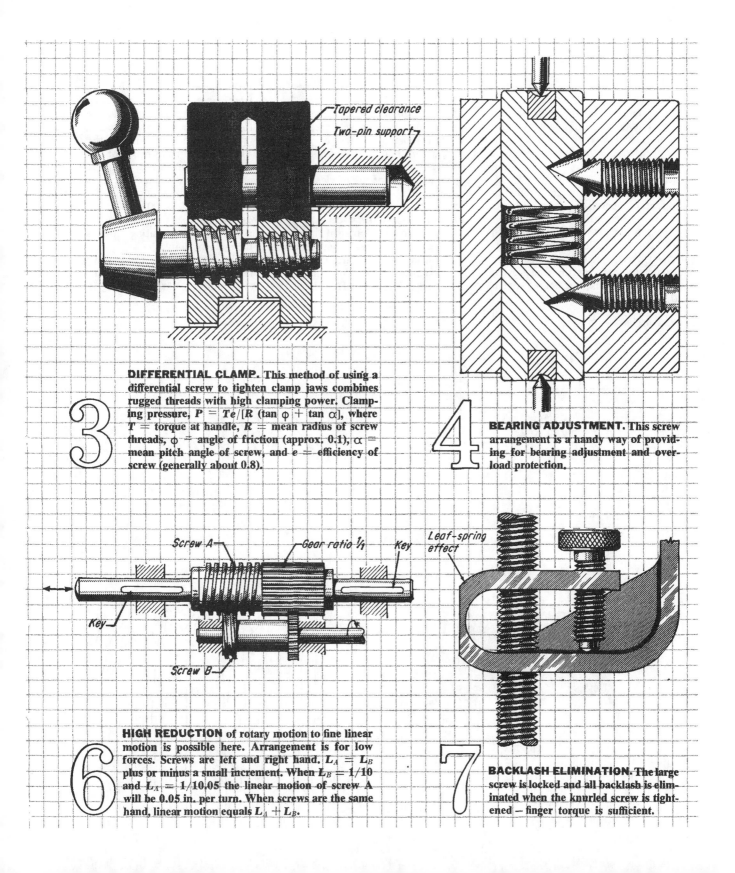

DIFFERENTIAL CLAMP. This method of using a differential screw to tighten clamp jaws combines rugged threads with high clamping power. Clamping pressure, $P = Te/[R (\tan \phi + \tan \alpha]$, where $T =$ torque at handle, $R =$ mean radius of screw threads, $\phi =$ angle of friction (approx. 0.1), $\alpha =$ mean pitch angle of screw, and $e =$ efficiency of screw (generally about 0.8).

Labels: *Tapered clearance*, *Two-pin support*

BEARING ADJUSTMENT. This screw arrangement is a handy way of providing for bearing adjustment and overload protection.

HIGH REDUCTION of rotary motion to fine linear motion is possible here. Arrangement is for low forces. Screws are left and right hand. $L_A = L_B$ plus or minus a small increment. When $L_B = 1/10$ and $L_A = 1/10.05$ the linear motion of screw A will be 0.05 in. per turn. When screws are the same hand, linear motion equals $L_A + L_B$.

Labels: *Screw A*, *Gear ratio* $\frac{1}{1}$, *Key*, *Key*, *Screw B*

BACKLASH ELIMINATION. The large screw is locked and all backlash is eliminated when the knurled screw is tightened — finger torque is sufficient.

Labels: *Leaf-spring effect*

World of Self-Locking Screws

Screws and bolts with self-locking ability are making the scene on more products because of today's emphasis on safety and reliability.

Frank Yeaple

Self-locking ability in fasteners is definitely in. The fear of costly lawsuits stemming from a fastener's loss of clamp load is making manufacturers of products spend a little more time and money to assure joint reliability. Hence design engineers have a freer hand in selecting locknuts and lockscrews.

Last month's issue contained a roundup of proprietary types of locknuts. Locknuts are generally cheaper than lockscrews, and fewer special parts (the locknuts) need to be stocked to handle original and replacement parts. But because lockscrews mate with a standard nut, they can provide a better guarantee that the locking system

will be retained even when the original nut is lost.

This roundup of self-locking screws and bolts includes many novel thread forms that are competing for the job of maintaining a tight joint under adverse conditions. Also included are special shank and head designs and the use of plastic inserts and anaerobics to improve self locking characteristics.

Variable pitch threads. A screw thread with an oscillating pitch distance can increase and decrease between maximum and minimum values to provide a high resistance to self loosening from shock and vibration. The Leed-Lok

prevailing-torque screw (Fig 1), available from National Lock Fasteners Division of Keystone Consolidated Industries (Rockford, Ill.), uses the varying pitch to induce interference on the flanks of the threads. The locking threads need not be produced along the entire length of the screw, but only at desired points. For example, the pitch for a ½-20 thread may vary as follows: 0.047, 0.050, 0.047. This significantly increases turning friction but does not deform the mating threads beyond their elastic limit.

Tilted threads. Another way to induce interference is by deflecting and slightly deforming the threads during the thread

1. Leed-lok threads

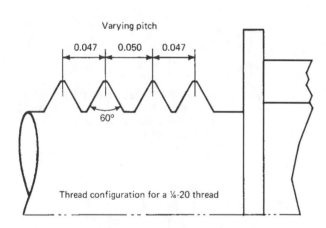

Varying pitch

0.047   0.050   0.047

60°

Thread configuration for a ¼-20 thread

2. E-Lok threads

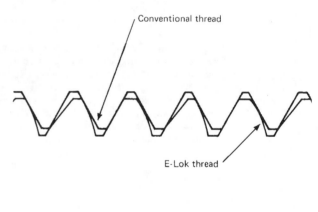

Conventional thread

E-Lok thread

3. Vibresist threads

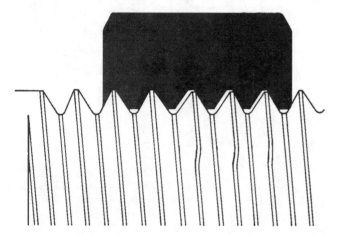

4. Tru-Flex threads

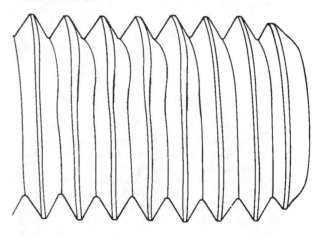

5. Orlo threads

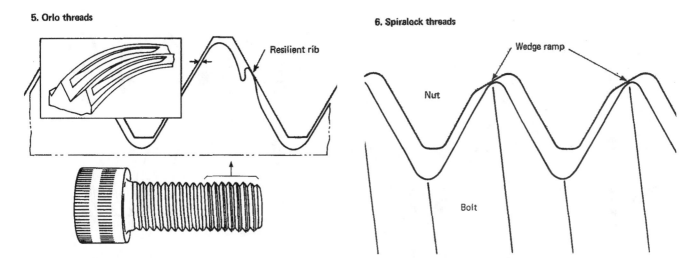

Resilient rib

6. Spiralock threads

Wedge ramp

Nut

Bolt

rolling process when the screws are made. Employed by the Everlock Division (Troy, Mich) of Microdot on its E-Lok screws (Fig 2), the process deflects the 60-deg thread form a certain amount (about 10 deg) from the normal to the screw axis. In general, every two tilted threads are followed by two straight (normal) threads over the chosen length of screw.

Partially offset threads. In this method, portion of each thread is deformed parallel to the thread helix axis (Fig 3). As an example, Vibresist screws produced by Russell, Burdsall, & Ward (Mentor, Ohio) provide a consistent prevailing torque, slight or substantial as required. The mating internal threads wedge between the offset screw threads. Spring action is reinforced, and because locking, here as with the other altered-thread types, depends on the elasticity of metal, the screw does not lose its grip at elevated temperatures. The locking threads can be specified adjacent to the head for locking in tapped holes, along the middle of the thread length for locking in tapped holes and near the screw end for locking with standard nuts.

Screws with somewhat similarly de-formed threads are offered by Cleveland Cap Screw (division of SPS Technologies, Cleveland, Ohio) on its Tru-Flex screws (Fig 4). In the process, threads near the end of the screw are specially deformed for a circumference of less than 180 deg of arc. When meshed with conventional female threads, the locking threads are repositioned from root to crest to induce a resisting torque. It was found that reshaping several threads within a 180-deg arc of the fastener circumference would provide ample prevailing torque with only minute changes in pitch diameter.

Resilient rib thread. The Orlo thread (Fig 5) has a cold-formed resilient rib on the non-pressure flank of the thread, either part way or continuously around. When a screw with Orlo threads is assembled into a threaded part, the ribs are compressed like springs to force the screw's pressure flanks against the mating threads, and this will increase resistance to rotational forces from vibration or shock. The spring-like action of the rib threads permits the screw to be reused effectively. Orlo threads are available on screws by Holo-Krome Co (West Hartford, Conn)., and Pioneer

Screw & Nut (Elk Grove Village, Ill).

Wedge-ramp roots. This internal thread form, called Spiralock (Fig 6), is applied to locknuts and tapped holes to provide locking characteristics to standard bolts. The key innovation is the addition of a 30-deg ramp to the roots of conventional 60-deg threads. When the bolt is seated, the crest of the bolt thread is pulled up tight against the ramp and is wedged firmly with positive metal-to-metal contact that runs the entire length of the nut or tapped hole. The special thread form, in fact, allows wide latitude in bolt tolerances. Spiralock™ locknuts are available from the Greer and the Kaynar divisions of Microdot (Greenwich, Conn.); taps for producing the thread may be obtained from Detroit Tap and Tool Co. (Warren, Mich.)

Threads with special wedge-ramp roots also have been applied to screws. In the Lok-thred form (Fig 7), available from Lock Thread Corporation and National Lock Fasteners (Rockford, Ill.), the thread of the screw itself performs the locking action. This male thread is shallow, with ample radii and a wide root pitched at a

7. Lok-Thred threads

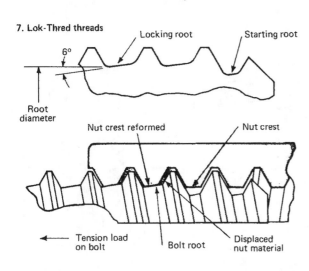

Locking root   Starting root

6°

Root diameter

Nut crest reformed   Nut crest

Tension load on bolt   Bolt root   Displaced nut material

8. Lamcolok threads

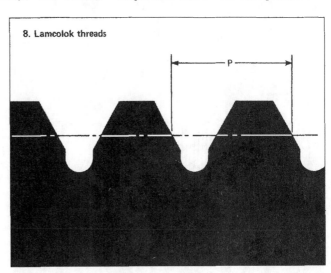

P

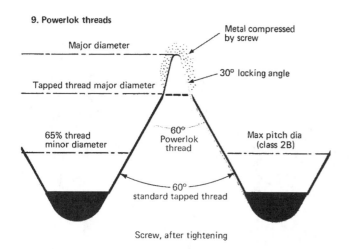

9. Powerlok threads

Major diameter

Metal compressed by screw

Tapped thread major diameter

30° locking angle

65% thread minor diameter

60° Powerlok thread

Max pitch dia (class 2B)

60° standard tapped thread

Screw, after tightening

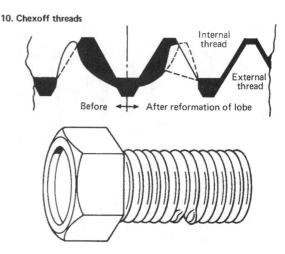

10. Chexoff threads

Internal thread

External thread

Before ← → After reformation of lobe

locking angle. These locking roots converge toward the head. A Lok-Thred bolt enters an ordinary tapped hole freely for a few turns, then meets resistance when the bolt root contacts and, by swaging, reforms the nut threads to a perfect fit about the wedge-like locking roots of the bolt. In service, much of the clamp load is carried on the tapered roots, which wedge against the nut crest to lock securely as the work load increases. Bolts which have this type of wedge root thread have a larger root diameter than ordinary threads, which gives strength in tension, torsion and shear, as well as increased endurance limits.

Locking roots. A self-locking thread based on flank interference with standard internal thread flanks (Fig 8) provides a clamping force and holding torque that remains constant. Developed by Lamson & Sessions (Cleveland, Ohio), the Lamcolock thread design can be rolled on almost any bolt or screw. It employs a recessed full-radius root and a decreased major diameter giving the thread a wide, squat look. The decreased major diameter provides room for material to flow into

the root recesses, thus averting possible galling.

Locking crests. The locking ability of Powerlok screws (Fig 9) is enhanced through the combination of a novel 60 deg - 30 deg thread form and a tri-lobular thread body section. Locking action is developed at the outermost radius of the torque arm of the screw body, whereas most locking screws develop their resistance at lesser radius points. The deeper thread form of the Powerlok geometry, along with a slight increase in the major diameter of the thread as compared to equivalent size conventional screws, also adds to the locking ability. Basically, the nut-thread metal is elastically deformed in the compressed areas created by the 30-deg portion of the thread. Powerlok screws are available from several fastener manufacturers, including Continental Screw Co (New Bedford, Mass.), Midland Screw Co. (Chicago, Ill.), Central Screw Division of Microdot, and Elco Industries (Rockford, Ill.).

Resilient bulges. By deforming several threads on one side of the Chexoff screw to form lobes (Fig 10), controlled thread

interference is induced when the screw is threaded. Available from Central Screw Co, (Des Plaines, Ill.), the special screws with lobes create a wedge-like effect by exerting pressures on the opposite side of the mating threads. The lobes may all be located on the same line or else staggered in order to provide considerable periphery pressures.

In another design, Deutsch Fasteners Corp. induces a resilient bulge on one side of the bolt (Fig 11) that increases in a similar manner the frictional contact between mating threads on the opposite side of the bolt. The bulge is formed by the interference action of a precision ball, pressed into a hole drilled close to the minor diameter of the threads during manufacture.

Sine-wave threads. In another approach to improving the locking characteristics of bolts, Valley-Todeco, Inc. (Sylmar, Calif), has developed its Sine-Lok interference-type thread (Fig 12) for use on the upper regions of a bolt, normally consisting merely of a straight shank. The lower portion of the bolt has conventional threads. During assembly, the bolt shank,

11. Resilient-bulge threads

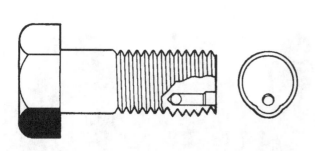

12. Sine-Lok bolt shank

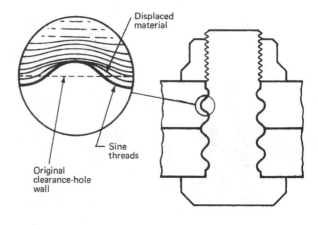

Displaced material

Original clearance-hole wall

Sine threads

13. Taper-Lok shank

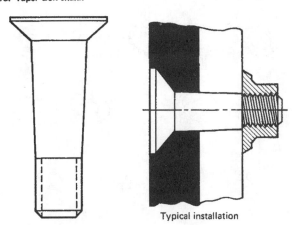

Typical installation

14. Uniflex head

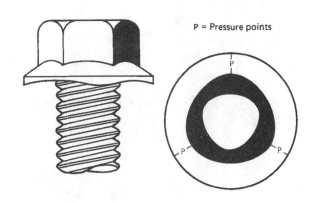

P = Pressure points

which has a series of modified sine waves, is threaded through the clearance holes of the two parts being assembled, instead of simply being pushed through. The sine wave threads displace the material of the clearance-hole wall into their roots. The protruding, threaded end of the bolt is then tightened by a standard nut. The bolt thus gets a double threading action that helps prevent loosening.

Tapered shank. Another way to increase the shank's grip of the parts being assembled together is to provide a slight taper to the shank, as shown in the Taper-Lok screw by Voi-Shan Division of VSI Corp. (Pasadena, Calif.). (Fig 13). Although only 0.25 in. per linear foot, the taper provides a controlled interference fit that compresses the material of the joint elastically around the hole to induce an excellent preload condition.

Head-grip screws. Many new head shapes for screws are designed to increase resistance to vibrational loosening in joints. The Uniflex head (Fig 14), developed by Continental Screw Co. (New

Bedford, Mass) to complement the company's line of trilobular thread-rolling screws, has a washer-like, undulating head-bearing surface. When the surface, which has three alternate high-and-low areas, is tightened against the joint being assembled, the relieved areas are aligned with potential stress points at the lobes of the trilobular screw thread. As a result of this bearing-area relief, thread engagement is increased and bolt loosening noticeably resisted.

High locking power and clamping force are provided with Tensilock screws available from Eaton Corporation (Massilon, Ohio) (Fig 15). The screw head has a concentric circle of 24 embedded, carburized teeth, with an outer concentric groove that permits flexing of the head to occur. The Durlok fastener available from SPS Technology's Cleveland Cap Screw (Cleveland, Ohio) also has ratchet-like teeth around the periphery of the bearing surface (Fig 16). To limit depth of penetration and marring of the mating surface, the serrations are encircled by a smooth

outer bearing area.

Lockscrews with the toothed portion of the head furnished in the form of a pre-assembled washer include the Melgrip screw by Elco Industries (Rockford, Ill) (Fig 17), and Sems screws available from a number of manufacturers, including Shakeproof Division of Illinois Tool (Elgin, Ill), National Lock Fasteners (Rockford, Ill.), and Central Screw (Fig 18). Melgrip's locking effectiveness results from the mating serrations on the underside of the bolt head and top of the washer surface, coupled by the bidirectional gripping teeth on the periphery of the washer which embed into the joint material. Thus, the washer cannot skid or score.

Sems screws are available in a vast variety of washer types to provide spring tensioning for improved loosening resistance, as well as to bridge oversized holes or insulate and protect material surfaces.

Nylon-pellet insert. Self-locking screws that use a resilient nylon insert in the threaded section to develop a

15. Tensilock head

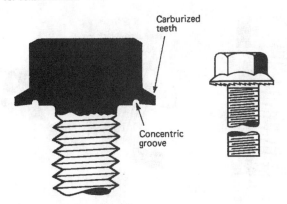

Carburized teeth

Concentric groove

16. Durlok head

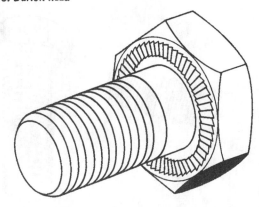

17. Melgrip washer-head

Mating serrations

Bi-directional gripping teeth

18. Sems washer types

prevailing-torque locking action (Fig 19) have been a familiar product for a number of years. The nylon pellet, press-fitted into the screw, projects slightly beyond the crest of the thread. Once the threads are engaged, the screw is held in position by lateral pressure. The pellet technique can be applied also to nuts and studs. Screws with nylon-pellet inserts are available from Nylok Fastener Division of USM Corp. (Paramus, N.J.) and ND Industries, Inc. (Troy, Mich.).

Nylon fused patch. Another effective and well-established way to help a screw resist loosening is by use of a plastic locking patch that is fused on a dimensionally controlled area of the screw threads, shown in Fig 20. This nylon patch is thickest in the center and feathers-out along the edge to provide a gradual engagement of the locking patch as it encounters mating threads. As the mating threads fully engage the patch, the nylon is compressed to build up a resistance to turning at the right in the figure, and a strong metal-to-metal contact between threads at the

left. This type of screw is available from the Esna Division of Amerace Corp. (Union, N.J.), Long-Lok Fasteners Corp. (Cincinnati, Ohio), Holo-Krome Co., (West Hartford, Conn.), and the Unbrako Division of SPS Industries (Jenkintown, Pa.)

Adhesive thread locking. Epoxy and anaerobic adhesives and stiffly viscous fluids have become popular for turning low-cost plain bolts and nuts into lock fasteners. Epoxy, which is a two-part adhesive, is applied in the form of alternating strips or microcapsules to the threaded fastener. Once applied, the epoxy (or one of the anaerobics) remains dormant on the fastener (Fig 21) until a cure is activated by engagement with a mating thread. The curing process for most of the chemical adhesives may continue for days, although by the twelfth hour a good bond has generally been achieved. The adhesive technique, however, does not offer reusability capabilities. Screws with pre-applied adhesives are available from ND Industries, (Troy, Mich.), Cleveland Cap

Screw (Cleveland, Ohio) and Camcar Division of Textron (Rockford, Ill). Or you can buy your own.

Highly viscous fluid coatings, such as Vibra-Tite available from ND Industries (Troy, Mich.) and Oakland Corp., offer a compromise by making the parts adjustable as well as improving self-locking. The user can apply it from a bottle with a brush applicator, like glue. Vibra-Tite, however, is not an adhesive, so its primary usage is not to provide a heavy-duty locking capability; but it does permit the fastener to be assembled, disassembled or adjusted.

Prevailing torque. Remember that most self-locking screws rely on friction in one form or another to hold fast against axial or transverse vibration. Transverse vibration is most difficult to protect against because accelerating forces can cause momentary slip at a microscopic level, eventually loosening the thread. Lab and field tests are recommended.

19. Nylon-pellet screw

20. Nylon locking-patch screw

Strong metal-to-metal contact

Compressed nylon

21. Adhesive-lock screw

SECTION 12

PINS

Slotted Spring Pins Find Many Jobs

Assembled under pressure, these fasteners provide powerful gripping action to locate and hold parts together.

Robert O. Parmley

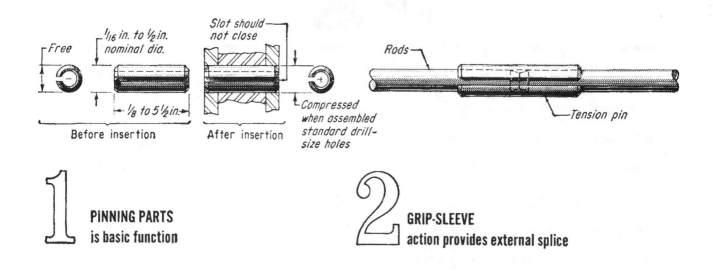

1 PINNING PARTS
is basic function

2 GRIP-SLEEVE
action provides external splice

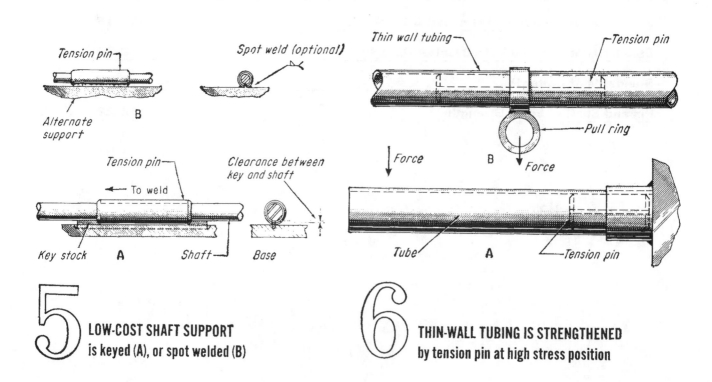

5 LOW-COST SHAFT SUPPORT
is keyed (A), or spot welded (B)

6 THIN-WALL TUBING IS STRENGTHENED
by tension pin at high stress position

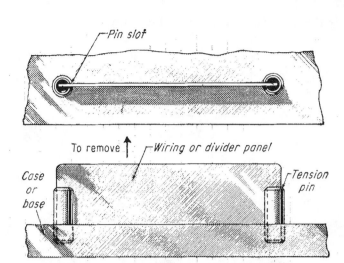

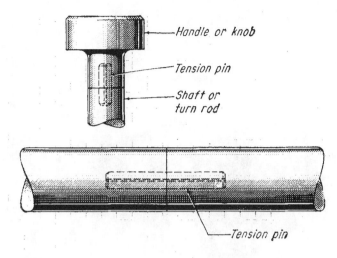

3
THIN PANELS
are inexpensively supported

4
LOW-TORQUE SHAFT
connection, or knob assembly

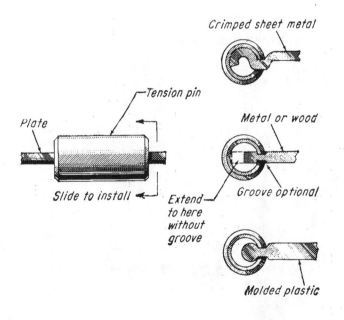

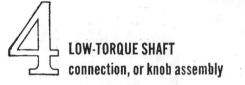

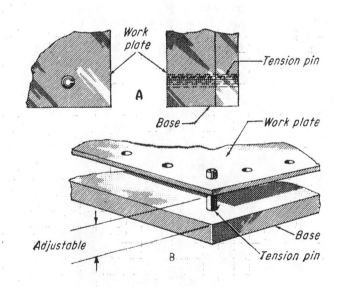

7
PROTECT PROJECTING LIPS

8
SUPPORT POSTS
locate (A), provide adjustment (B)

8 Unusual Jobs for Spring Pins

Be sure you get top value from these versatile assembly devices. These examples show how.

Andrew J. Turner

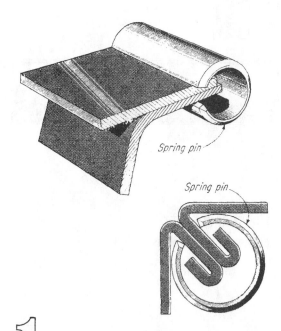

Spring pin

Spring pin

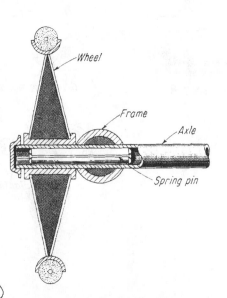

1⅛-in.–square opening in sheet metal

Rubber hose, 1in. O.D.

4 – ³/₁₆-in. dia. x 1in. rollpins

Section "A-A"

1 SLOT IN PIN does duty as anchoring device, holding two pieces together. Fastening can be either permanent or temporary. Parts can be metal or non-metal.

2 PROTECT HYDRAULIC tubing or electrical wires touching sharp edges of casings by clipping pins over the edges of the hole. Its size is only slightly reduced.

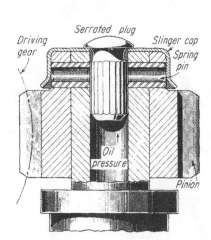

Serrated plug

Driving gear

Slinger cap

Spring pin

Oil pressure

Pinion

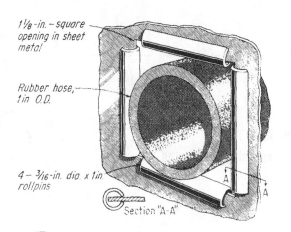

Wheel

Frame

Axle

Spring pin

5 LUBRICANT PASSAGE is combined with retaining pin for gear. Also, slinger ring not only performs functionally but improves appearance too.

6 STIFFEN LIGHT-DUTY structures such as tubular axles with spring pins; they are simple to install and add considerable strength to the assembly.

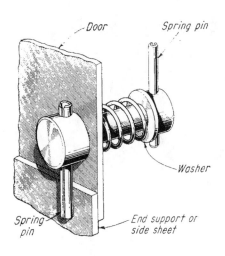

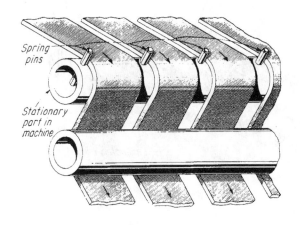

TWO PINS SERVE AS HANDLE and latch. This low-cost assembly replaces an expensive forged handle and a fabricated-metal latch-piece.

AS BELT GUIDES, spring pins eliminate molded spacers, or costly machined grooves for spacer rings which would otherwise be needed.

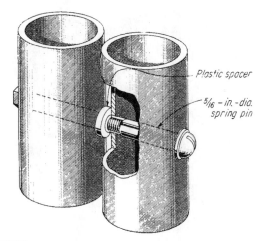

HARDENED STEEL SLEEVE for pivot-screw gives durability to legs of folding table illustrated here, while keeping costs competitively low.

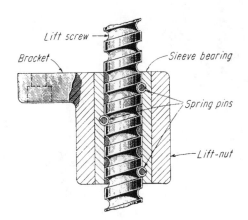

LOW-COST THREAD in lift-nut can be made by fitting spring pins at correct pitch-positions as shown. Rotate the pins to reduce wear.

8 Electrical Jobs for Spring Pins

Put these handy assembly devices to work as terminals, connectors, actuators, etc.

Andrew J. Turner

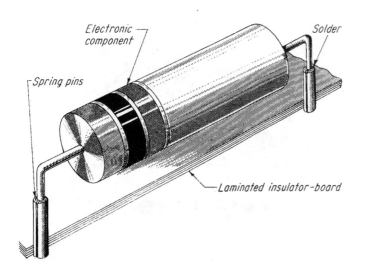

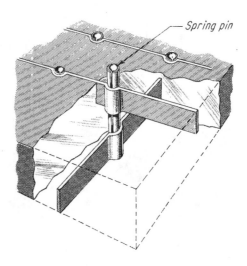

1 **LOW-COST TERMINALS** are made by assembling two 1/16-in.-dia tin-dipped Rollpins into phenolic board. The board should be about 3/32 in. thick.

2 **AS ELECTRICAL CONNECTORS** in "patchboard" circuits, spring pins have ample conductivity. Select various circuits by removing or inserting pins.

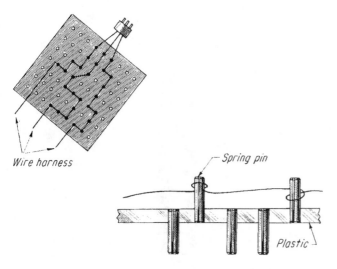

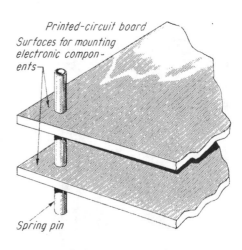

5 **FORMING FIXTURE** for wire harnesses is quickly adjusted when different harness-shapes are needed. Plastic sheet has pin holes on ¼-in. centers.

6 **STAND-OFFS** for printed-circuit boards can be spring pins. Select a pin long enough to ensure adequate spacing between the boards.

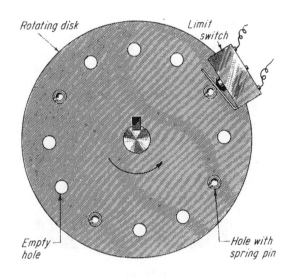

Rotating disk   Limit switch

Empty hole

Hole with spring pin

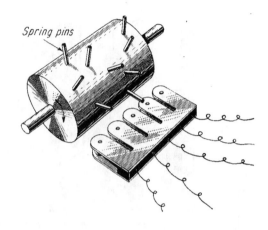

Spring pins

SWITCH ACTUATORS can be quickly relocated in rotating disk if spring pins are employed. Hard steel of pin gives excellent wear resistance.

DRUM-MOUNTED ACTUATORS function in similar way to spring-pin actuators in Fig. 3. Protruding length of pins may be critical, but is easily adjusted.

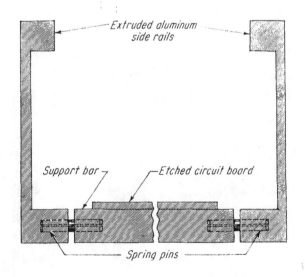

Extruded aluminum side rails

Support bar   Etched circuit board

Spring pins

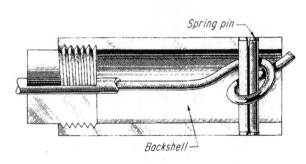

Spring pin

Backshell

SUPPORT BARS in electronic units can be easily and quickly installed into the sliding chassis with spring pins. Close tolerances are not needed.

STRAIN RELIEF for wire in electrical connectors will not slip during assembly. Loop wire then fill shell with potting compound to seal wire in place.

8 More Spring Pin Applications

Some additional ways that these fasteners, assembled under pressure, can grip and locate parts. They can even valve fluids.

Robert O. Parmley

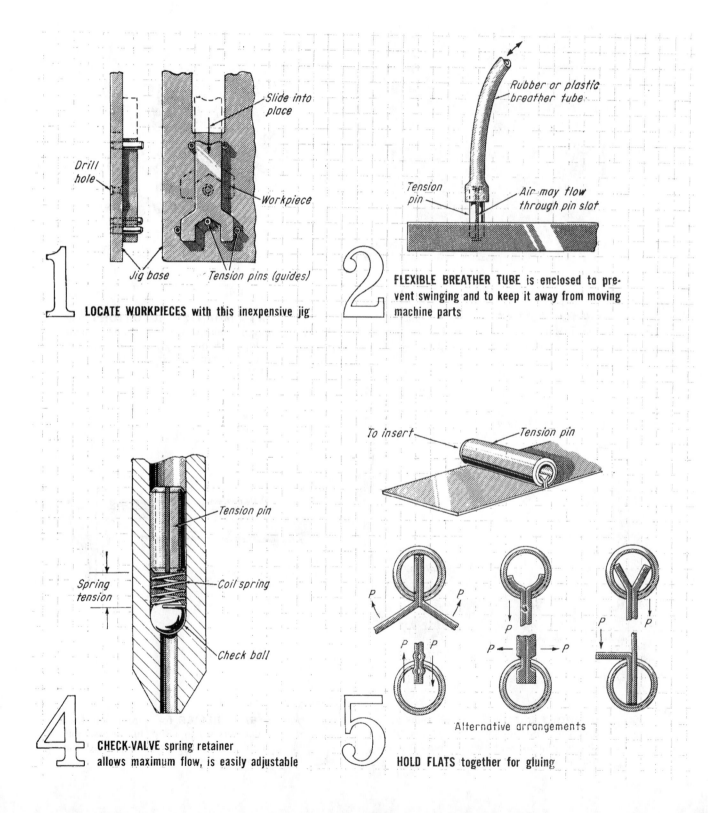

1 **LOCATE WORKPIECES** with this inexpensive jig

Slide into place
Drill hole
Workpiece
Jig base
Tension pins (guides)

2 **FLEXIBLE BREATHER TUBE** is enclosed to prevent swinging and to keep it away from moving machine parts

Rubber or plastic breather tube
Tension pin
Air may flow through pin slot

4 **CHECK-VALVE** spring retainer allows maximum flow, is easily adjustable

Tension pin
Spring tension
Coil spring
Check ball

5 **HOLD FLATS** together for gluing

To insert
Tension pin
P
Alternative arrangements

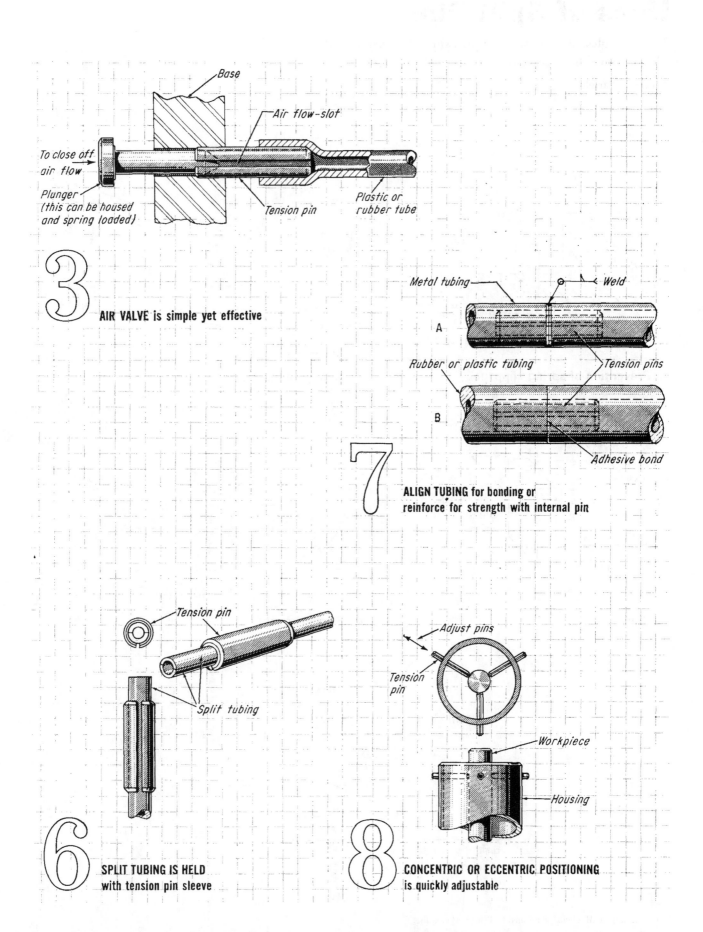

3 AIR VALVE is simple yet effective

Base
Air flow-slot
To close off air flow
Plunger (this can be housed and spring loaded)
Tension pin
Plastic or rubber tube

7 ALIGN TUBING for bonding or reinforce for strength with internal pin

Metal tubing
Weld
A
Rubber or plastic tubing
Tension pins
B
Adhesive bond

6 SPLIT TUBING IS HELD with tension pin sleeve

Tension pin
Split tubing

8 CONCENTRIC OR ECCENTRIC POSITIONING is quickly adjustable

Adjust pins
Tension pin
Workpiece
Housing

Uses of Split Pins

Ten examples show how these pins simplify assembly of jigs and fixtures.
The pins are easily removed.

Robert O. Parmley

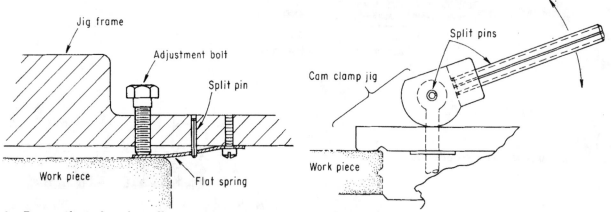

1. Prevention of spring slippage

2. Cam pivot and handle

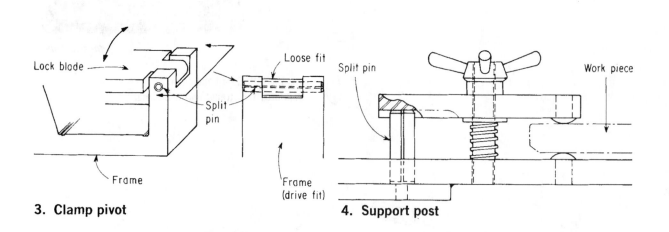

3. Clamp pivot

4. Support post

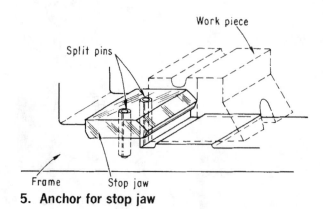

5. Anchor for stop jaw

Source: *American Machinist*, Published by Penton Media, Inc.

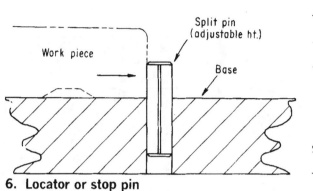

6. Locator or stop pin

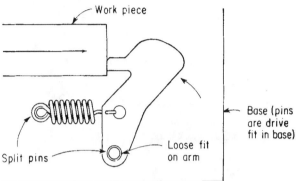

7. Spring anchor and arm pivot

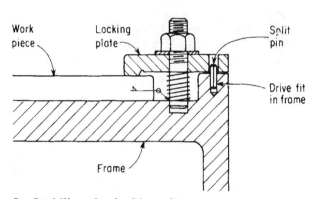

8. Stabilizer for locking plate

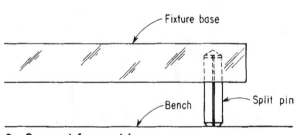

9. Support for post leg

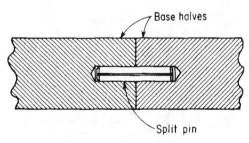

10. Dowels for fixture base

Slotted tubular pins are intended to be forced into their locations; free diameter should be larger than hole diameter so the pin exerts radial force all along its mounting hole to resist axial motion when properly mounted. Maximum compression is controlled by the amount of gap when the pin is free. When it acts as a pivot, the hole through the pivoting member should be a free fit (see figures 7 and 8) so the pin will not be worked loose from its anchor hole. These pins may be made of heat-treated carbon steel, corrosion-resistant steel, or beryllium-copper.

Design Around Spiral Wrapped Pins

Coil, rolled, or spiral wrapped pins come in a wide range of lengths and diameters.
Their applications are limitless; here are eight.

Robert O. Parmley

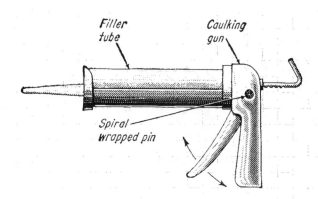

2 Pivot. Pin is a drive fit in the handle housing, and acts as a pivot for trigger

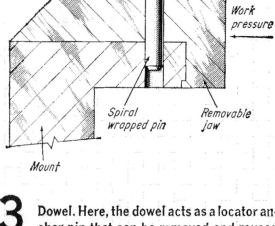

3 Dowel. Here, the dowel acts as a locator anchor pin that can be removed and reused

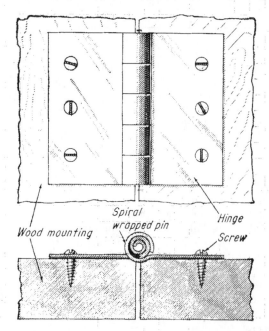

6 Hinge pins. If hole size in both members is different hinge will be free moving; if the hole size is the same in both members a friction hinge is the result

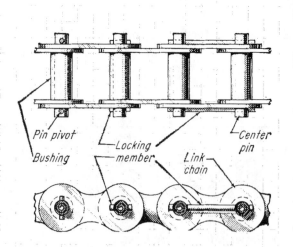

7 Link chain connection. Pins are used as pivot or locking members. An advantage: Both types are removable and reusable

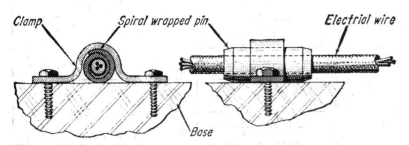

1 Wire gripper. As clamp is tightened in place, pin coils, and secures the wire

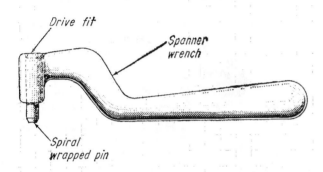

4 Wrench pin. Coil construction permits pin to fit holes with large tolerances

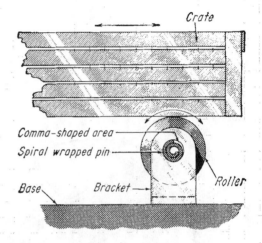

5 Lubricated shaft for work roller. Comma-shaped area of the spiral wrapped pin forms an oil reservoir for the roller

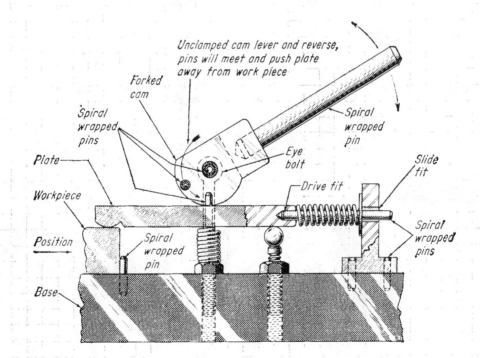

8 Pivot, stop, locator, handle, and anchor are typical applications in the design of a clamp. The pins are drive or slide fits, and can be removed and reused if the clamp position must be changed

A Penny-Wise Connector: The Cotter Pin

They're simple, inexpensive and make excellent electrical connectors.
Why not consider cotter pins the next time you need one?

Robert O. Parmley

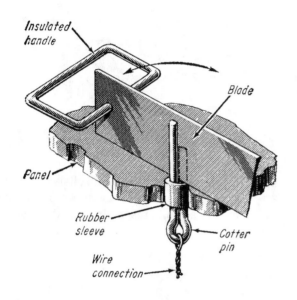

1 Knife blade connector

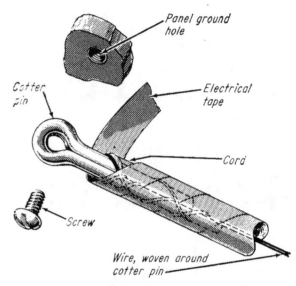

3 End mounting connection

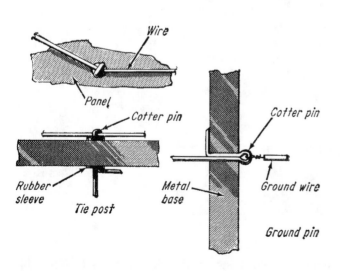

2 Tie post and ground connections

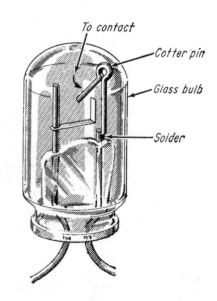

4 Glow switch

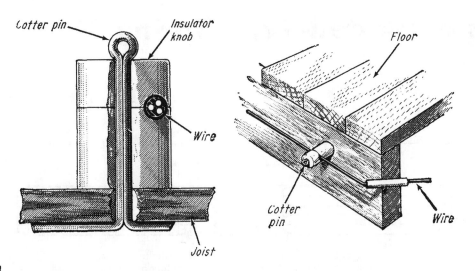

5   Knob anchor pin

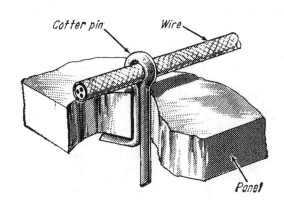

6   Wire eyelet

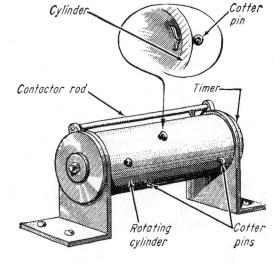

8   Cylinder contactor

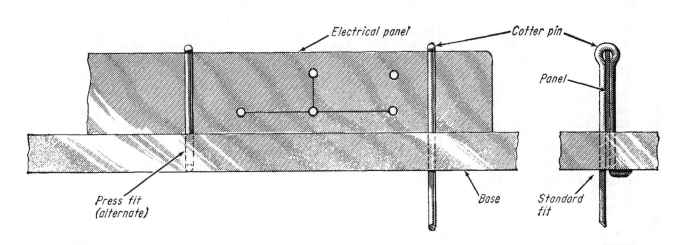

7   Electrical panel stabilizer

Standards of Slotted–Type Spring Pins

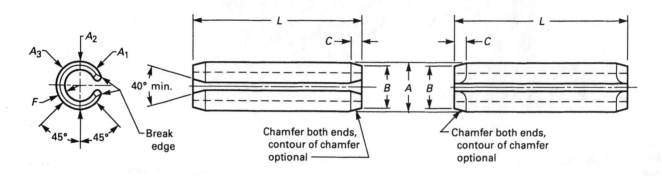

Style 1   Style 2

Optional Constructions

DIMENSIONS OF SLOTTED-TYPE SPRING PINS

Nominal Size of Basic Pin Diameter [Note (1)]	A Pin Diameter Max. [Note (2)]	A Pin Diameter Min. [Note (3)]	B Chamfer Diameter Max.	C Chamfer Length Max.	C Chamfer Length Min.	F Stock Thickness Basic	Recommended Hole Size Max.	Recommended Hole Size Min.	Double Shear Load, Min., lb SAE 1070–1095 and SAE 51420 [Note (4)]	Double Shear Load, Min., lb SAE 30302 and 30304	Double Shear Load, Min., lb Beryllium Copper
1/16 0.062	0.069	0.066	0.059	0.028	0.007	0.012	0.065	0.062	430	250	270
5/64 0.078	0.086	0.083	0.075	0.032	0.008	0.018	0.081	0.078	800	460	500
3/32 0.094	0.103	0.099	0.091	0.038	0.008	0.022	0.097	0.094	1,150	670	710
1/8 0.125	0.135	0.131	0.122	0.044	0.008	0.028	0.129	0.125	1,875	1,090	1,170
9/64 0.141	0.149	0.145	0.137	0.044	0.008	0.028	0.144	0.140	2,175	1,260	1,350
5/32 0.156	0.167	0.162	0.151	0.048	0.010	0.032	0.160	0.156	2,750	1,600	1,725
3/16 0.188	0.199	0.194	0.182	0.055	0.011	0.040	0.192	0.187	4,150	2,425	2,600
7/32 0.219	0.232	0.226	0.214	0.065	0.011	0.048	0.224	0.219	5,850	3,400	3,650
1/4 0.250	0.264	0.258	0.245	0.065	0.012	0.048	0.256	0.250	7,050	4,100	4,400
5/16 0.312	0.330	0.321	0.306	0.080	0.014	0.062	0.318	0.312	10,800	6,300	6,750
3/8 0.375	0.395	0.385	0.368	0.095	0.016	0.077	0.382	0.375	16,300	9,500	10,200
7/16 0.438	0.459	0.448	0.430	0.095	0.017	0.077	0.445	0.437	19,800	11,500	12,300
1/2 0.500	0.524	0.513	0.485	0.110	0.025	0.094	0.510	0.500	27,100	15,800	17,000
5/8 0.625	0.653	0.640	0.608	0.125	0.030	0.125	0.636	0.625	46,000	18,800	. . .
3/4 0.750	0.784	0.769	0.730	0.150	0.030	0.150	0.764	0.750	66,000	23,200	. . .

GENERAL NOTE: For additional requirements refer to General Data for Spring Pins on pages 27, 29, and 30 from original document.

CHARACTERISTICS	SYMBOL
Straightness	−
Diameter	ø

NOTES:

(1) Where specifying nominal size in decimals, zeros preceding the decimal shall be omitted.

(2) Maximum diameter shall be checked by GO ring gage.

(3) Minimum diameter shall be average of three diameters measured at points illustrated A min $= \dfrac{A_1 + A_2 + A_3}{3}$.

(4) Sizes 5/8 in. (0.625) and larger are produced from SAE 6150H alloy steel, not SAE 1070-1095.

Standards of Coiled–Type Spring Pins

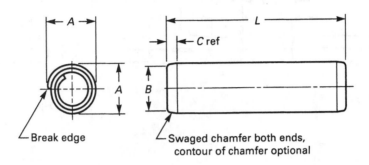

Break edge

Swaged chamfer both ends, contour of chamfer optional

DIMENSIONS OF COILED-TYPE SPRING PINS

Nominal Size or Basic Pin Diameter [Note (1)]	A Pin Diameter — Standard Duty Max. [Note (2)]	A Pin Diameter — Standard Duty Min. [Note (3)]	A Pin Diameter — Heavy Duty Max. [Note (2)]	A Pin Diameter — Heavy Duty Min. [Note (3)]	A Pin Diameter — Light Duty Max. [Note (2)]	A Pin Diameter — Light Duty Min. [Note (3)]	B Chamfer Dia. Max.	C Chamfer Length Ref.	Recommended Hole Size Max.	Recommended Hole Size Min.	Standard Duty SAE 1070–1095 and SAE 51420 [Notes (4), (5)]	Standard Duty SAE 30302 and 30304	Heavy Duty SAE 1070–1095 and SAE 51420 [Note (5)]	Heavy Duty SAE 30302 and 30304	Light Duty SAE 1070–1095 and SAE 51420	Light Duty SAE 30302 and 30304
1/32 0.031	0.035	0.033	...	...	...	...	0.029	0.024	0.032	0.031	90	65	...	...	...	...
0.039	0.044	0.041	...	...	...	...	0.037	0.024	0.040	0.039	135	100	...	...	...	...
3/64 0.047	0.052	0.049	...	...	...	...	0.045	0.024	0.048	0.047	190	145	...	...	...	...
0.052	0.057	0.054	...	...	...	...	0.050	0.024	0.053	0.051	250	190	...	...	...	...
1/16 0.062	0.072	0.067	0.070	0.066	0.073	0.067	0.059	0.028	0.065	0.061	330	265	475	360	205	160
5/64 0.078	0.088	0.083	0.086	0.082	0.089	0.083	0.075	0.032	0.081	0.077	550	425	800	575	325	250
3/32 0.094	0.105	0.099	0.103	0.098	0.106	0.099	0.091	0.038	0.097	0.093	775	600	1,150	825	475	360
7/64 0.109	0.120	0.114	0.118	0.113	0.121	0.114	0.106	0.038	0.112	0.108	1,050	825	1,500	1,150	650	500
1/8 0.125	0.138	0.131	0.136	0.130	0.139	0.131	0.121	0.044	0.129	0.124	1,400	1,100	2,000	1,700	825	650
5/32 0.156	0.171	0.163	0.168	0.161	0.172	0.163	0.152	0.048	0.160	0.155	2,200	1,700	3,100	2,400	1,300	1,000
3/16 0.188	0.205	0.196	0.202	0.194	0.207	0.196	0.182	0.055	0.192	0.185	3,150	2,400	4,500	3,500	1,900	1,450
7/32 0.219	0.238	0.228	0.235	0.226	0.240	0.228	0.214	0.065	0.224	0.217	4,200	3,300	5,900	4,600	2,600	2,000
1/4 0.250	0.271	0.260	0.268	0.258	0.273	0.260	0.243	0.065	0.256	0.247	5,500	4,300	7,800	6,200	3,300	2,600
5/16 0.312	0.337	0.324	0.334	0.322	0.339	0.324	0.304	0.080	0.319	0.308	8,700	6,700	12,000	9,300	5,200	4,000
3/8 0.375	0.403	0.388	0.400	0.386	0.405	0.388	0.366	0.095	0.383	0.370	12,600	9,600	18,000	14,000	...	...
7/16 0.438	0.469	0.452	0.466	0.450	0.471	0.452	0.427	0.095	0.446	0.431	17,000	13,300	23,500	18,000	...	...
1/2 0.500	0.535	0.516	0.532	0.514	0.537	0.516	0.488	0.110	0.510	0.493	22,500	17,500	32,000	25,000	...	...
5/8 0.625	0.661	0.642	0.658	0.640	...	...	0.613	0.125	0.635	0.618	35,000	...	48,000	...	...	...
3/4 0.750	0.787	0.768	0.784	0.766	...	...	0.738	0.150	0.760	0.743	50,000	...	70,000	...	...	...

GENERAL NOTES:
(a) For additional requirements refer to General Data for Spring Pins on pages 27, 29, and 30 from original document.
(b) Light-duty SAE 1070 and 1075 pins are not produced in diameters smaller than 3/32 in.

CHARACTERISTICS	SYMBOL
Straightness	–
Diameter	ø

NOTES:
(1) Where specifying nominal size in decimals, zeros preceding the decimal shall be omitted.
(2) Maximum diameter shall be checked by GO ring gage.
(3) Minimum diameter shall be checked by NO GO ring gage.
(4) Sizes 1/32 in. (0.031) through 0.052 in. are not available in SAE 1070-1095 carbon steel.
(5) Sizes 5/8 in. (0.625) and larger are produced from SAE 6150H alloy steel, not SAE 1070-1095 carbon steel.

Standards of Grooved Pins

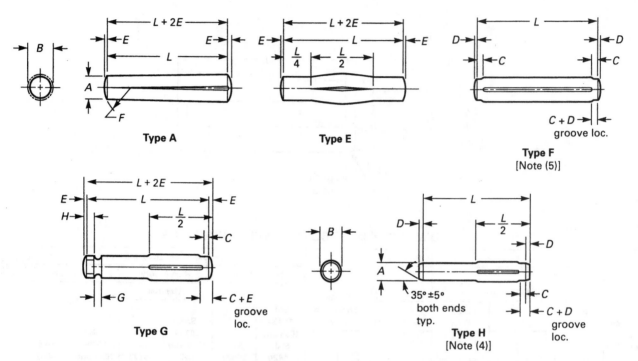

Type A

Type E

Type F
[Note (5)]

Type G

Type H
[Note (4)]

DIMENSIONS OF GROOVED PINS [Note (1)]

Nominal Size or Basic Pin Diameter [Note (2)]	A Pin Diameter		C Pilot Length	D [Note (3)] Chamfer Length	E [Note (3)] Crown Height	F [Note (3)] Crown Radius		G Neck Width		H Shoulder Length		J Neck Radius	K Neck Diameter	
	Max.	Min.	Ref	Min.	Nom.	Max.	Min.	Max.	Min.	Max.	Min.	Ref	Max.	Min.
1/32 [6]	0.0312	0.0312	0.0297	0.015	...	...	...	...	...	...	...	...	...	...	...
3/64 [6]	0.0469	0.0469	0.0454	0.031	...	...	...	...	...	...	...	...	...	...	...
1/16	0.0625	0.0625	0.0610	0.031	0.005	0.0065	0.088	0.068	...	...	...	...	...	...	...
5/64 [6]	0.0781	0.0781	0.0766	0.031	0.005	0.0087	0.104	0.084	...	...	...	...	...	...	...
3/32	0.0938	0.0938	0.0923	0.031	0.005	0.0091	0.135	0.115	0.038	0.028	0.041	0.031	0.016	0.067	0.057
7/64 [6]	0.1094	0.1094	0.1074	0.031	0.005	0.0110	0.150	0.130	0.038	0.028	0.041	0.031	0.016	0.082	0.072
1/8	0.1250	0.1250	0.1230	0.031	0.005	0.0130	0.166	0.146	0.069	0.059	0.041	0.031	0.031	0.088	0.078
5/32	0.1563	0.1563	0.1543	0.062	0.005	0.0170	0.198	0.178	0.069	0.059	0.057	0.047	0.031	0.109	0.099
3/16	0.1875	0.1875	0.1855	0.062	0.016	0.0180	0.260	0.240	0.069	0.059	0.057	0.047	0.031	0.130	0.120
7/32	0.2188	0.2188	0.2168	0.062	0.016	0.0220	0.291	0.271	0.101	0.091	0.072	0.062	0.047	0.151	0.141
1/4	0.2500	0.2500	0.2480	0.062	0.016	0.0260	0.322	0.302	0.101	0.091	0.072	0.062	0.047	0.172	0.162
5/16	0.3125	0.3125	0.3105	0.094	0.031	0.0340	0.385	0.365	0.132	0.122	0.104	0.094	0.062	0.214	0.204
3/8	0.3750	0.3750	0.3730	0.094	0.031	0.0390	0.479	0.459	0.132	0.122	0.135	0.125	0.062	0.255	0.245
7/16	0.4375	0.4375	0.4355	0.094	0.031	0.0470	0.541	0.521	0.195	0.185	0.135	0.125	0.094	0.298	0.288
1/2	0.5000	0.5000	0.4980	0.094	0.031	0.0520	0.635	0.615	0.195	0.185	0.135	0.125	0.094	0.317	0.307

GENERAL NOTE: For additional requirements and recommended hole sizes see Section 7 from original document.

NOTES:
(1) For expanded diameters applicable to pins made from corrosion resistant steel or monel, see Table 6B; and for pins made from other materials, see Table 6A.
(2) Where specifying nominal size in decimals, zeros preceding decimal and in the fourth decimal place shall be omitted.
(3) Pins in 1/32 and 3/64 in. sizes of any length and all sizes 1/4 in. nominal length, or shorter, are not crowned or chamfered. See para. 7.4 of General Data. Alloy steel pins of all types shall have chamfered ends conforming with Type F pins, included within the pin length.
(4) Type H replaces Types B and D as previously used in ANSI B18.8.2-1978 (see Appendix C).
(5) Type F replaces Type C as previously used in ANSI B18.8.2-1978 (see Appendix C).
(6) Non-stock items — not recommended for new design.

Standards of Round–Head Grooved Drive Studs

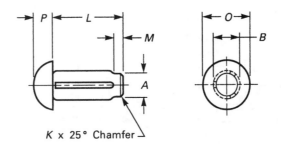

K x 25° Chamfer

DIMENSIONS OF ROUND-HEAD GROOVED DRIVE STUDS

Stud Size Number and Basic Shank Diameter [Note (1)]	A Shank Diameter		O Head Diameter		P Head Height		B Expanded Diameter ± 0.002 Nominal Stud Length [Note (2)]								K Chamfer
	Max.	Min.	Max.	Min.	Max.	Min.	$\frac{1}{8}$	$\frac{3}{16}$	$\frac{1}{4}$	$\frac{5}{16}$	$\frac{3}{8}$	$\frac{1}{2}$	$\frac{5}{8}$	$\frac{3}{4}$	Min.
0 0.067	0.067	0.065	0.130	0.120	0.050	0.040	0.074	0.074	0.074						0.005
2 0.086	0.086	0.084	0.162	0.146	0.070	0.059	0.096	0.096	0.095						0.005
4 0.104	0.104	0.102	0.211	0.193	0.086	0.075		0.115	0.113	0.113					0.005
6 0.120	0.120	0.118	0.260	0.240	0.103	0.091			0.132	0.130	0.130				0.005
7 0.136	0.136	0.134	0.309	0.287	0.119	0.107				0.147	0.147	0.144			0.005
8 0.144	0.144	0.142	0.309	0.287	0.119	0.107					0.155	0.153	0.153		0.005
10 0.161	0.161	0.159	0.359	0.334	0.136	0.124					0.173	0.171	0.171		0.016
12 0.196	0.196	0.194	0.408	0.382	0.152	0.140						0.206	0.204	0.204	0.016
14 0.221	0.221	0.219	0.457	0.429	0.169	0.156						0.234	0.232	0.232	0.016
16 0.250	0.250	0.248	0.472	0.443	0.174	0.161						0.263			0.016

GENERAL NOTE: For additional requirements and recommended hole sizes refer to General Data for Grooved Pins, also Grooved Drive Studs and Grooved T-Head Cotter Pins on pages 18, 19, 26, and 27 from original document.

NOTES:
(1) Where specifying stud size in decimals, zeros preceding decimal and in the fourth decimal place shall be omitted.
(2) Sizes and length, for which B values are tabulated are normally readily available in carbon steel. For other size-length combinations or materials, manufacturers should be consulted.

PILOT LENGTH DIMENSIONS FOR ROUND-HEAD GROOVED DRIVE STUDS

Nominal Length	Nominal Size																			
	0		2		4		6		7		8		10		12		14		16	
	M, Pilot Length																			
	Max.	Min.	Max.	Min.	Max.	Min.	Max.	Min.	Max.	Min.	Max.	Min.	Max.	Min.	Max.	Min.	Max.	Min.	Max.	Min.
$\frac{1}{8}$	0.051	0.031	0.051	0.031																
$\frac{3}{16}$	0.067	0.047	0.067	0.047	0.067	0.047														
$\frac{1}{4}$	0.082	0.062	0.082	0.062	0.082	0.062	0.082	0.062												
$\frac{5}{16}$					0.098	0.078	0.098	0.078	0.098	0.078										
$\frac{3}{8}$					0.114	0.094	0.114	0.094	0.114	0.094	0.114	0.094	0.114	0.094						
$\frac{1}{2}$									0.14	0.12	0.14	0.12	0.14	0.12	0.14	0.12	0.14	0.12	0.14	0.12
$\frac{5}{8}$											0.18	0.16	0.18	0.16	0.18	0.16	0.18	0.16		
$\frac{3}{4}$															0.20	0.18	0.20	0.18		

GENERAL NOTE: To find total pilot length of lengths (L) not shown above, use the next shorter length.

Standards of Grooved T–Head Cotter Pins

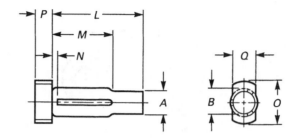

DIMENSIONS OF GROOVED T-HEAD COTTER PINS [Note (1)]

Nominal Size or Basic Shank Diameter [Note (2)]		A Shank Diameter		B Expanded Shank Diameter		N Length	O Head Diameter		P Head Height		Q Head Width		Recommended Hole Size	
		Max.	Min.	Max.	Min.	Max.	Max.	Min.	Max.	Min.	Max.	Min.	Max.	Min.
$5/32$	0.156	0.154	0.150	0.168	0.163	0.08	0.26	0.24	0.11	0.09	0.18	0.15	0.161	0.156
$3/16$	0.187	0.186	0.182	0.201	0.195	0.09	0.30	0.28	0.13	0.11	0.22	0.18	0.193	0.187
$1/4$	0.250	0.248	0.244	0.265	0.258	0.12	0.40	0.38	0.17	0.15	0.28	0.24	0.257	0.250
$5/16$	0.312	0.310	0.305	0.326	0.320	0.16	0.51	0.48	0.21	0.19	0.34	0.30	0.319	0.312
$23/64$	0.359	0.358	0.353	0.375	0.369	0.18	0.57	0.54	0.24	0.22	0.38	0.35	0.366	0.359
$1/2$	0.500	0.498	0.493	0.520	0.514	0.25	0.79	0.76	0.32	0.30	0.54	0.49	0.508	0.500

GENERAL NOTE: For additional requirements refer to General Data for Grooved Pins, also Grooved Drive Studs and Grooved T-Head Cotter Pins on pages 18, 19, 26, and 27 from original document.

NOTES:
(1) For groove lengths, M, which vary with pin length, see Table 8A.
(2) Where specifying nominal size in decimals, zeros preceding decimal and in the fourth decimal place shall be omitted.

GROOVE LENGTH DIMENSIONS FOR GROOVED T-HEAD COTTER PINS

Nominal Length	Nominal Size											
	$5/32$		$3/16$		$1/4$		$5/16$		$23/64$		$1/2$	
	M, Pilot Length [Note (1)]											
	Max.	Min.	Max.	Min.	Max.	Min.	Max.	Min.	Max.	Min.	Max.	Min.
$3/4$	0.50	0.48	0.50	0.48								
$7/8$	0.50	0.48	0.50	0.48								
1	0.62	0.60	0.62	0.60	0.62	0.60						
$1 1/8$	0.68	0.66	0.68	0.66	0.68	0.66	0.68	0.66				
$1 1/4$			0.75	0.73	0.75	0.73	0.75	0.73	0.75	0.73		
$1 1/2$					0.88	0.86	0.88	0.86	0.88	0.86		
$1 3/4$							1.00	0.98	1.00	0.98		
2							1.25	1.23	1.25	1.23	1.25	1.23
$2 1/4$											1.31	1.29
$2 1/2$											1.50	1.48
$2 3/4$											1.62	1.60
3											1.85	1.83

NOTE:
(1) Sizes and lengths for which M values are tabulated are normally readily avilable. For other size-length combinations, manufacturers should be consulted.

Standards of Cotter Pins

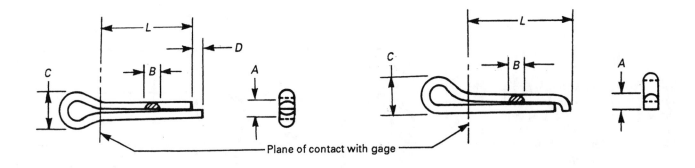

Extended Prong
Square Cut Type

Hammer Lock Type

— Plane of contact with gage

DIMENSIONS OF COTTER PINS

Nominal Size[1,2] or Basic Pin Dia.		Total Shank Diameter A[3]		Wire Width B		Head Diameter C	Extended Prong Length D	Gage Hole Diameter ±0.001
		Max.	Min.	Max.	Min.	Min.	Min.	
1/32	0.031	0.032	0.028	0.032	0.022	0.06	0.01	0.047
3/64	0.047	0.048	0.044	0.048	0.035	0.09	0.02	0.062
1/16	0.062	0.060	0.056	0.060	0.044	0.12	0.03	0.078
5/64	0.078	0.076	0.072	0.076	0.057	0.16	0.04	0.094
3/32	0.094	0.090	0.086	0.090	0.069	0.19	0.04	0.109
7/64	0.109	0.104	0.100	0.104	0.080	0.22	0.05	0.125
1/8	0.125	0.120	0.116	0.120	0.093	0.25	0.06	0.141
9/64	0.141	0.134	0.130	0.134	0.104	0.28	0.06	0.156
5/32	0.156	0.150	0.146	0.150	0.116	0.31	0.07	0.172
3/16	0.188	0.176	0.172	0.176	0.137	0.38	0.09	0.203
7/32	0.219	0.207	0.202	0.207	0.161	0.44	0.10	0.234
1/4	0.250	0.225	0.220	0.225	0.176	0.50	0.11	0.266
5/16	0.312	0.280	0.275	0.280	0.220	0.62	0.14	0.312
3/8	0.375	0.335	0.329	0.335	0.263	0.75	0.16	0.375
7/16	0.438	0.406	0.400	0.406	0.320	0.88	0.20	0.438
1/2	0.500	0.473	0.467	0.473	0.373	1.00	0.23	0.500
5/8	0.625	0.598	0.590	0.598	0.472	1.25	0.30	0.625
3/4	0.750	0.723	0.715	0.723	0.572	1.50	0.36	0.750

GENERAL NOTE: For additional requirements, refer to General Data for Cotter Pins in Sections 1 and 3 from original document.
NOTES:
(1) Where specifying nominal size in decimals, zero preceding decimal shall be omitted.
(2) 5/64, 7/32, 7/16 and 3/4 not preferred for new design.
(3) Total shank diameter, A dimension, is two times wire thickness. A is measured at end of pin where no gap is permitted.

Pin and Shaft of Equal Strength

Herman J. Scholtze

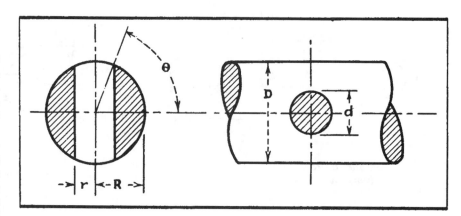

THE ACCOMPANYING TABLE gives the sizes of round driving pins and round shafts drilled to receive the pin in which both parts are equally strong in shear, for the condition that the shaft and pin are made of the same material.

The author has discovered that when the pin diameter equals 40 percent of the shaft diameter, the shearing stress in the pin equals the shearing stress in the shaft, also that the polar moment of inertia of the drilled shaft equals the shaft radius to the fourth power.

Values given in the table in columns headed "Torque," and "Load on One End of Pin" have been computed for a shear stress of 12,000 lb. per sq. in. For other values of shear stress, the load on one end of pin equals the cross-section area of the pin multiplied by the allowable shear stress, and the torque equals the load on the pin multiplied by the shaft diameter.

R = radius of shaft, in.
r = radius of pin, in.
S_s = shearing stress in shaft, lb. per sq. in.
S_p = shearing stress in pin, lb. per sq. in.
J = polar moment of shaft cross-section through the axis of pin bore
T_s = torque on shaft, in. lb.
T_p = torque delivered by pin, in. lb.
θ = central angle subtended by one half the chord of circular segment section or drilled shaft, radians

$$T_p = 2\pi r^2 R S_p$$

$$T_s = J S_s / R$$

$$r = \sqrt{J}/(R\sqrt{2\pi})$$

$$J = R^4\theta - r^4\left(\frac{\sin\theta}{3\cos^3\theta} + \frac{2\tan\theta}{3}\right)$$

$$r = \frac{1}{R\sqrt{2\pi}}\left[R^4\theta - r^4\left(\frac{\sin\theta}{3\cos^3\theta} + \frac{2\tan\theta}{3}\right)\right]^{1/4}$$

Equal Strength Shafts and Pins of Similar Material

Dia. of Shaft, in. D	Dia. of Pin, in. d	Polar Moment of Inertia, J	Polar Section Modulus, J/R	Torque, in. lb. at 12,000 lb. per sq. in. Shear Stress T	Load on One End of Pin, lb. $P = T/D$	Cross-Section Area of One End of Pin, sq. in.
1/4	0.100	0.000244	0.001952	23.5	94	0.00785
5/16	0.125	0.000597	0.003820	45.8	146	0.01277
3/8	0.150	0.001236	0.006579	79	210	0.01767
7/16	0.175	0.002290	0.01047	125	286	0.02405
1/2	0.200	0.003906	0.01562	187	374	0.03142
5/8	0.250	0.009537	0.03051	366	590	0.04909
3/4	0.300	0.01977	0.05273	635	845	0.07069
7/8	0.350	0.03663	0.08374	1,010	1,160	0.09621
1	0.400	0.06250	0.1250	1,500	1,500	0.1257
1-1/4	0.500	0.1526	0.2442	2,940	2,350	0.1963
1-1/2	0.600	0.3164	0.4218	5,100	3,400	0.2827
1-3/4	0.700	0.5862	0.6700	8,000	4,570	0.3848
2	0.800	1.0000	1.0000	12,000	6,000	0.5027
2-1/4	0.900	1.6018	1.4238	17,000	7,550	0.6362
2-1/2	1.00	2.4414	1.9531	23,400	9,350	0.7854
2-3/4	1.10	3.5745	2.6000	31,200	11,350	0.9500
3	1.20	5.0625	3.3750	40,500	13,500	1.131
3-1/2	1.40	9.3789	5.3593	64,000	18,200	1.539
4	1.60	16.000	8.0000	96,000	24,000	2.011
4-1/2	1.80	25.629	11.390	125,000	27,700	2.545
5	2.00	39.062	15.625	187,000	37,500	3.142
5-1/2	2.20	57.191	20.797	240,000	43,750	3.801
6	2.40	81.000	27.000	324,000	54,000	4.524
7	2.80	150.062	42.875	515,000	73,500	6.158
8	3.20	256.000	64.000	770,000	96,000	8.042
9	3.60	410.062	91.125	1,090,000	121,000	10.18
10	4.00	625.000	125.000	1,500,000	150,000	12.57
11	4.40	915.062	166.375	2,000,000	182,000	15.21
12	4.80	1,296.000	216.000	2,600,000	216,000	18.10

SECTION 13

SPRINGS

12 Ways to Put Springs to Work

Variable-rate arrangements, roller positioning, space saving, and other ingenious ways to get the most from springs.

L. Kasper

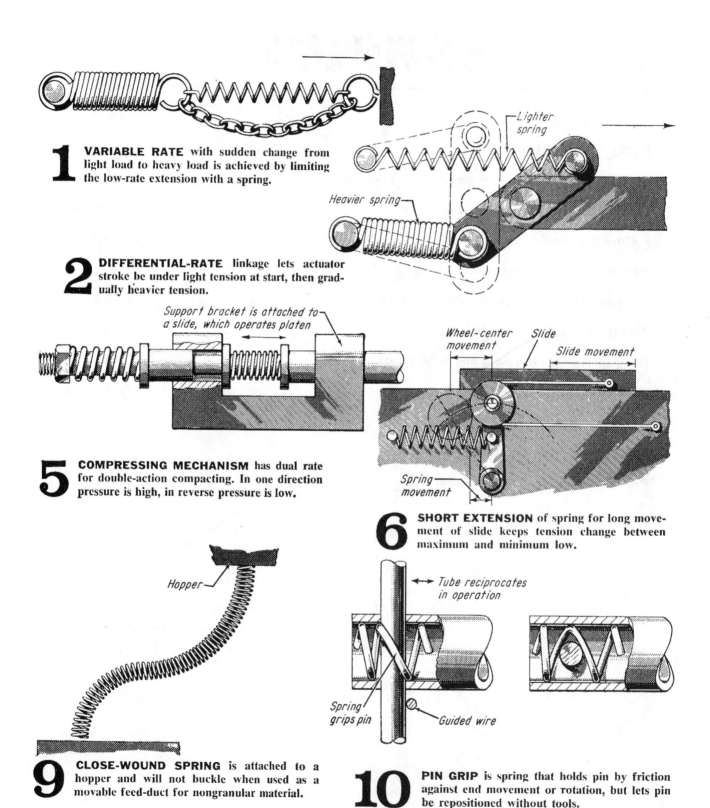

1 **VARIABLE RATE** with sudden change from light load to heavy load is achieved by limiting the low-rate extension with a spring.

2 **DIFFERENTIAL-RATE** linkage lets actuator stroke be under light tension at start, then gradually heavier tension.

Lighter spring

Heavier spring

5 **COMPRESSING MECHANISM** has dual rate for double-action compacting. In one direction pressure is high, in reverse pressure is low.

Support bracket is attached to a slide, which operates platen

Wheel-center movement Slide Slide movement

Spring movement

6 **SHORT EXTENSION** of spring for long movement of slide keeps tension change between maximum and minimum low.

Hopper

Tube reciprocates in operation

Spring grips pin Guided wire

9 **CLOSE-WOUND SPRING** is attached to a hopper and will not buckle when used as a movable feed-duct for nongranular material.

10 **PIN GRIP** is spring that holds pin by friction against end movement or rotation, but lets pin be repositioned without tools.

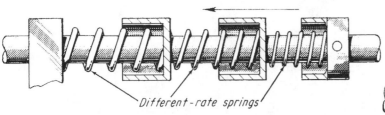

Different-rate springs

3 **THREE-STEP RATE** change at predetermined positions. The lighter springs will always compress first regardless of their position.

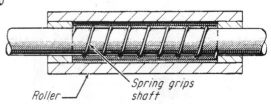

Roller   Spring grips shaft

4 **ROLLER POSITIONING** by tight-wound spring on shaft obviates necessity for collars. Roller will slide under excess end thrust.

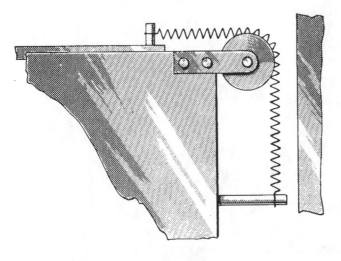

7 **SPRING WHEEL** helps distribute deflection over more coils than if spring rested on corner. Less fatigue and longer life result.

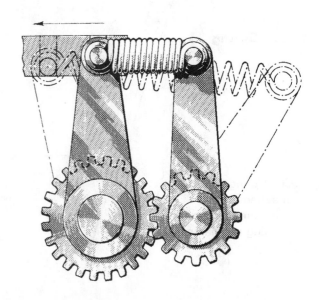

8 **INCREASED TENSION** for same movement is gained by providing a movable spring mount and gearing it to the other movable lever.

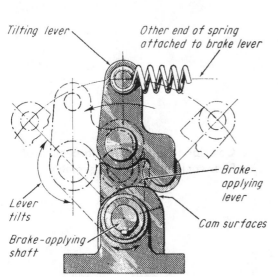

Tilting lever

Other end of spring attached to brake lever

Brake-applying lever

Cam surfaces

Lever tilts

Brake-applying shaft

11 **TENSION VARIES** at different rate when brake-applying lever reaches the position shown. Rate is reduced when tilting lever tilts.

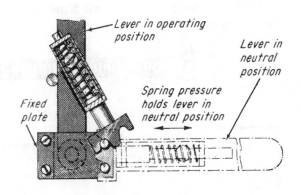

Lever in operating position

Lever in neutral position

Spring pressure holds lever in neutral position

Fixed plate

12 **TOGGLE ACTION** here is used to make sure the gear-shift lever will not inadvertently be thrown past neutral.

Multiple Uses of Coil Springs

R. O. Parmley

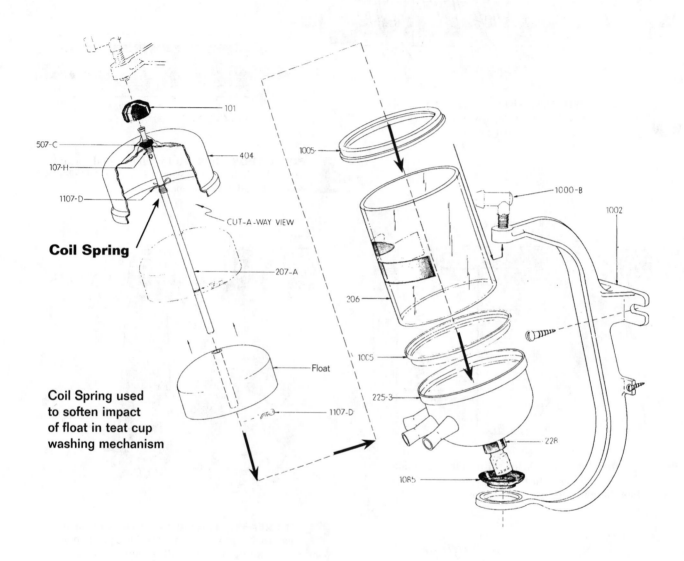

Coil Spring

Coil Spring used to soften impact of float in teat cup washing mechanism

101

507-C

107-H

1107-D

404

CUT-A-WAY VIEW

207-A

Float

1107-D

1005

1000-B

1002

206

1005

225-3

228

1085

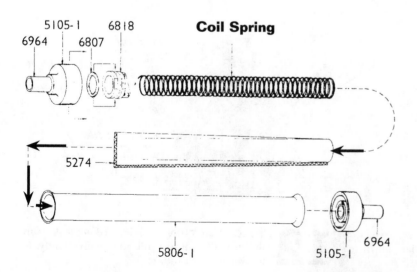

5105-1   6818   **Coil Spring**

6964   6807

Coil Spring used as reinforcement of filter sock in milk filter assembly

5274

5806-1   5105-1   6964

Source: Bender Machine Works, Inc.

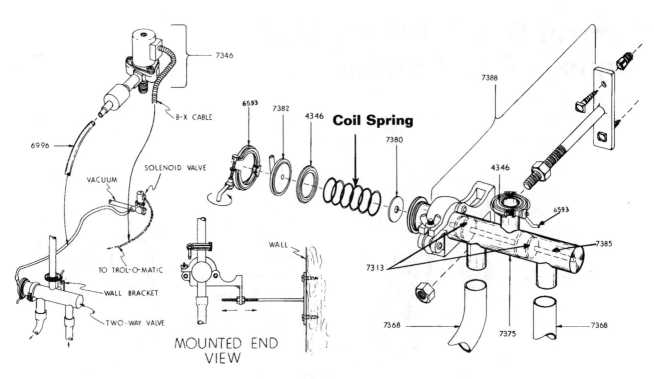

Coil Spring used stabilizing component in two-way valve assembly

Source: Bender Machine Works, Inc.

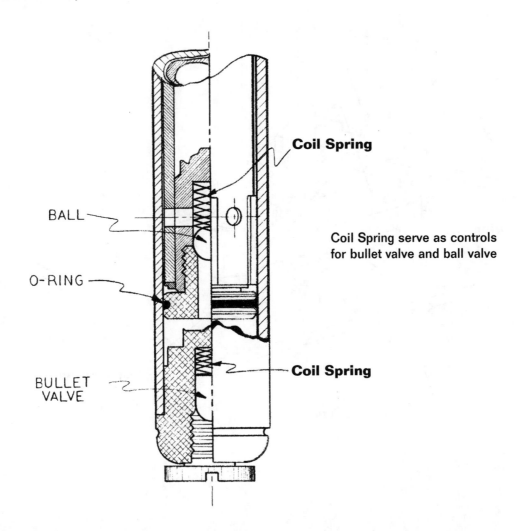

Coil Spring serve as controls for bullet valve and ball valve

Control Depth Primer Tool Employs Coil Springs

E. E. Lawrence, Inventor
R. O. Parmley, Draftsman

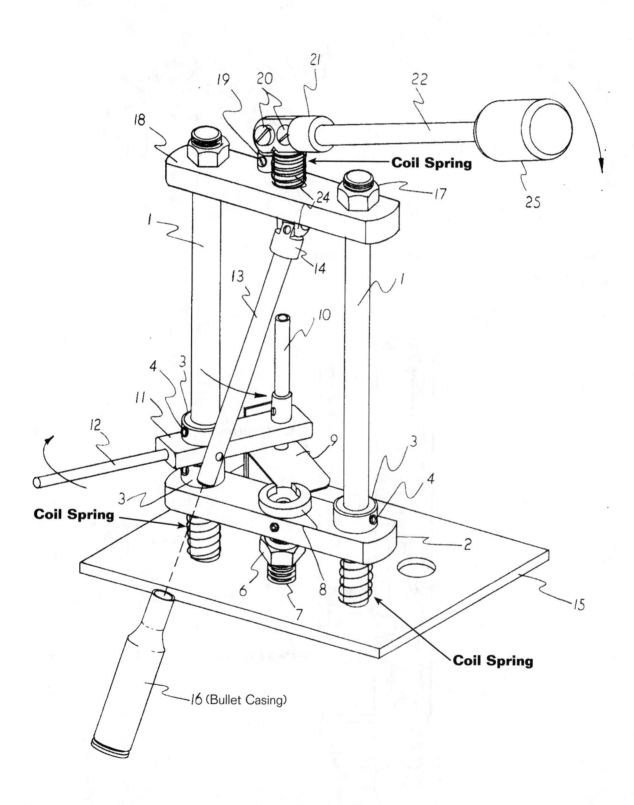

Coil Spring

Coil Spring

Coil Spring

16 (Bullet Casing)

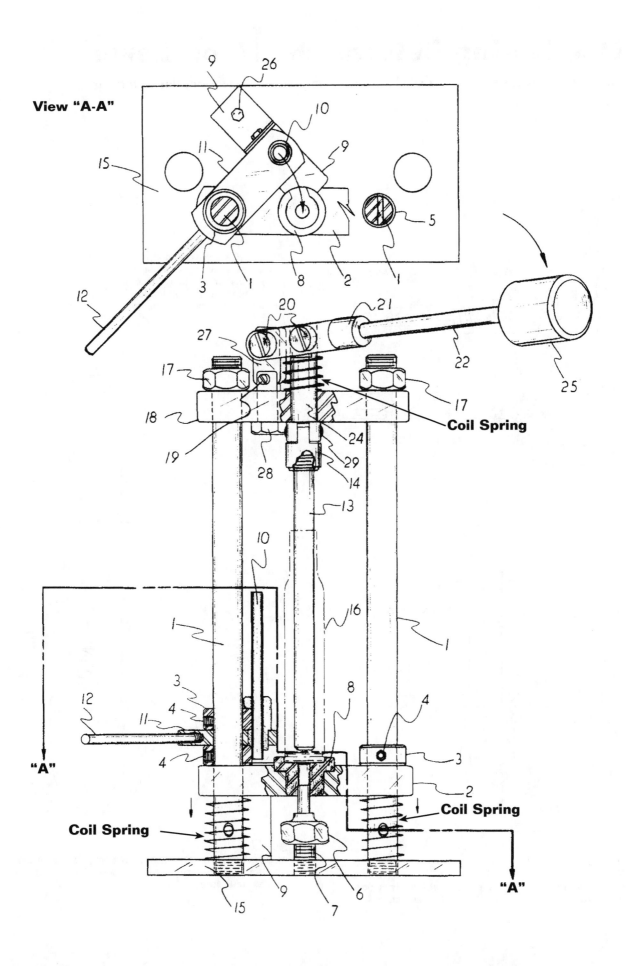

View "A-A"

Coil Spring

Coil Spring

Coil Spring

One Spring Returns the Hand Lever

These seven designs need only a single spring—compression, extension, flat or torsion.

L. Kasper

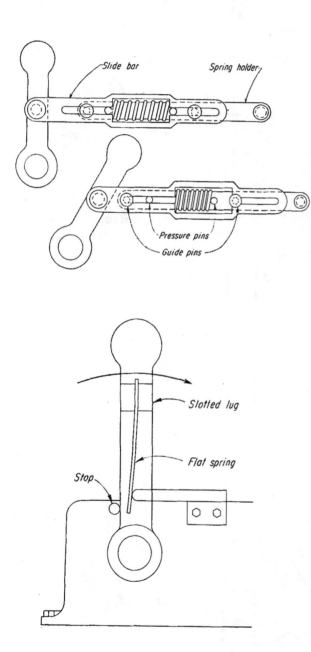

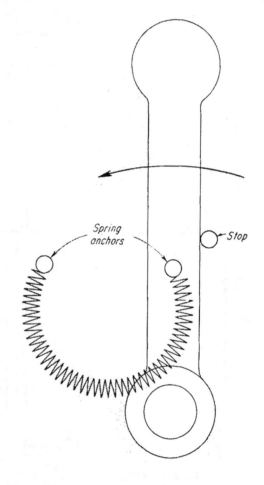

1
SLIDE BAR attached to lever compresses spring against pressure pins in either direction. Guide pins in spring holder hit end of slot to limit movement.

2
FLAT SPRING has initial tension which gives positive return for even a small lever-movement.

3
CLOSE-WOUND HELICAL SPRING gives almost constant return force. Anchor post for spring also acts as limit stop.

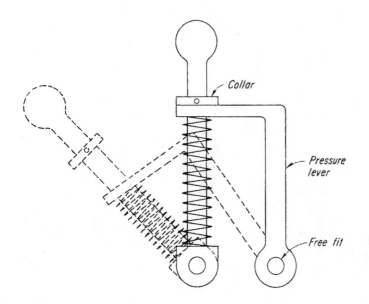

4
PRESSURE LEVER returns hand lever because it rotates on a different center. Collar sets starting position.

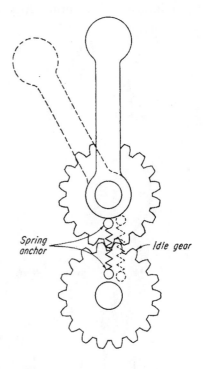

5
GEARS extend spring when lever moves up to 180° in either direction.

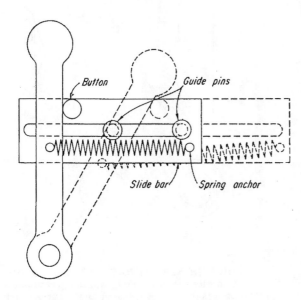

6
SLIDE BAR rides on guide pins as lever pushes it to right. Stretched spring pulls slide bar against lever to return lever to vertical position.

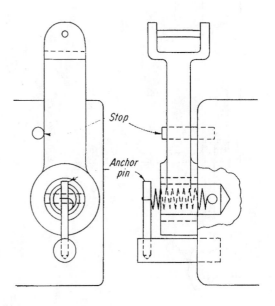

7
OPEN-WOUND HELICAL SPRING extends inside shaft of handle. Coils must be wound in direction of movement so that spring tightens instead of unwinds as lever turns.

6 More One Spring Lever Return Designs

A flat, torsion or helical spring does the job alone.

L. Kasper

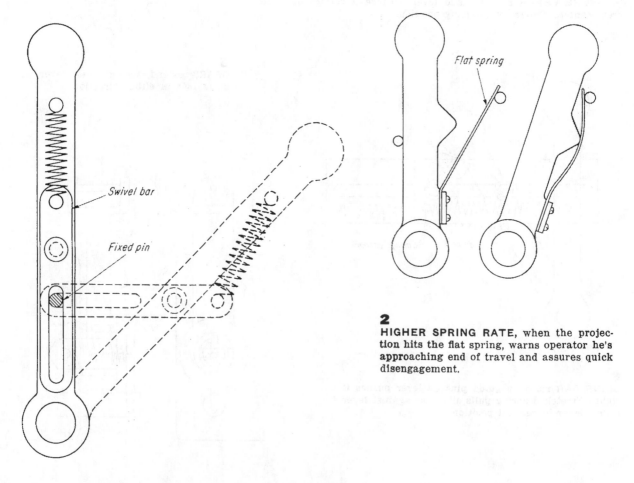

1
SWIVEL BAR, which slides on fixed pin, returns hand lever. Slot in swivel bar is limit stop for movement either way.

Swivel bar

Fixed pin

Flat spring

2
HIGHER SPRING RATE, when the projection hits the flat spring, warns operator he's **approaching** end of travel and assures quick disengagement.

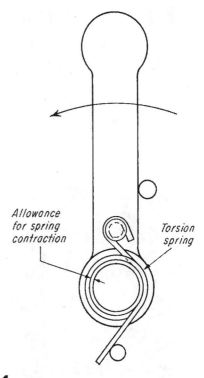

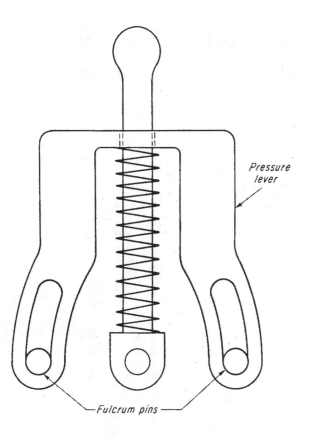

3
DOUBLE PRESSURE-LEVER returns handle to center from either direction by compressing spring. Lever pivots on one pin and comes to stop against the other pin.

4
TORSION SPRING must have coil diameter larger than shaft diameter to allow for spring contraction during windup.

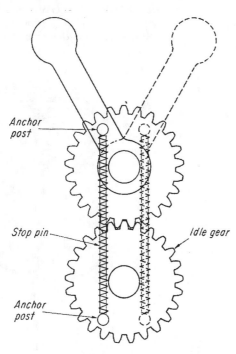

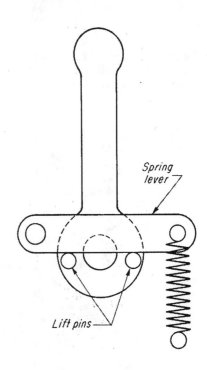

5
LEVER flops to stop because of spring pull. Stop-pins inside springs limit movement.

6
SELF-CENTERING HAND LEVER returns to vertical as soon as it's released. Any movement lifts spring lever and creates a righting force.

How to Stiffen Bellows with Springs

Rubber bellows are an essential part of many products. Here are eight ways to strengthen, cushion, and stabilize bellows with springs.

Robert O. Parmley

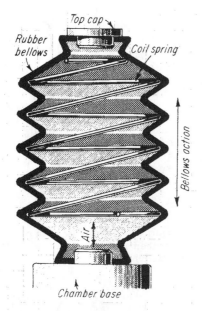

1 INTERNAL COIL SPRING strengthens and adds vertical stability. To install spring, just "corkscrew" it into place.

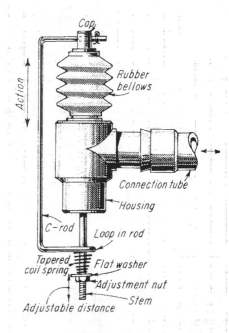

2 CUSHION BELLOWS SUPPORT-ROD with coil spring. Adjustment is provided and bellows are strengthened by this arrangement.

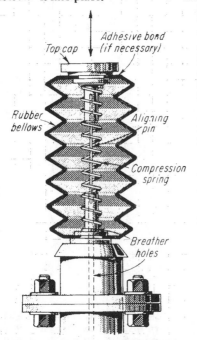

5 COMPRESSION STRENGTH for bellows is best obtained with a coil spring, mounted internally as shown.

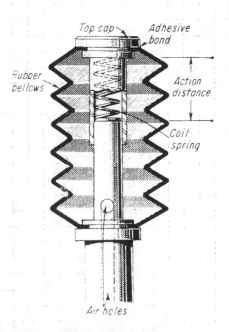

6 INTERNAL RIGIDITY of bellows is here provided by a mating rod and sleeve in which a compression spring is fitted.

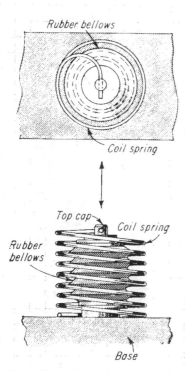

3 **EXTERNAL STABILITY** is provided here, with the added advantage of simple assembly that strengthens bellows, too.

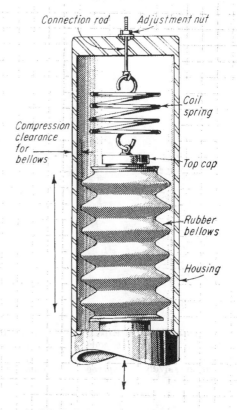

4 **ADJUSTMENT WITH TENSION SPRING** lets bellows be enclosed in casting while adjustment is provided externally.

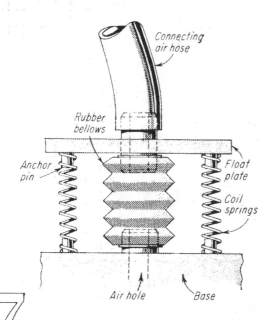

7 **BELLOWS STIFFENER AND STABILIZER** are sometimes combined by means of a platform and four mounting springs.

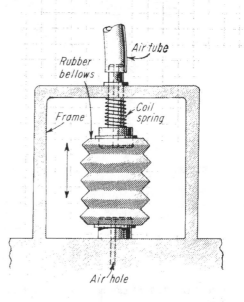

8 **HOUSED STIFFENING UNIT** gives solid mount for hose connection, together with spring action for bellows.

Springs: How to Design for Variable Rate

Eighteen diagrams show how stops, cams, linkages and other arrangements can vary the load/deflection ratio during extension or compression.

James F. Machen

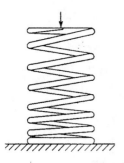

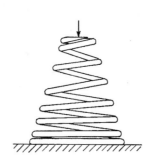

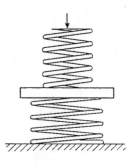

1
WITH TAPERED-PITCH SPRINGS (1), the number of effective coils changes with deflection—the coils "bottom" progressively. Tapered

2
O.D. and pitch (2) combine to produce similar effect except spring with tapered O.D. will have shorter solid height.

3
IN DUAL SPRINGS one spring closes solid before the other.

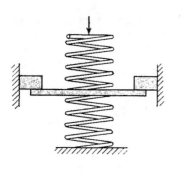

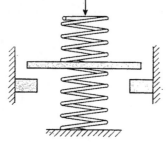

STOPS (4, 5) can be used with either compression or extension springs.

4

5

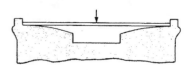

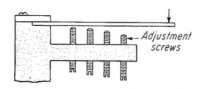

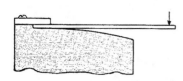

6

7

8

LEAF SPRINGS (6, 7, 8) can be arranged so that their effective lengths change with deflection.

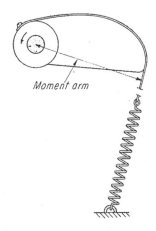

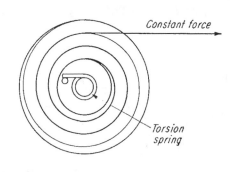

10
TORSION SPRING combined with variable-radius pulley gives constant force.

9
CAM-AND-SPRING DEVICE causes torque relationship to vary during rotation as moment arm changes.

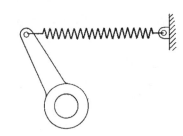

LINKAGE-TYPE ARRANGEMENTS (11, 12) are often used in instruments where torque control or anti-vibration suspension is required.

11

12

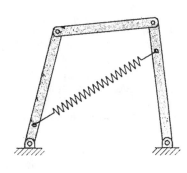

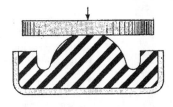

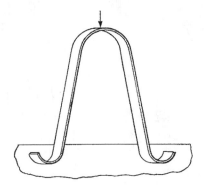

14
MOLDED-RUBBER SPRING has deflection characteristics that vary with its shape.

13
4-BAR MECHANISM in conjunction with a spring has a great variety of load/deflection characteristics.

15
ARCHED LEAF-SPRING gives almost constant force when shaped like the one illustrated.

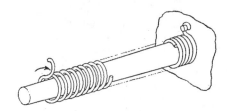

16
TAPERED MANDREL AND TORSION SPRING. Effective number of coils decreases with torsional deflection.

Adjustable Extension Springs

Henry Martin

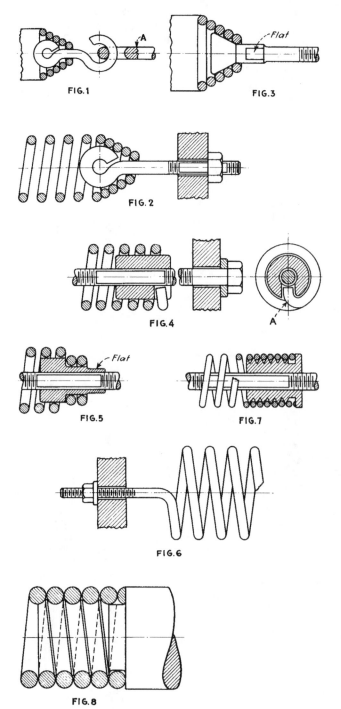

FIG. 1

FIG. 3

FIG. 2

FIG. 4

A

FIG. 5

FIG. 7

FIG. 6

FIG. 8

FIG. 9

Design of the end of a tension or extension spring using some form of loop integral with the spring is often unsatisfactory, since many spring failures occur somewhere in the loop, most often at the base of the loop adjacent to the spring body. Use of the accompanying tested methods has reduced breakage and therefore down-time of machinery, especially where adjustability of tension and length is required

FIG. 1—Spring-end is tapered about a loop made of larger diameter and somewhat softer wire than that used for the spring. Upper end of wire is also formed into a loop, larger and left open to engage a rod-end or eye-bolt *A*.

FIG. 2—A loop is formed at the end of a soft steel rod threaded at the opposite end for a hex adjusting nut. Ordinary threaded rod-end may be substituted if desired.

FIG. 3—End of adjusting screw is upset in shape of a conical head to coincide with taper of spring-end. Unless initial tension of spring is sufficiently great a wrench flat on stem is provided to facilitate adjustment.

FIG. 4—The last coil of spring is bent inwardly to form hoop *A* which engages slot in nut. Although a neat and simple design, all spring tension is exerted on hook at one point, somewhat off-center of spring axis. Not recommended for heavy loads.

FIG. 5—An improved method over Fig. 4. The nut is shouldered to accommodate two end coils which are wound smaller than the body of spring. Flats are provided for use of wrench during adjustment.

FIG. 6—When wire size permits, the spring end can be left straight and threaded for adjustment. Because of the small size of nut a washer must also be used as shown.

FIG. 7—The shouldered nut is threaded with a coarse V-thread and is screwed into the end of the spring. The point of tangency between the 30-deg. side of thread and wire diameter should be such that the coils cannot pull off. The end of the spring is squared for sufficient friction so that nut need not be held when turning the adjusting screw.

FIG. 8—For close-wound extension springs, end of rod may be threaded with a shallow thread the root of which is the same curvature as that of the spring wire. This form of thread cut with the crests left sharp provides greater engagement contact.

FIG. 9—For more severe duty, the thread is cut deeper than that shown in Fig. 8. The whole spring is close-wound, but when screwed on adjusting rod, the coils are spread, thereby creating greater friction for better holding ability. Spring is screwed against the relieved shoulder of rod.

Fig. 10—When design requires housed spring, adjusting rod is threaded internally. Here also, the close-wound coils are spread when assembled. Unless housing bore is considerably larger than shouldered diameter of adjusting rod, or sufficient space is available for a covered spring, methods shown in Figs. 8 or 9 will be less expensive.

Fig. 11—A thin piece of cold-drawn steel is drilled to exact pitch of the coils with a series of holes slightly larger than spring wire. Three or four coils are screwed into the piece which has additional holes for further adjustment. It will be seen that all coils so engaged are inactive or dead coils.

Fig. 12—A similar design to that shown in Fig. 11, except that a smaller spring lies inside the larger one. Both springs are wound to the same pitch for ease of adjustment. By staggering the holes as shown, the outer diameter of the inner spring may approach closely that of the inner diameter of the outer spring, thereby leaving sufficient space for a third internal spring if necessary.

Fig. 13—When the spring is to be guarded, and to prevent binding of the spring attachment in the housing, the end is cross-shaped as shown in the section. The two extra vanes are welded to the solid vane. The location of the series of holes in each successive vane is such as to advance spring at one quarter the pitch.

Fig. 14—This spring end has three vanes and is turned, bored and milled from solid round stock where welding facilities are not convenient. In sufficient quantities, the use of a steel casting precludes machining bar stock. The end with the hole is milled approximately ¼ in. thick for the adjusting member.

Fig. 15—A simple means of adjusting tension and length of spring. The spring anchor slides on a plain round rod and is fastened in any position by a square head setscrew and brass clamping shoe. The eye in the end of the spring engages a hole in the anchor.

Fig. 16—A block of cold-drawn steel is slotted to accommodate the eye of the spring by means of a straight pin. The block is drilled slightly larger than the threaded rod and adjustment and positioning is by the two hex nuts.

Fig. 17—A similar arrangement to that shown in Fig. 16. The spring finger is notched at the outer end for the spring-eye as illustrated in the sectioned end view. In these last three methods, the adjustable member can be made to accommodate 2 or 3 springs if necessary.

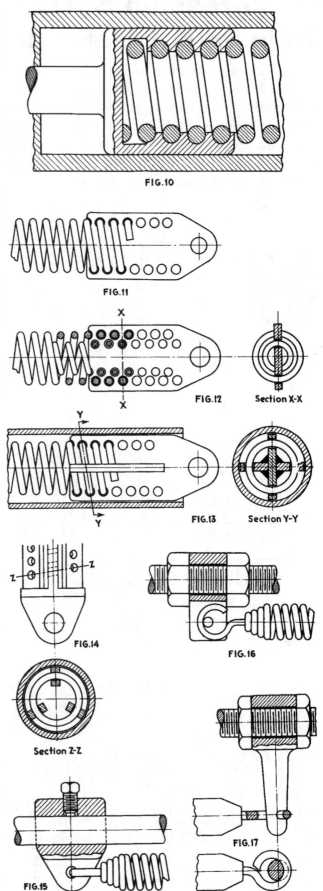

FIG.10

FIG.11

FIG.12   Section X-X

FIG.13   Section Y-Y

FIG.14

Section Z-Z

FIG.16

FIG.17

FIG.15

Compression Spring Adjustment Methods I

In many installations where compression springs are used, adjustability of the spring tension is frequently required. The methods shown incorporate various designs of screw and nut adjustment with numerous types of spring-centering means to guard against buckling. Some designs incorporate frictional reducing members to facilitate adjustment especially for springs of large diameter and heavy wire.

Henry Martin

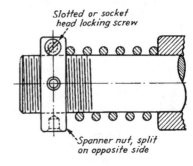

Slotted or socket head locking screw

Spanner nut, split on opposite side

FIG.3

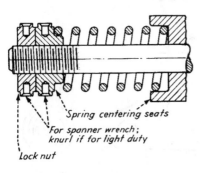

Spring centering seats

For spanner wrench; knurl if for light duty

Lock nut

FIG.1

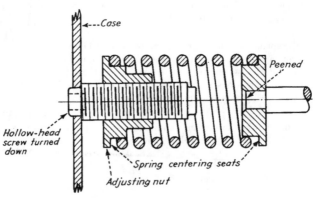

Case

Peened

Hollow-head screw turned down

Spring centering seats

Adjusting nut

FIG.2

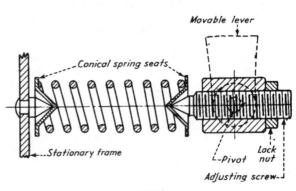

Spring retaining washer

Adjustment nut

Spring centering seats

FIG.4

Movable lever

Conical spring seats

Stationary frame

Pivot

Lock nut

Adjusting screw

FIG.5

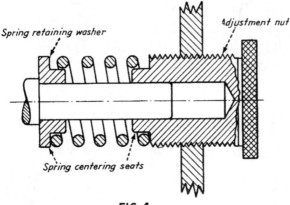

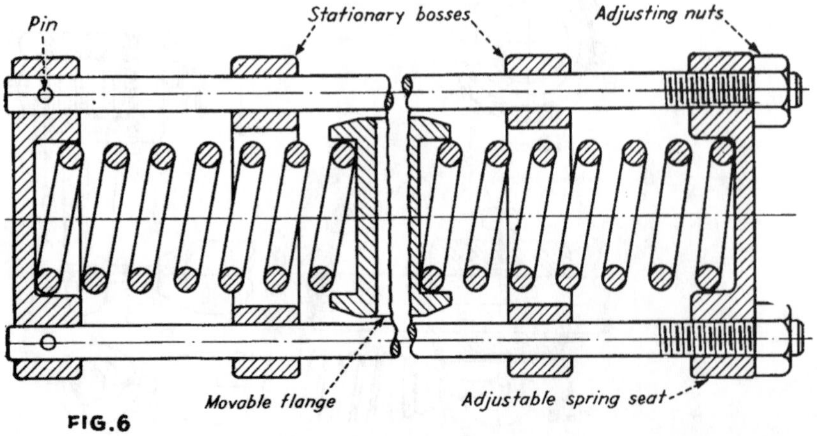

Pin

Stationary bosses

Adjusting nuts

Movable flange

Adjustable spring seat

FIG.6

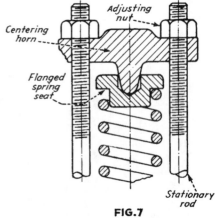

Adjusting nut

Centering horn

Flanged spring seat

Stationary rod

FIG.7

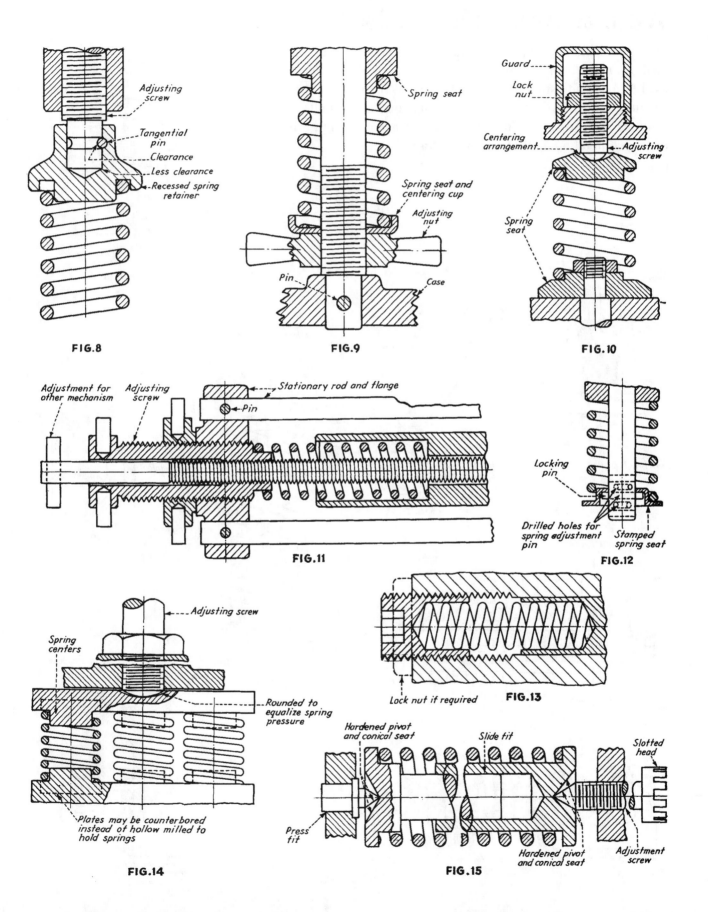

FIG. 8

FIG. 9

FIG. 10

FIG. 11

FIG. 12

FIG. 13

FIG. 14

FIG. 15

Compression Spring Adjustment Methods II

In this concluding group of adjustable compression springs, several methods are shown in which some form of anti-friction device is used to make adjustment easier. Thrust is taken against either single or multiple steel balls, the latter including commercial ball thrust bearings. Adjustments of double spring arrangements and other unconventional methods are also illustrated.

Henry Martin

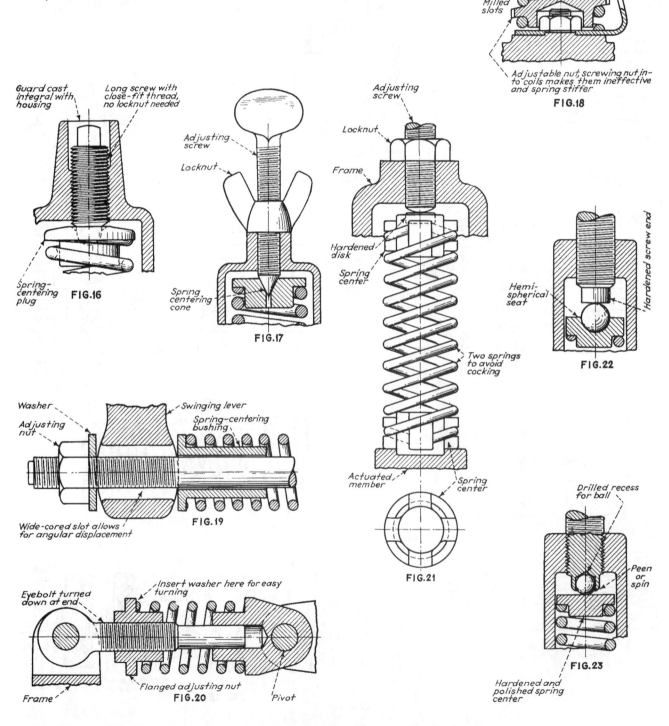

Adjustable nut, screwing nut into coils makes them ineffective and spring stiffer
Locking key
Milled slots
FIG.18

Guard cast integral with housing
Long screw with close-fit thread, no locknut needed
Spring-centering plug
FIG.16

Adjusting screw
Locknut
Spring centering cone
FIG.17

Adjusting screw
Locknut
Frame
Hardened disk
Spring center
Two springs to avoid cocking
Actuated member
Spring center
FIG.21

Hardened screw end
Hemispherical seat
FIG.22

Washer
Adjusting nut
Swinging lever
Spring-centering bushing
Wide-cored slot allows for angular displacement
FIG.19

Eyebolt turned down at end
Insert washer here for easy turning
Flanged adjusting nut
FIG.20
Frame
Pivot

Drilled recess for ball
Peen or spin
Hardened and polished spring center
FIG.23

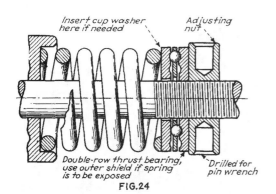

Insert cup washer here if needed

Adjusting nut

Double-row thrust bearing, use outer shield if spring is to be exposed

Drilled for pin wrench

FIG.24

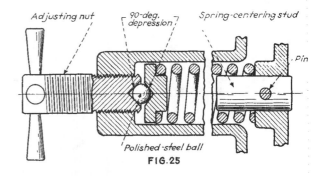

Adjusting nut

90-deg. depression

Spring-centering stud

Pin

Polished-steel ball

FIG.25

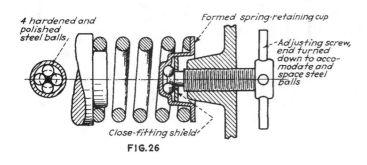

4 hardened and polished steel balls

Formed spring-retaining cup

Adjusting screw, end turned down to accomodate and space steel balls

Close-fitting shield

FIG.26

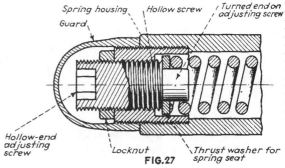

Spring housing

Guard

Hollow screw

Turned end on adjusting screw

Hollow-end adjusting screw

Locknut

Thrust washer for spring seat

FIG.27

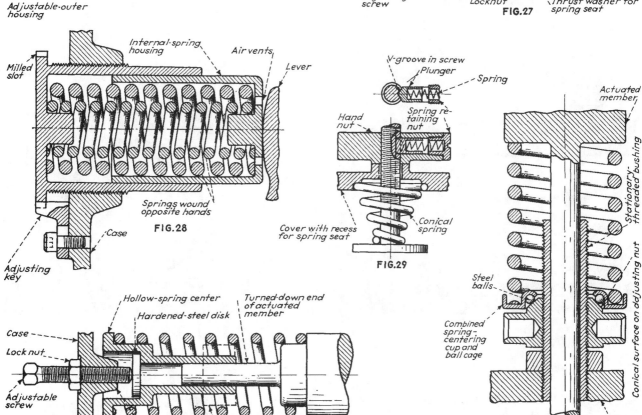

Adjustable-outer housing

Internal-spring housing

Air vents

Lever

Milled slot

Springs wound opposite hands

FIG.28

Case

Adjusting key

V-groove in screw

Plunger

Spring

Hand nut

Spring retaining nut

Cover with recess for spring seat

Conical spring

FIG.29

Actuated member

Stationary threaded bushing

Steel balls

Combined spring-centering cup and ball cage

Conical surface on adjusting nut

Frame

FIG.31

Hollow-spring center

Hardened-steel disk

Turned-down end of actuated member

Case

Lock nut

Adjustable screw

Boss projecting into counterbored hole saves space

FIG.30

Support, if spring is long

Flat Springs in Mechanisms

These devices all rely on a flat spring for their efficient actions,
which would otherwise need more complex configurations.

L. Kasper

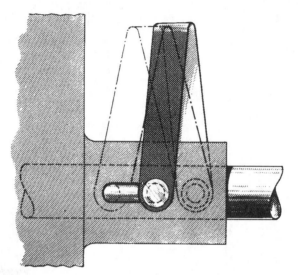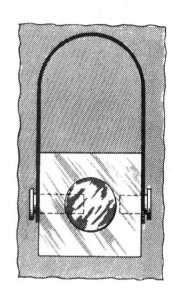

1   **CONSTANT FORCE** is approached be-cause of the length of this U-spring. Don't align studs or spring will fall.

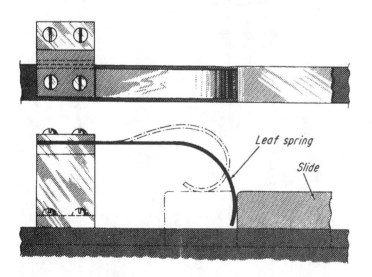

Leaf spring

Slide

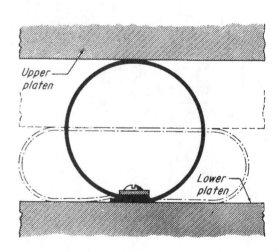

Upper platen

Lower platen

4   **SPRING-LOADED SLIDE** will always re-turn to its original position unless it is pushed until the spring kicks out.

5   **INCREASING SUPPORT AREA** as the load increases on both upper and lower platens is provided by a circular spring.

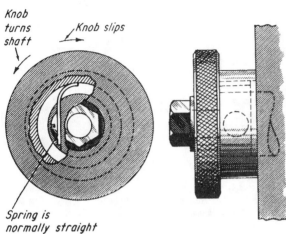

Knob turns shaft

Knob slips

Spring is normally straight

Grip springs have preloaded tension

Slide   *Anchor bar*   *Handle*

2

FLAT-WIRE SPRAG is straight until the knob is assembled; thus tension helps the sprag to grip for one-way clutching.

3

EASY POSITIONING of the slide is possible when the handle pins move a grip spring out of contact with the anchor bar.

6

CONSTANT TENSION in the spring, and thus force required to activate slide, is (almost) provided by this single coil.

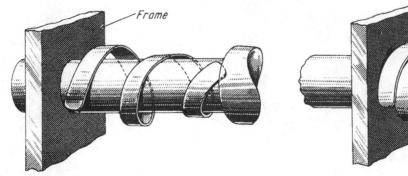

Frame

7

VOLUTE SPRING here lets the shaft be moved closer to the frame, thus allowing maximum axial movement.

Flat Springs Find More Work

Five additional examples for the way flat springs perform important jobs in mechanical devices.

L. Kasper

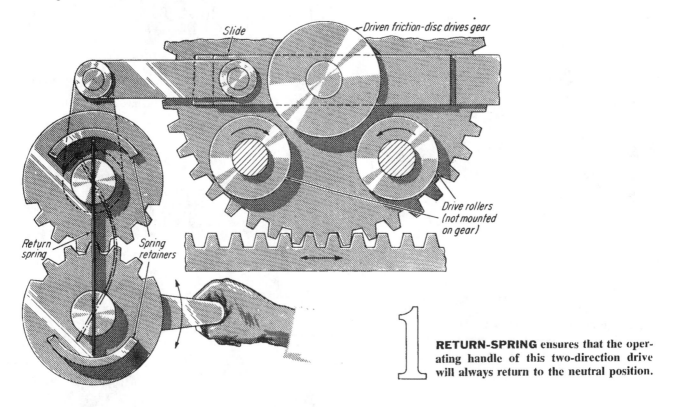

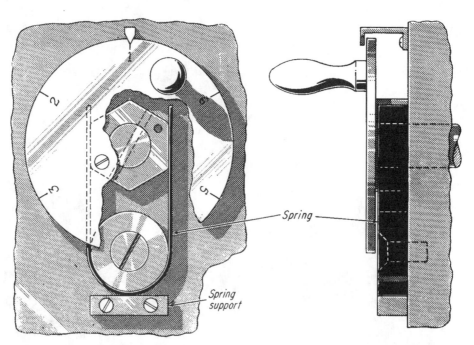

RETURN-SPRING ensures that the operating handle of this two-direction drive will always return to the neutral position.

INDEXING is accomplished simply, efficiently, and at low cost by the flat-spring arrangement shown here.

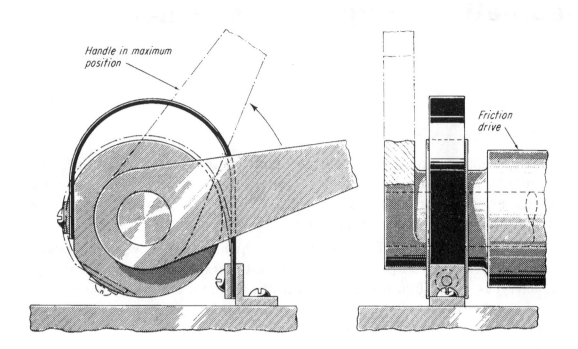

Handle in maximum position

Friction drive

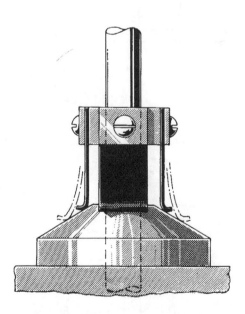

2 **SPRING-MOUNTED DISK** changes center position as handle is rotated to move friction drive, also acts as built-in limit stop.

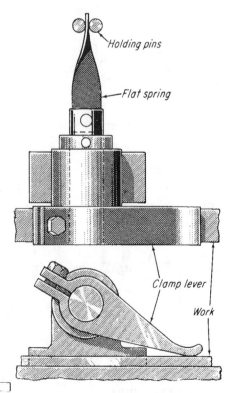

Holding pins

Flat spring

Clamp lever

Work

CUSHIONING device features rapid increase of spring tension because of the small pyramid angle. Rebound is minimum, too.

HOLD-DOWN CLAMP has flat spring assembled with initial twist to provide clamping force for thin material.

Pneumatic Spring Reinforcement

Robert O. Parmley, P.E.

A typical pneumatic spring is basically a column of trapped air or gas which is configured within a designed chamber to utilize the pressure of said air (or gas) for the unit's spring support action. The compressibility of the confined air provides the elasticity or flexibility of the pneumatic spring.

There are many designs of pneumatic springs which include: hydro-pneumatic, pneumatic spring/shock absorber, cylinder, piston, constant-volume, constant mass and bladder types. The latter, bladder type, is one of the most basic designs. This type of pneumatic spring is generally composed or rubber or plastic membranes without any integral reinforcement. See Figure 1.

A cost-effective method to reinforce the bladder membrane is to utilize a steel coil spring for external support. Figure 2 illustrates the conceptual design. Proper sizing of the coil spring is necessary to avoid undue stress and pinching of the membrane during both the flexing action and rest phase.

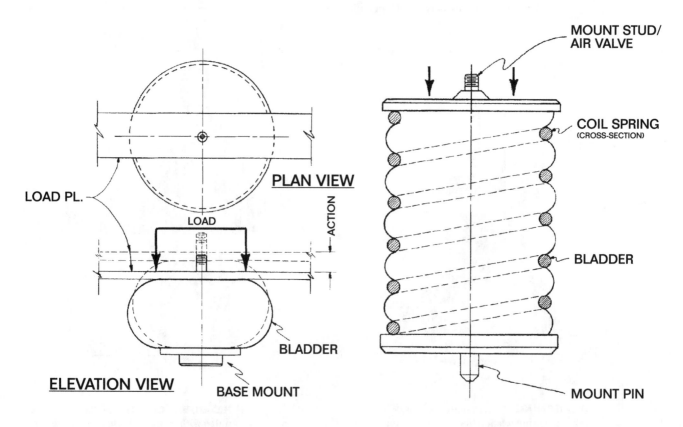

Figure 1   **Figure 2**

SECTION 14

CAMS

Generating Cam Curves

It usually doesn't pay to design a complex cam curve if it can't be easily machined—so check these mechanisms before starting your cam design.

Preben W. Jensen

I F you have to machine a cam curve into the metal blank without using a master cam, how accurate can you expect it to be? That depends primarily on how precisely the mechanism you use can feed the cutter into the cam blank. The mechanisms described here have been carefully selected for their practicability. They can be employed directly to machine the cams, or to make master cams for producing others.

The cam curves are those frequently employed in automatic-feed mechanisms and screw machines. They are the circular, constant-velocity, simple-harmonic, cycloidal, modified cycloidal, and circular-arc cam curve, presented in that order.

Circular cams

This is popular among machinists because of the ease in cutting the groove. The cam (Fig 1A) has a circular groove whose center, A, is displaced a distance a from the cam-plate center, A_0, or it may simply be a plate cam with a spring-loaded follower (Fig 1B).

Interestingly, with this cam you can easily duplicate the motion of a four-bar linkage (Fig 1C). Rocker BB_0 in Fig 1C, therefore, is equivalent to the motion of the swinging follower in Fig 1A.

The cam is machined by mounting the plate eccentrically on a lathe. The circular groove thus can be cut to close tolerances with an excellent surface finish.

If the cam is to operate at low speeds you can replace the roller with an arc-formed slide. This permits the transmission of high forces. The optimum design of such "power cams" usually requires time-consuming computations, but charts were published re-

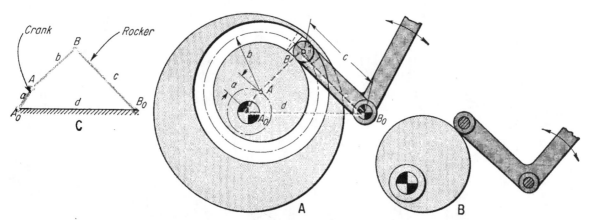

1. **Circular cam groove** is easily machined on turret lathe by mounting the plate eccentrically onto the truck. Plate cam in **(B)** with spring load follower produces same output motion. Many designers are unaware that this type of cam has same output motion as four-bar linkage **(C)** with the indicated equivalent link lengths. Hence it's the easiest curve to pick when substituting a cam for an existing linkage.

cently (see Editor's Note at end of article) which simplify this aspect of design.

The disadvantage (or sometimes, the advantage) of the circular-arc cam is that, when traveling from one given point, its follower reaches higher speed accelerations than with other equivalent cam curves.

Constant-velocity cams

A constant-velocity cam profile can be generated by rotating the cam plate and feeding the cutter linearly, both with uniform velocity, along the path the translating roller follower will travel later (Fig 2A). In the case of a swinging follower, the tracer (cutter) point is placed on an arm equal to the length of the actual swinging roller follower, and the arm is rotated with uniform velocity (Fig 2B).

Simple-harmonic cams

The cam is generated by rotating it with uniform velocity and moving the cutter with a scotch yoke geared to the rotary motion of the cam. Fig 3A shows the principle for a radial translating follower; the same principle is, of course, applicable for offset translating and swinging roller follower. The gear ratios and length of the crank working in the scotch yoke control the pressure angles (the angles for the rise or return strokes).

For barrel cams with harmonic motion the jig in Fig 3B can easily be set up to do the machining. Here, the barrel cam is shifted axially by means of the rotating, weight-loaded (or spring-loaded) truncated cylinder.

The scotch-yoke inversion linkage (Fig 3C) replaces the gearing called for in Fig 3A. It will cut an approximate simple-harmonic motion curve when the cam has a swinging roller follower, and an exact curve when the cam has a radial or offset translating roller follower. The slotted member is fixed to the machine frame *1*. Crank *2* is driven around the center *0*. This causes link *4* to oscillate back and forward in simple harmonic motion. The sliding piece *5* carries the cam to be cut, and the cam is rotated around the center of *5* with uniform velocity. The length of arm *6* is made equal to the length of the swinging roller follower of the actual cam mechanism and the device adjusted so that the extreme positions of the center of *5* lie on the center line of *4*.

The cutter is placed in a stationary spot somewhere along the centerline of member *4*. In case a radial or offset translating roller follower is used, the sliding piece *5* is fastened to *4*.

The deviation from simple harmonic motion when the cam has a swinging follower causes an increase in acceleration ranging from 0 to 18% (Fig 3D), which depends on the total angle of oscillation of the follower. Note that for a typical total oscillating angle of 45 deg, the increase in acceleration is about 5%.

Cycloidal motion

This curve is perhaps the most desirable from a designer's viewpoint because of its excellent acceleration characteristic. Luckily, this curve is comparatively easy to generate. Before selecting the mechanism it is worthwhile looking at the underlying theory of the cycloids because it is possible to generate not only cycloidal motion but a whole family of similar curves.

The cycloids are based on an offset sinusoidal wave (Fig 4). Because the radii of curvatures in points *C*, *V*, and *D* are infinite (the curve is "flat" at these points), if this curve was a cam groove and moved in the direction of line *CVD*, a translating roller follower, actu-

ated by this cam, would have zero acceleration at points *C*, *V*, and *D* no matter in what direction the follower is pointed.

Now, if the cam is moved in the direction of *CE* and the direction of motion of the translating follower is lined perpendicular to *CE*, the acceleration of the follower in points *C*, *V*, and *D* would still be zero.

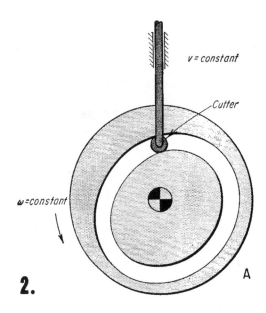

2.

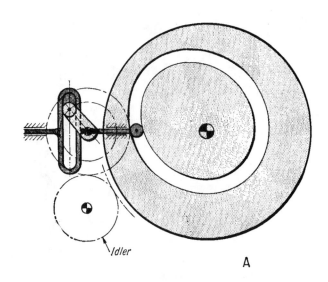

3. For producing simple harmonic curves: (A) Scotch yoke device feeds cutter while gearing arrangement rotates cam; (B) truncated-cylinder slider for

2. **Constant-velocity** cam is machined by feeding the cutter and rotating the cam at constant velocity. Cutter is fed linearly **(A)** or circularly **(B)**, depending on type of follower.

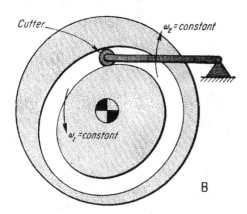

B

This has now become the basic cycloidal curve, and it can be considered as a sinusoidal curve of a certain amplitude (with the amplitude measured perpendicular to the straight line) superimposed on a straight (constant-velocity) line.

The cycloidal is considered the best *standard* cam contour because of its low dynamic loads and low

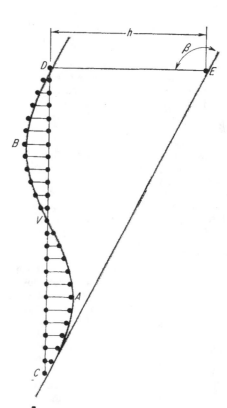

4. **Layout of a cycloidal curve.**

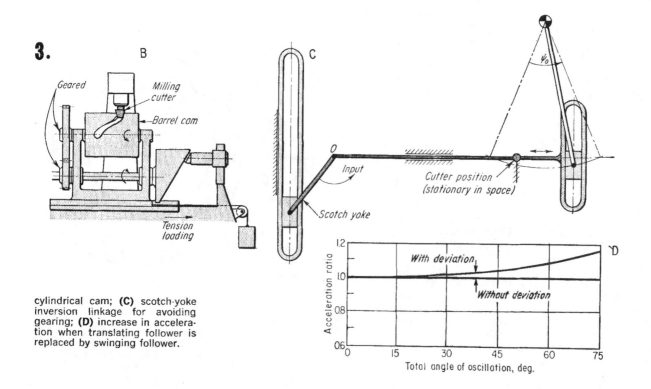

3.

B

cylindrical cam; **(C)** scotch-yoke inversion linkage for avoiding gearing; **(D)** increase in acceleration when translating follower is replaced by swinging follower.

shock and vibration characteristics. One reason for these outstanding attributes is that it avoids any sudden change in acceleration during the cam cycle. But improved performances are obtainable with certain modified cycloidals.

Modified cycloids

To get a modified cycloid, you need only change the direction and magnitude of the amplitude, while keeping the radius of curvature infinite at points C, V, and D.

Comparisons are made in Fig 5 of some of the modified curves used in industry. The true cycloidal is shown in the cam diagram of A. Note that the sine amplitudes to be added to the constant-velocity line are perpendicular to the base. In the Alt modification shown in B (after Hermann Alt, German kinematician, who first analyzed it), the sine amplitudes are perpendicular to the constant-velocity line. This results in improved (lower) velocity characteristics (see D), but higher acceleration magnitudes (see E).

The Wildt modified cycloidal (after Paul Wildt) is constructed by selecting a point w which is 0.57 the distance $T/2$, and then drawing line wp through yp which is midway along OP. The base of the sine curve is then constructed perpendicular to yw. This modification results in a maximum acceleration of 5.88 h/T^2, whereas the standard cycloidal curve has a maximum acceleration of 6.28 h/T^2. This is a 6.8% reduction in acceleration.

(It's quite a trick to construct a cycloidal curve to go through a particular point P—where P may be anywhere within the limits of the box in C—and with a specific slope at P. There is a growing demand for this type of modification, and a new, simple, graphic technique developed for meeting such requirements will be shown in the next issue.)

Generating the modified cycloidals

One of the few devices capable of generating the family of modified cycloidals consists of a double carriage and rack arrangement (Fig 6A).

The cam blank can pivot around the spindle, which in turn is on the movable carriage I. The cutter center is stationary. If the carriage is now driven at constant speed by the lead screw, in the direction of the arrow, the steel bands 1 and 2 will also cause the cam blank to rotate. This rotation-and-translation motion to the cam will cause a spiral type of groove.

For the modified cycloidals, a second motion must be imposed on the cam to compensate for the deviations from the true cycloidal. This is done by a second steel band arrangement. As carriage I moves, the bands 3 and 4 cause the eccentric to rotate. Because of the stationary frame, the slide surrounding the eccentric is actuated horizontally. This slide is part of carriage II, with the result that a sinusoidal motion is imposed on to the cam.

Carriage I can be set at various angles β to match angle β in Fig 5B and C. The mechanism can also be modified to cut cams with swinging followers.

Circular-arc cams

Although in recent years it has become the custom to turn to the cycloidal and other similar curves even when speeds are low, there are many purposes for which

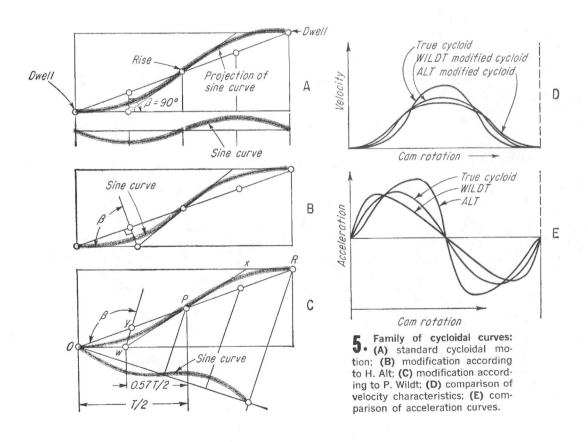

5. Family of cycloidal curves: (A) standard cycloidal motion; (B) modification according to H. Alt; (C) modification according to P. Wildt; (D) comparison of velocity characteristics; (E) comparison of acceleration curves.

circular-arc cams suffice. Such cams are composed of circular arcs, or circular arcs and straight lines. For comparatively small cams the cutting technique illustrated in Fig 7 produces good accuracy.

Assume that the contour is composed of circular arc *1-2* with center at O_2, arc *3-4* with center at O_3, arc *4-5* with center at O_1, arc *5-6* with center at O_4, arc *7-1* with center at O_1, and the straight lines *2-3* and *6-7*. The method involves a combination of drilling, lathe turning, and template filing.

First, small holes about 0.1 in diameter are drilled at O_1, O_3, and O_4, then a hole is drilled with the center at O_2 and radius of r_2. Next the cam is fixed in a turret lathe with the center of rotation at O_1, and the steel plate is cut until it has a diameter of $2r_5$. This takes care of the larger convex radius. The straight lines *6-7* and *2-3* are now milled on a milling machine.

Finally, for the smaller convex arcs, hardened pieces are turned with radii r_1, r_3, and r_4. One such piece is shown in Fig 7B. The templates have hubs which fit into the drilled holes at O_1, O_3, and O_4. Now the arc *7-1*, *3-4*, and *5-6* are filed, using the hardened templates as a guide. Final operation is to drill the enlarged hole at O_1 to a size that a hub can be fastened to the cam.

This method is frequently better than copying from a drawing or filing the scallops away from a cam where a great number of points have been calculated to determine the cam profile.

Compensating for dwells

One disadvantage with the previous generating devices is that, with the exception of the circular cam, they cannot include a dwell period within the rise-and-fall cam

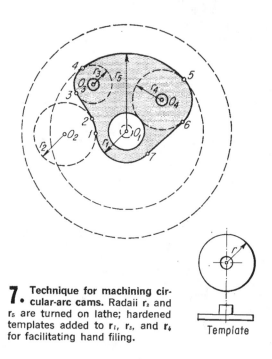

7. **Technique for machining circular-arc cams.** Radaii r_2 and r_5 are turned on lathe; hardened templates added to r_1, r_3, and r_4 for facilitating hand filing.

Template

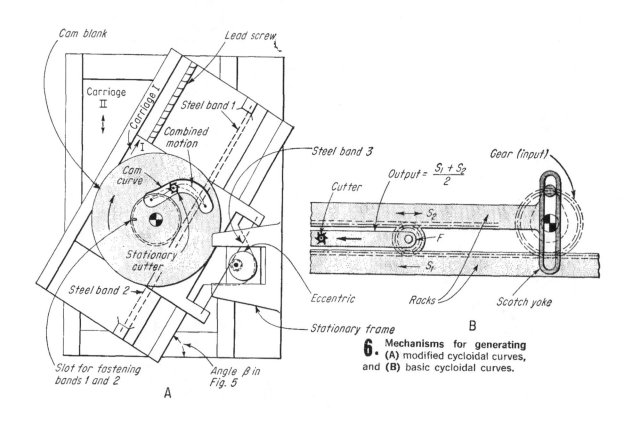

6. **Mechanisms for generating** **(A)** modified cycloidal curves, and **(B)** basic cycloidal curves.

cycle. The mechanisms must be disengaged at the end of rise and the cam rotated in the exact number of degrees to where the fall cycle begins. This increases the inaccuracies and slows down production.

There are two devices, however, that permit automatic machining through a specific dwell period: the double-geneva drive and the double eccentric mechanism.

Double-genevas with differential

Assume that the desired output contains dwells (of specific duration) at both the rise and fall portions, as shown in Fig 8A. The output of a geneva that is being rotated clockwise will produce an intermittent motion similar to the one shown in Fig 8B—a rise-dwell-rise-dwell . . . etc, motion. These rise portions are distorted simple-harmonic curves, but are sufficiently close to the pure harmonic to warrant use in many applications.

If the motion of another geneva, rotating counter-clockwise as shown in (C), is added to that of the clockwise geneva by means of a differential (D), then the sum will be the desired output shown in (A).

The dwell period of this mechanism is varied by shifting the relative position between the two input cranks of the genevas.

The mechanical arrangement of the mechanism is shown in Fig 8D. The two driving shafts are driven by gearing (not shown). Input from the four-star geneva to the differential is through shaft 3; input from the eight-station geneva is through the spider. The output from the differential, which adds the two inputs, is through shaft 4.

The actual device is shown in Fig 8E. The cutter is fixed in space. Output is from the gear segment which rides on a fixed rack. The cam is driven by the motor which also drives the enclosed genevas. Thus, the entire device reciprocates back and forth on the slide to feed the cam properly into the cutter.

Genevas driven by couplers

When a geneva is driven by a constant-speed crank, as shown in Fig 8D, it has a sudden change in acceleration at the beginning and end of the indexing cycle (as the crank enters or leaves a slot). These abrupt changes can be avoided by employing a four-bar linkage with coupler in place of the crank. The motion of the coupler point C (Fig 9) permits smooth entry into the geneva slot.

Double eccentric drive

This is another device for automatically cutting cams with dwells. Rotation of crank A (Fig 10) imparts an oscillating motion to the rocker C with a prolonged dwell at both extreme positions. The cam, mounted on the rocker, is rotated by means of the chain drive and thus is fed into the cutter with the proper motion. During the dwells of the rocker, for example, a dwell is cut into the cam.

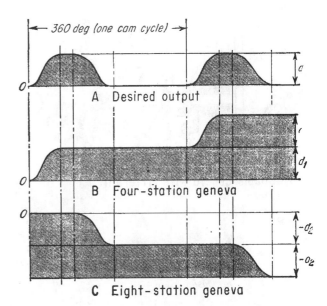

A Desired output

B Four-station geneva

C Eight-station geneva

8. Double genevas with differential for obtaining long dwells. Desired output characteristic (A) of cam is obtained by adding the motion (B) of a four-station geneva to that of (C) eight-station geneva. The mechanical arrangement of genevas with a differential is shown in (D); actual device is shown in (E). A wide variety of output dwells (F) are obtained by varying the angle between the driving cranks of the genevas.

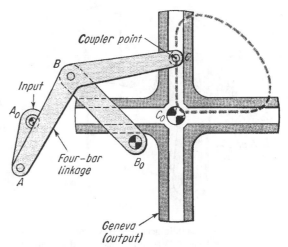

9. Four-bar coupler mechanism for replacing the cranks in genevas to obtain smoother acceleration characteristics.

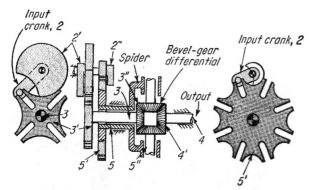

D   Double geneva with differential

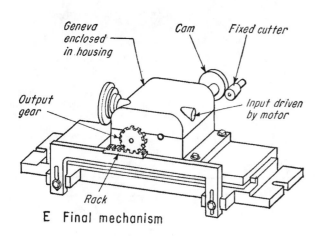

E   Final mechanism

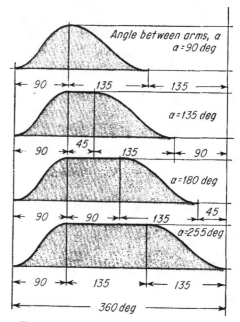

F   Various dwell resultants

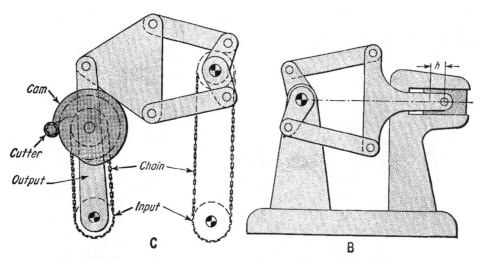

10. **Double eccentric drive** for automatically cutting cams with dwells. Cam is rotated and oscillated, with dwell periods at extreme ends of oscillation corresponding to desired dwell periods in cam.

Cams and Gears Team Up in Programmed Motion

Pawls and ratchets are eliminated in this design, which is adaptable to the smallest or largest requirements; it provides a multitude of outputs to choose from at low cost.

Theodore Simpson

A new and extremely versatile mechanism provides a programmed rotary output motion simply and inexpensively. It has been sought widely for filling, weighing, cutting, and drilling in automatic and vending machines.

The mechanism, which uses overlapping gears and cams (drawing below), is the brainchild of mechanical designer Theodore Simpson of Nashua, N. H.

Based on a patented concept that could be transformed into a number of configurations , PRIM (Programmed Rotary Intermittent Motion), as the mechanism is called, satisfies the need for smaller devices for instrumentation without using spring pawls or ratchets.

It can be made small enough for a wristwatch or as large as required.

Versatile output. Simpson reports the following major advantages:

• Input and output motions are on a concentric axis.

• Any number of output motions of varied degrees of motion or dwell time per input revolution can be provided.

• Output motions and dwells are variable during several consecutive input revolutions.

• Multiple units can be assembled on a single shaft to provide an almost limitless series of output motions and dwells.

• The output can dwell, then snap around.

How it works. The basic model

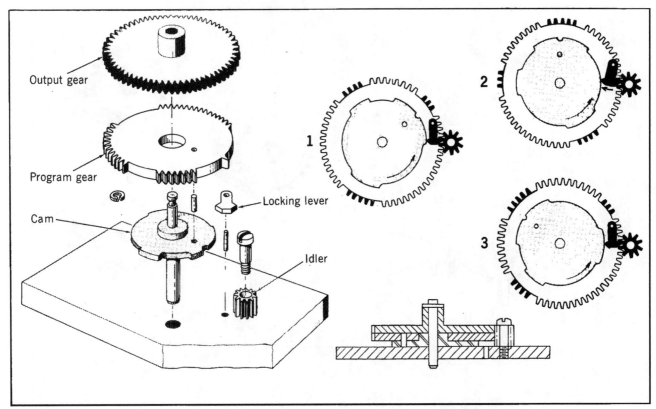

Output gear

Program gear

Cam

Locking lever

Idler

1

2

3

Basic intermittent-motion mechanism, at left in drawings, goes through the rotation sequence as numbered above.

(drawing, below left) repeats the output pattern, which can be made complex, during every revolution of the input.

Cutouts around the periphery of the cam give the number of motions, degrees of motion, and dwell times desired. Tooth sectors in the program gear match the cam cutouts.

Simpson designed the locking lever so one edge follows the cam and the other edge engages or disengages, locking or unlocking the idler gear and output. Both program gear and cam are lined up, tooth segments to cam cutouts, and fixed to the input shaft. The output gear rotates freely on the same shaft, and an idler gear meshes with both output gear and segments of the program gear.

As the input shaft rotates, the teeth of the program gear engage the idler. Simultaneously, the cam releases the locking lever and allows the idler to rotate freely, thus driving the output gear.

Reaching a dwell portion, the teeth of the program gear disengage from the idler, the cam kicks in the lever to lock the idler, and the output gear stops until the next program-gear segment engages the idler.

Dwell time is determined by the space between the gear segments. The number of output revolutions does not have to be the same as the number of input revolutions. An idler of a different size would not affect the output, but a cluster idler with a matching output gear can increase or decrease the degrees of motion to meet design needs.

For example, a step-down cluster with output gear to match could reduce motions to fractions of a degree, or a step-up cluster with matching output gear could increase motions to several complete output revolutions.

Snap action. A second cam and a spring are used in the snap-action version (drawing below). Here, the cams have identical cutouts.

One cam is fixed to the input and the other is lined up with and fixed to the program gear. Each cam has a pin in the proper position to retain a spring; the pin of the input cam extends through a slot in the program gear cam that serves the function of a stop pin.

Both cams rotate with the input shaft until a tooth of the program gear engages the idler, which is locked and stops the gear. At this point, the program cam is in position to release the lock, but misalignment of the peripheral cutouts prevents it from doing so.

As the input cam continues to rotate, it increases the torque on the spring until both cam cutouts line up. This positioning unlocks the idler and output, and the built-up spring torque is suddenly released. It spins the program gear with a snap as far as the stop pin allows; this action spins the output.

Although both cams are required to release the locking lever and output, the program cam alone will relock the output—a feature of convenience and efficient use.

After snap action is complete and the output is relocked, the program gear and cam continue to rotate with the input cam and shaft until they are stopped again when a succeeding tooth of the segmented program gear engages the idler and starts the cycle over again.

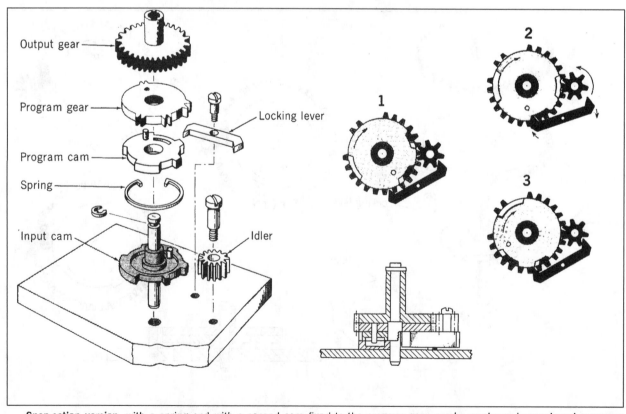

Snap-action version, with a spring and with a second cam fixed to the program gear, works as shown in numbered sequence.

Spherical Cams: Linking Up Shafts

European design is widely used abroad but little-known in the U.S. Now a German engineering professor is telling the story in this country, stirring much interest.

Anthony Hannavy

Problem: to transmit motion between two shafts in a machine when, because of space limitations, the shaft axes may intersect each other. One answer is to use a spherical-cam mechanism, unfamiliar to most American designers but used in Europe to provide many types of motion in agricultural, textile, and printing machinery.

Recently, Prof. W. Meyer zur Cappellen of the Institute of Technology, Aachen, Germany, visited the U. S. to show designers how spherical-cam mechanisms work and how to design and make them. He and his assistant kinematician at Aachen, Dr. G. Dittrich, are in the midst of experiments with complex spherical-cam shapes and with the problems of manufacturing them.

Fundamentals. Key elements of spherical-cam mechanism (above Fig. 1) can be considered as being posi-

3   Spherical mechanism with radial follower

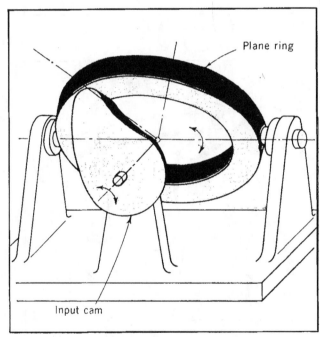

4   Cam mechanism with flat-faced follower

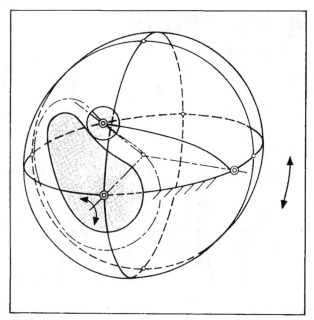

Radial roller follower shown on a sphere

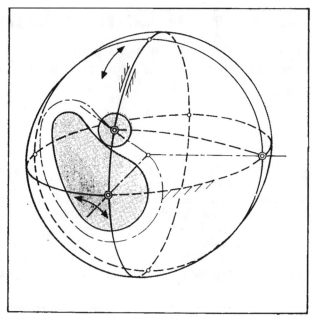

Mechanism with radial roller follower shown on a sphere

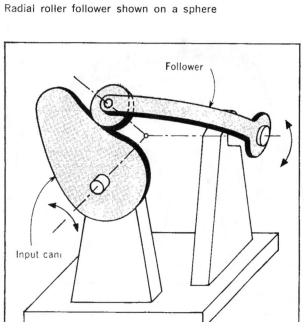

1 Spherical cam mechanism with radial follower

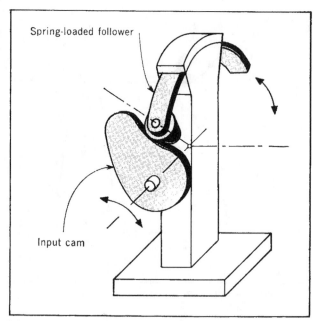

2 Cam mechanism with rocking roller follower

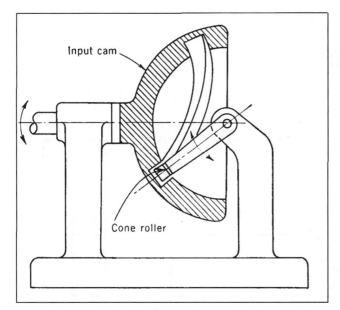

5 Hollow-sphere cam mechanism

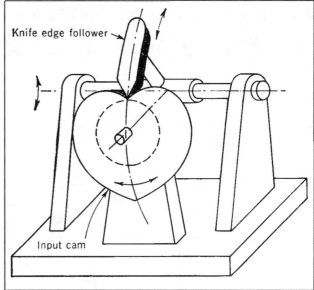

6 Mechanism with Archimedean spiral; knife-edge follower

tioned on a sphere. The center of this sphere is the point where the axes of rotation of the input and follower cams intersect.

In a typical configuration in an application (Fig. 1), the input and follower cams are shown with depth added to give them a conical roller surface. The roller is guided along the conical surface of the input cam by a rocker, or follower.

A schematic view of a spherical-cam mechanism (above Fig. 2) shows how the follower will rise and fall along a linear axis. In the same type of design (Fig. 2), the follower is spring-loaded. The designer can also use a rocking roller follower (Fig. 3) that oscillates about an axis that, in turn, intersects with another shaft.

These spherical-cam mechanisms using a cone roller have the same output motion characteristics as spherical-cam designs with non-rotating circular cone followers or spherically-shaped followers. The flat-faced follower in Fig. 4 rotates about an axis that is the contact face rather than the center of the plane ring. The plane ring follower corresponds to the flat-faced follower in plane kinematics.

Closed-form guides. Besides having the follower contained as in Fig. 2, spherical-cam mechanisms can be designed so the cone roller on the follower is guided along the body of the input cam. For example, in Fig.

5, the cone roller moves along a groove that has been machined on the spherical inside surface of the input cam. However, this type of guide encounters difficulties unless the guide is carefully machined. The cone roller tends to seize.

Although cone rollers are recommended for better motion transfer between the input and output, there are some types of motion where their use is prohibited.

For instance, to obtain the motion diagram shown in Fig. 6, a cone roller would have to roll along a surface where any change in the concave section would be limited to the diameter of the roller. Otherwise there would be a point where the output motion would be interrupted. In contrast, the use of a knife-edge follower theoretically imposes no

limit on the shape of the cam. However, one disadvantage with knife-edge followers is that they. unlike cone followers, slide and hence wear faster.

Manufacturing methods. Spherical cams are usually made by copying from a stencil. In turn, the cam-shaped tools can be copied from a stencil. Normally the cams are milled, but in special cases they are ground.

Three methods for manufacture are used to make the stencils:
- Electronically controlled point-by-point milling.
- Guided-motion machining.
- Manufacture by hand.

However, this last method is not recommended, because it isn't as accurate as the other two.

Modifications and Uses
for Basic Types of Cams

Edward Rahn

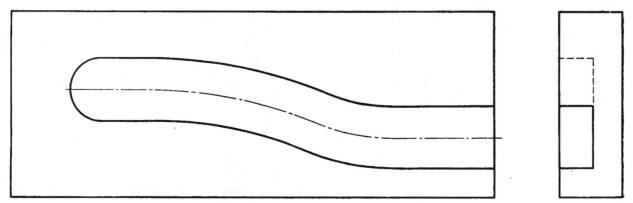

FLAT PLATE CAM—Essentially a displacement cam. With it, movement can be made from one point to another along any desired profile. Often used in place of taper attachments on lathes for form turning. Some have been built in sections up to 15 ft. long for turning the outside profile on gun barrels. Such cams can be made either on milling machines or profiling machines.

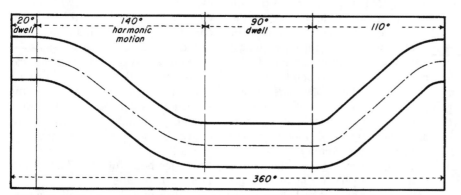

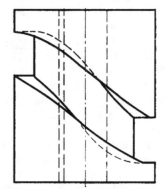

BARREL CAM—Sometimes called a cylindrical cam. The follower moves in a direction parallel to the cam axis and lever movement is reciprocating. As with other types of cam, the base curve can be varied to give any desired movement. Internal as well as external barrel cams are practical. A limitation: internal cams less than 11 in. in diam. are difficult to make on cam millers.

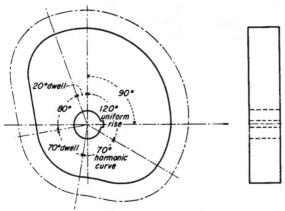

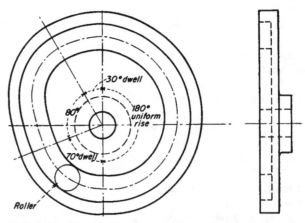

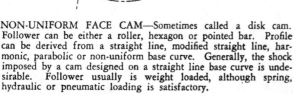

NON-UNIFORM FACE CAM—Sometimes called a disk cam. Follower can be either a roller, hexagon or pointed bar. Profile can be derived from a straight line, modified straight line, harmonic, parabolic or non-uniform base curve. Generally, the shock imposed by a cam designed on a straight line base curve is undesirable. Follower usually is weight loaded, although spring, hydraulic or pneumatic loading is satisfactory.

BOX CAM—Gives positive movement in two directions. A profile can be based on any desired base curve, as with face cams, but a cam miller is needed to cut it; whereas with face cams, a band saw and disk grinder could conceivably be used. No spring, pneumatic or hydraulic loading is needed for the followers. This type cam requires more material than for a face cam, but is no more expensive to mill.

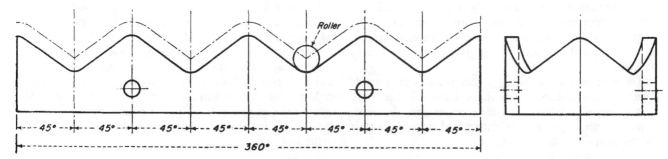

SIDE CAM—Essentially a barrel cam having only one side. Can be designed for any type of motion, depending on requirements and speed of operation. Spring or weight loaded followers of either the pointed or roller type can be used. Either vertical or horizontal mounting is permissible. Cutting of the profile is usually done on a shaper or a cam miller equipped with a small diameter cutter, although large cams 24 in. in diameter are made with 7-in. cutters.

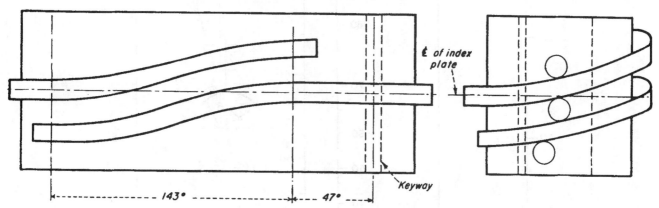

INDEX CAM—Within limits, such cams can be designed for any desired acceleration, deceleration and dwell period. A relatively short period for acceleration can be alloted on high speed cams such as those used on zipper-making equipment on which indexing occurs 1,200 to 1,500 times per minute. Cams of this sort can also be designed with four or more index stations.

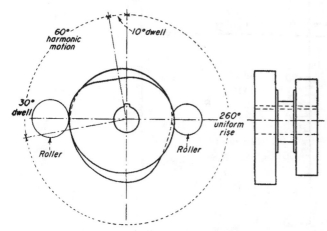

DOUBLE FACE CAM—Similar to single face cam except that it provides positive straight line movement in two directions. The supporting fork for the rollers can be mounted separately or between the faces. If the fork fulcrum is extended beyond the pivot point, the cam can be used for oscillatory movement. With this cam, the return stroke on a machine can be run faster than the feed stroke. Cost is more than that for a box cam

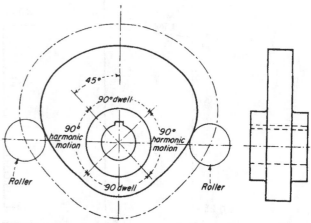

SINGLE-FACE CAM WITH TWO FOLLOWERS—Similar in action to a box or double face cam except flexibility is less than that for the latter type. Cam action for feed and return motions must be the same to prevent looseness of cam action. Used in place of box cams or double face cams to conserve space, and instead of single face cams to provide more positive movement for the roller followers.

Nomogram for Parabolic Cam with Radically Moving Follower

Rudolph Gruenberg

THE DEVELOPMENT of theoretically correct cam profiles is often complex and time consuming. In applications having neither high speeds nor forces, such efforts are unwarranted. In these applications, parabolic or gravity cams are usually adequate.

The efficiency of operating a cam-mechanism depends predominantly on the pressure angle. Since it is a measure of the greatest side thrust on the follower arm, the maximum pressure angle must be determined since it controls the physical dimensions of the cam.

The useful work transferred from the cam-shaft to the follower increases as the pressure angle decreases because the force component in the direction of the follower is proportional to the cosine of the pressure angle.

Designing for maximum efficiency, therefore, involves a trial and error balance of least maximum cam pressure angle against the mechanical limitations of the cam and its adjacent components. The nomogram below reduces this to a minimum and affords a quick check of an already established design.

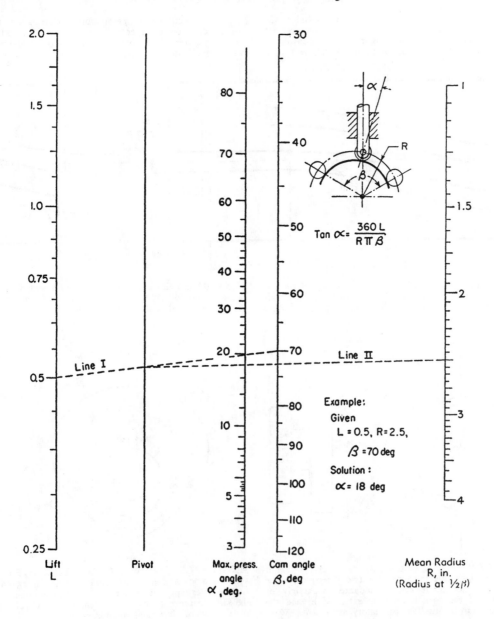

$$\text{Tan } \alpha = \frac{360\,L}{R\,\pi\,\beta}$$

Example:
Given
L = 0.5, R = 2.5,
β = 70 deg

Solution:
α = 18 deg

Lift
L

Pivot

Max. press.
angle
α , deg.

Cam angle
β, deg

Mean Radius
R, in.
(Radius at ½β)

SECTION 15

GROMMETS, SPACERS & INSERTS

A Fresh Look at Rubber Grommets

A small component that's often neglected in the details of a design. Here are eight unusual applications.

Robert O. Parmley

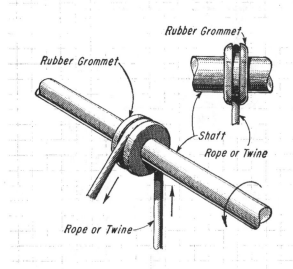

1 Pulley for slow rotation

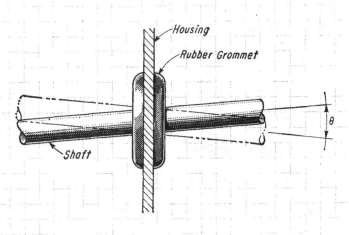

2 Handle shaft misalignment

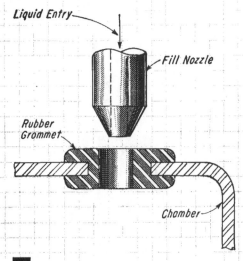

5 Seal for liquid filling

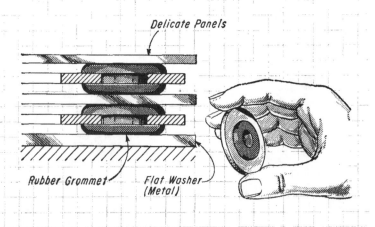

6 Cushioned spacers

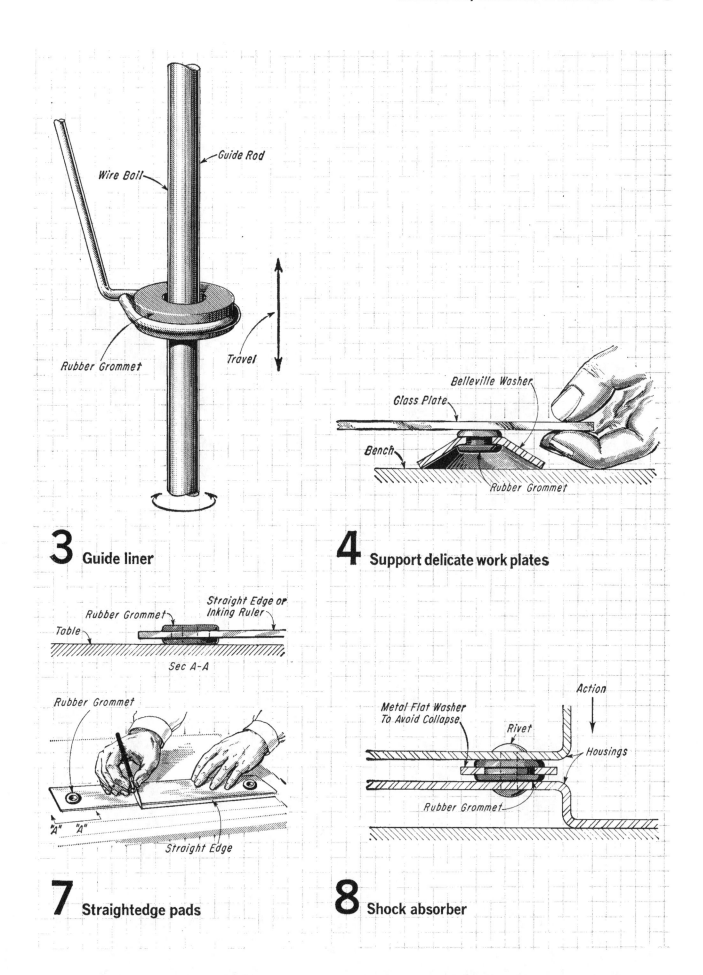

Guide Rod

Wire Bail

Rubber Grommet

Travel

3 Guide liner

Belleville Washer

Glass Plate

Bench

Rubber Grommet

4 Support delicate work plates

Rubber Grommet

Straight Edge or
Inking Ruler

Table

Sec A-A

Rubber Grommet

"A" "A"

Straight Edge

7 Straightedge pads

Metal Flat Washer
To Avoid Collapse

Action

Rivet

Housings

Rubber Grommet

8 Shock absorber

These Spacers Are Adjustable

Rubber, metal springs, jackscrews, pivoting bars and sliding wedges allow adjustment after assembly.

Richard A. Cooper

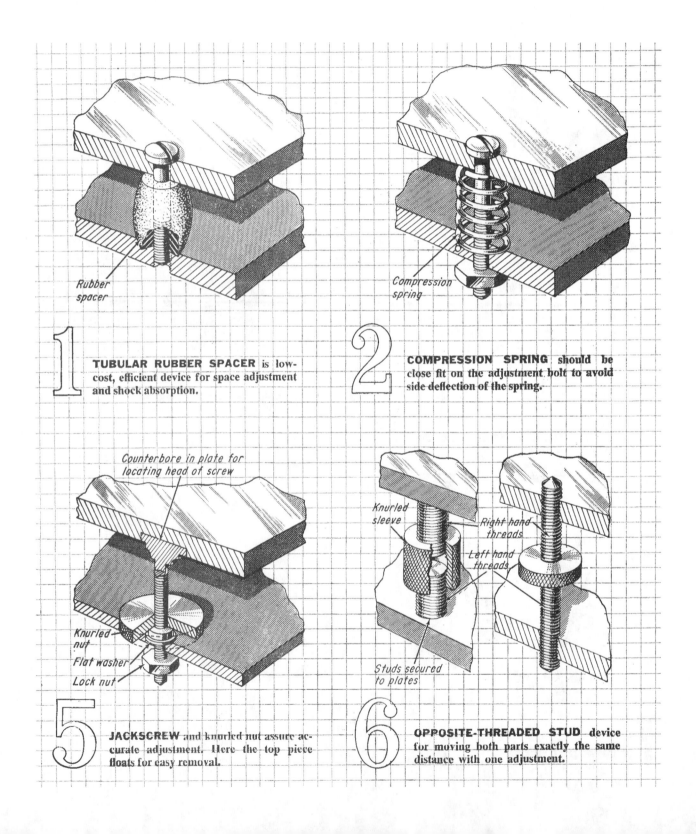

Rubber spacer

1 TUBULAR RUBBER SPACER is low-cost, efficient device for space adjustment and shock absorption.

Compression spring

2 COMPRESSION SPRING should be close fit on the adjustment bolt to avoid side deflection of the spring.

Counterbore in plate for locating head of screw

Knurled nut
Flat washer
Lock nut

5 JACKSCREW and knurled nut assure accurate adjustment. Here the top piece floats for easy removal.

Knurled sleeve
Right hand threads
Left hand threads
Studs secured to plates

6 OPPOSITE-THREADED STUD device for moving both parts exactly the same distance with one adjustment.

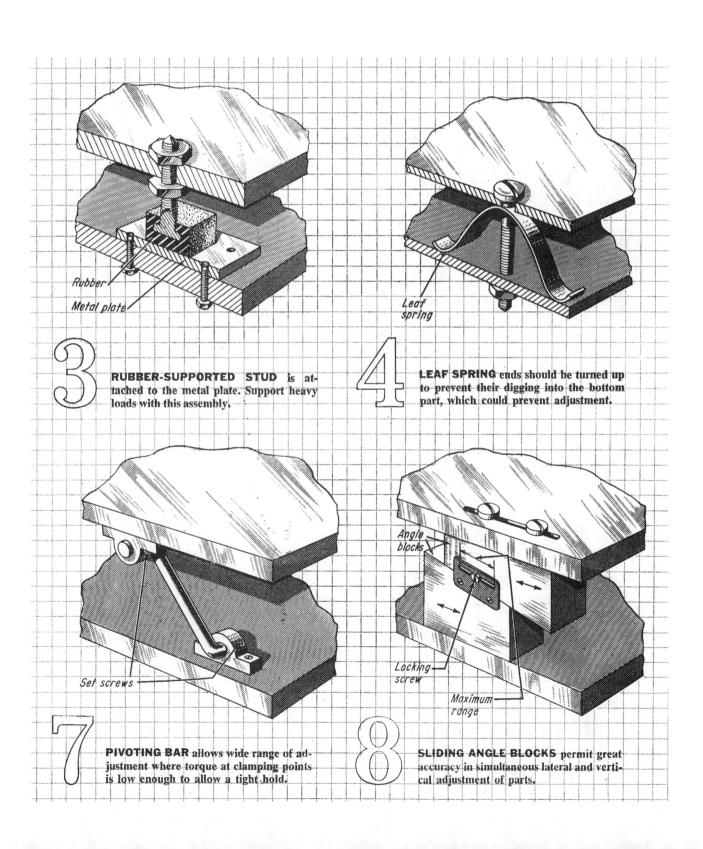

3 **RUBBER-SUPPORTED STUD** is attached to the metal plate. Support heavy loads with this assembly.

Rubber

Metal plate

4 **LEAF SPRING** ends should be turned up to prevent their digging into the bottom part, which could prevent adjustment.

Leaf spring

7 **PIVOTING BAR** allows wide range of adjustment where torque at clamping points is low enough to allow a tight hold.

Set screws

8 **SLIDING ANGLE BLOCKS** permit great accuracy in simultaneous lateral and vertical adjustment of parts.

Angle blocks

Locking screw

Maximum range

Odd Jobs for Rubber Mushroom Bumpers

High energy absorption at low cost is the way mushroom bumpers are usually billed.
But they have other uses; here are seven that are rather unconventional.

Robert O. Parmley

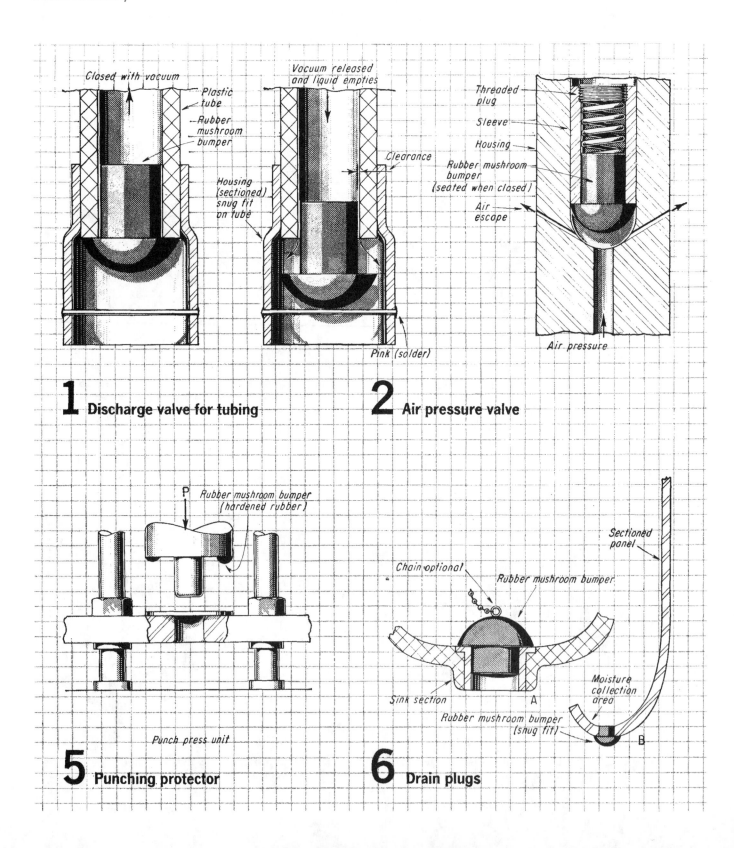

1 Discharge valve for tubing

Closed with vacuum
Plastic tube
Rubber mushroom bumper
Housing (sectioned) snug fit on tube

2 Air pressure valve

Vacuum released and liquid empties
Clearance
Pink (solder)

Threaded plug
Sleeve
Housing
Rubber mushroom bumper (seated when closed)
Air escape
Air pressure

5 Punching protector

P
Rubber mushroom bumper (hardened rubber)
Punch press unit

6 Drain plugs

Chain optional
Rubber mushroom bumper
Sectioned panel
Sink section
A
Moisture collection area
Rubber mushroom bumper (snug fit)
B

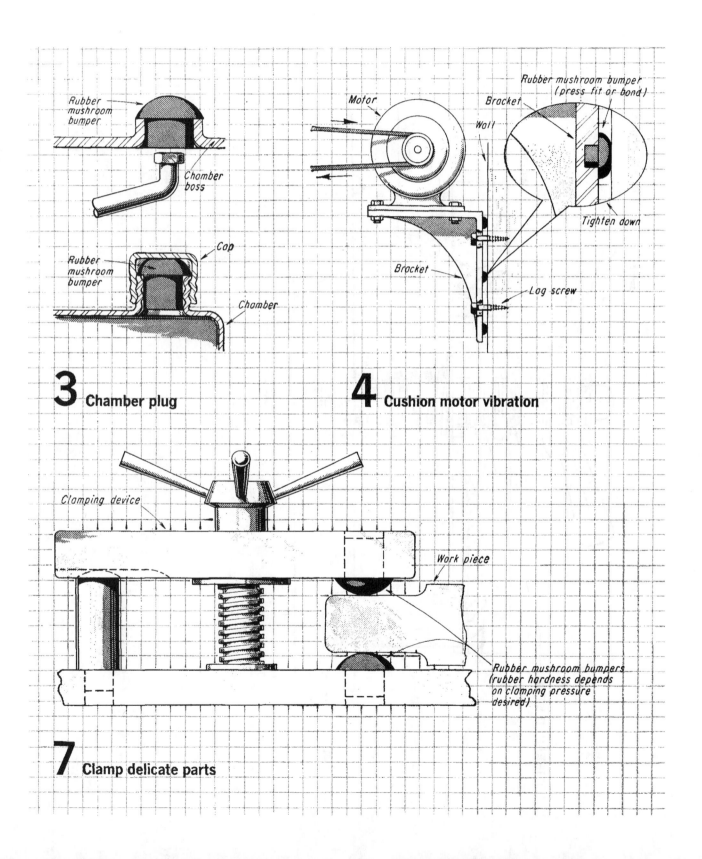

3 Chamber plug

Rubber mushroom bumper

Chamber boss

Rubber mushroom bumper

Cap

Chamber

4 Cushion motor vibration

Motor

Wall

Bracket

Rubber mushroom bumper (press fit or bond)

Tighten down

Bracket

Lag screw

7 Clamp delicate parts

Clamping device

Work piece

Rubber mushroom bumpers (rubber hardness depends on clamping pressure desired)

Spacers Used in Jigs & Fixtures

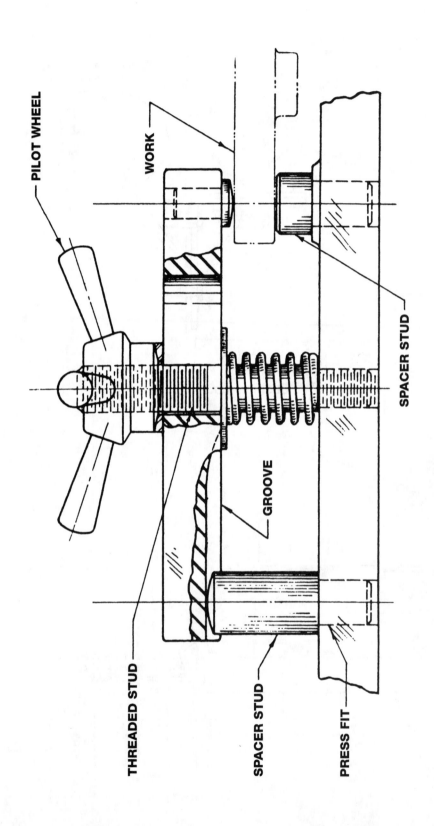

Flanged Inserts Stabilize Multi-Stroke Reloading Press

E. E. Lawrence, Inventor
Robert O. Parmley, Draftsman

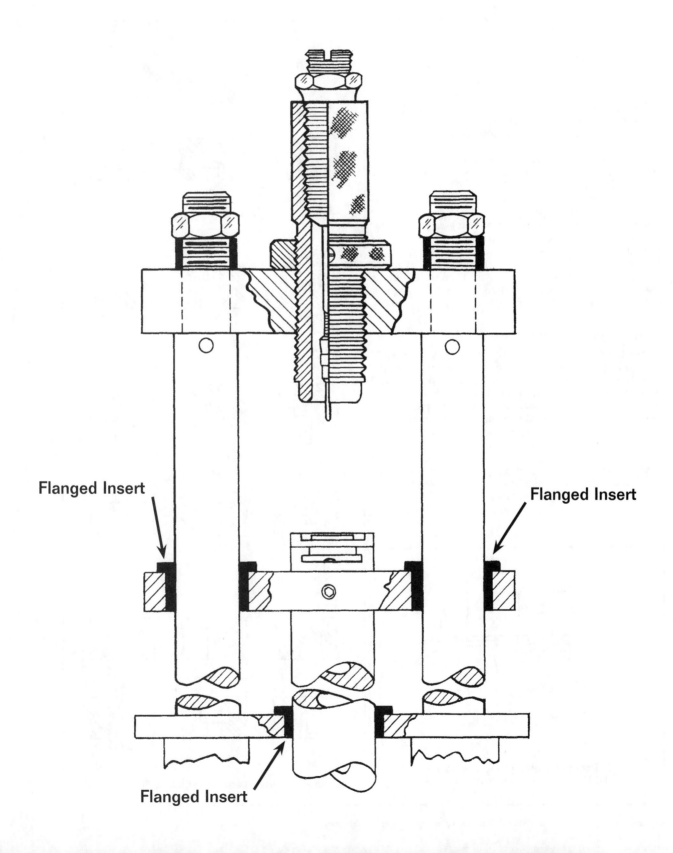

Flanged Insert

Flanged Insert

Flanged Insert

Metal Inserts for Plastic Parts

Plastics are increasingly used in automobiles and appliances and thus
a major company compiled these data.

Molded parts should be designed around any inserts
that are required. This work is done after the type of
compound is selected. Inserts are used for two basic
reasons:

1. To add strength to the plastic part or to control
shrinkage. Sometimes the purpose is to be decorative
or to avoid injuries.

2. To provide an attachment means for the con-
ductance of heat or electricity.

Special means of retention are not necessary for
inserts not subject to movement with relation to the
molding material. Inserts of round bar stock, coarse
diamond knurled and grooved, provide the strongest
anchorage under torque and tension. A large single
groove, as in Fig. 1, is better than two or more grooves
and smaller knurled areas.

Inserts secured by press or shrink fits

Inserts can be secured in the plastic by a press fit
or shrink fit. Both methods rely upon shrinkage of the
plastic, which is greatest immediately after removal
from the mold. The part should be made so that the
required tightness is obtained after the plastic has
cooled and shrinkage has occurred. The amount of
interference required depends on the size, rigidity
and stresses to be encountered in service. The hole to
receive the mating part should be round and countersunk. In addition, the mating part should be chamfered and filleted to facilitate proper assembly and
to eliminate stress concentration.

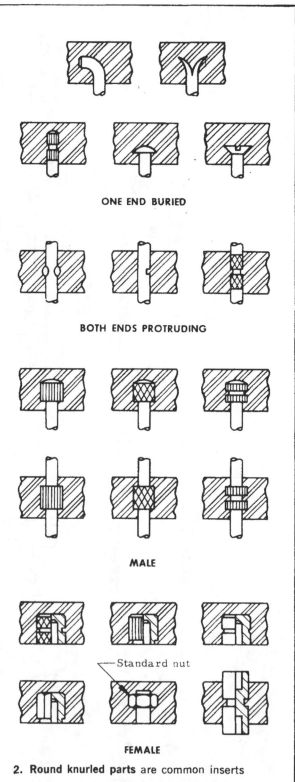

ONE END BURIED

BOTH ENDS PROTRUDING

MALE

Standard nut

FEMALE

2. Round knurled parts are common inserts

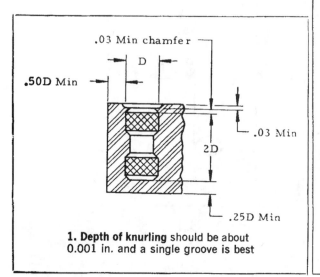

.03 Min chamfer

D

.50D Min

.03 Min

2D

.25D Min

1. Depth of knurling should be about
0.001 in. and a single groove is best

Reprinted with permission from *American Machinist*, A Penton Media Publication

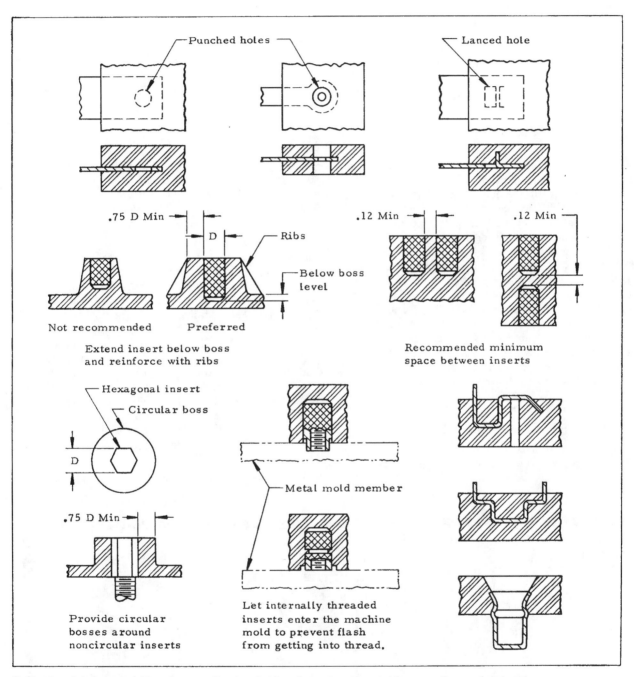

3. Sheet metal and special inserts, as well as knurled inserts, are used to provide connections and closures

How to Select Threaded Inserts

Why use threaded inserts? What types are available? What factors govern their selection? How do you determine whether a particular insert meets the given strength requirements and which one is most economical?

S. H. Davison

When the design calls for lightweight materials like aluminum, magnesium and plastics, threaded holes become a problem because of the low shear strength of these materials. To bolt such components to other machine parts, the threaded holes must be quite deep to develop the required pull-out strength, or they must have a coarse thread to increase shear area.

Further complications arise when the components require frequent disassembly for overhaul or repair; this may give excessive thread wear that will prevent satisfactory thread engagement with the bolt, particularly for high-strength bolts in the fine-thread series; and the depth of engagement needed will require a part so thick as to nullify the lightweight advantage of the material. To overcome these problems, thread inserts of high-strength material are often used.

Inserts available

Three main types of inserts are available:

1. Wire thread inserts are precision coils of diamond-shape stainless steel wire. They line a tapped hole and present a strong, accurate, standard internal thread for the bolt or stud. The OD on these inserts is from 11 to 30% greater than the OD of the internal thread—the lower limit is normally used on the fine-thread inserts specified in aircraft engines. The pull-out strength of wire thread insert depends on shear area of its OD thread.

2. Solid self-tapping inserts require only a drilled hole and counterbore in the part, and are made in two basic forms. (1) A solid self-tapping, self-locking bushing has a 60% external thread approximating the American National form with a uniform OD extending to the first 2 to 2½ threads. These leading threads are slightly truncated to provide the cutting action necessary for self-tapping. Chips flow out through the side hole drilled in the insert wall at the truncated lead threads. (2) A second type has lead threads similar to the pilot threads of a standard tap. These threads progressively flow into truncated threads extending the entire length of the insert—with the exception of the last three threads, which are standard.

In both types, two or three angular slots, depending on the particular design, provide the cutting action for self-tapping.

3. Solid bushings for pre-tapped holes are also made in

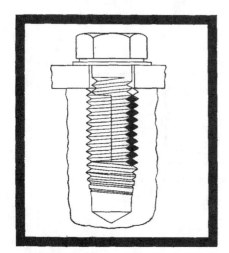

WIRE THREAD INSERTS made of diamond-shape stainless steel wire. Pull-out strength of material depends on ratio of internal to external thread shear area.

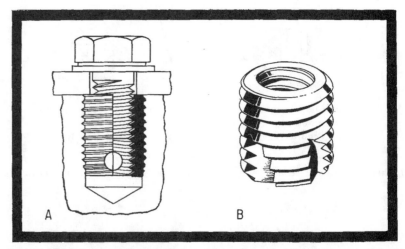

SOLID SELF-TAPPING INSERTS are of two types—(A) chips are removed through the side hole; or (B) with truncated lead threads and two or three angular slots which provide cutting action.

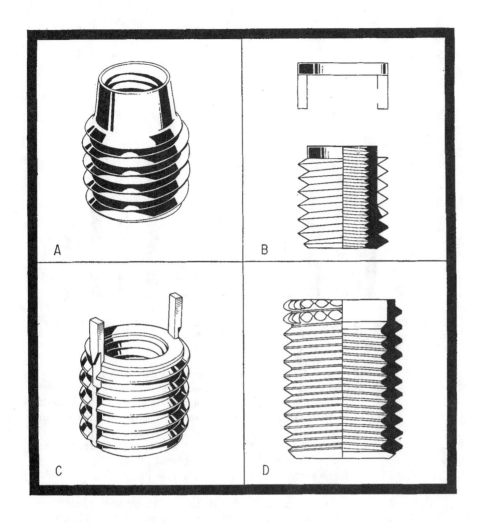

two general types. The first uses modified external threads that form an interference with the parent material, and provide locking action. The second type has many variations, but is characterized by standard external and internal threads, with various types of pins or keys to lock the bushing to the parent material. Some of the most widely used variations are:

A two-piece insert with a locking ring and two keys fits into mating grooves in upper external threads. The ring is pressed into place after the insert is screwed into tapped hole; it cuts through enough threads of parent material to provide a positive lock. A counterbore in the tapped hole is required for the ring, but assembly and replacement can be made with standard tools.

Another solid bushing insert has two integral keys which act as a broaching tool when insert is installed flush with the parent material. Locking pins are pressed into the base of the tapped hole through the grooves in the external thread.

Still another, a solid bushing, has standard internal and external threads and an expandable upper collar with serrations in the outer surface to lock the insert in the parent material.

Factors that affect selection

These factors must be considered in selecting the best type:

- Shear strength of parent material
- Operating temperature
- Load requirement
- Vibratory loads
- Assembly tooling—serviceability and ease of installation
- Relative cost

Shear strength of parent material below 40,000 psi generally calls for threaded inserts. This includes most of the aluminum alloys, all magnesium alloys and plastic materials. But other factors must be considered.

High operating temperature effects the shear strength by reducing strength of the parent material; an insert with a larger shear area may be required.

Bolt loading frequently makes it necessary to use threaded inserts. For example, if the full pull-out strength of a 125,000-psi bolt is required, it is probable that the parent material will need a threaded insert to increase the shear area and thus reduce the effective shear stress.

Vibratory loads may reduce bolt preload, and require a threaded insert to increase the effective shear area. Or vibration may cause creep, galling, and excessive wear, and inserts with both external and internal thread-locking features will be needed.

The pullout capacity of an insert is a function of projected shear area, and should equal the tensile strength of the bolt. This means pull-out strength should be greater than torque-applied tensile strength of the bolt.

In wire thread inserts the projected shear area per coil

RELATIVE EVALUATION—5 TYPES OF THREAD INSERTS

(A—self-tapping insert; B—wire thread insert; C—solid bushing for pre-tapped holes; D—solid bushings for pre-tapped holes and external interference threads; E—self-tapping insert)

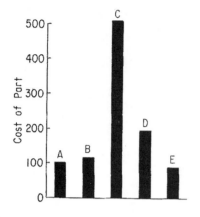

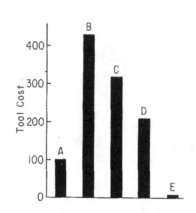

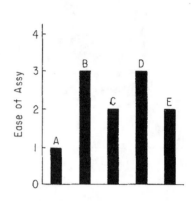

COST OF PART is price quoted for lots of 1000.

TOOL COST for each type is based on manufacturer's prices for tooling a standard tapping head.

EASE OF ASSEMBLY is a qualitative evaluation.

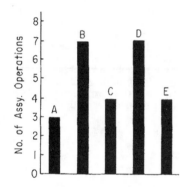

NUMBER OF ASSEMBLY OPERATIONS covers complete installation of an insert, including drill, counterbore, tap, ream, install and reinspect.

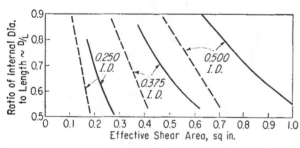

A USEFUL RELATION is effective shear area to D/L ratio. It determines required insert length or pull-out strength. Solid curves are for self-tapping inserts; dotted curves for wire thread inserts.

is relatively small; only way to increase the total projected shear area is to increase the number of coils. On the other hand, in solid and self-tapping inserts the projected shear area can be increased by a larger OD as well as by more threads, while maintaining the same bolt diameter.

One way to determine adequacy of pull-out capacity is

to plot the ratio of the internal diameter vs insert length as a function of the effective shear area developed in the parent material. The accompanying curves for three sizes of self-tapping and wire thread inserts were derived from tests in which the insert was pulled out of the parent material. Similar curves could be developed to determine the length needed for any other type of insert.

For example, assume that a $\frac{1}{4}$-28 bolt with an ultimate strength of 5000 lb is to be used in a material with a shear strength of 20,000 psi. The required shear area is 5000 lb/20,000 psi = 0.25 sq in. From the accompanying curves, the D/L ratio is 0.57; insert length, $L = 0.25/0.57 = 0.438$ in.

Similar calculations, using the same curves, can determine whether length of the insert is sufficient to give a required amount of creep resistance: The creep strength of the parent material is substituted for shear strength in the above calculation.

Also, if the insert length is limited, these calculations will give the available pull-out strength, which will vary with shear area of the insert. This analysis can be used to determine either the required length or pull-out strength, and from this, the thickness of the parent material for minimum weight and maximum economy.

Solid threaded bushings often permit using a shorter bolt than for the wire thread insert with limited shear area. With a large number of fasteners in an assembly, weight saving in reduction of parent material is much greater than the small extra weight added by the solid insert.

Other important factors in selecting inserts are assembly tooling, serviceability, relative cost, and ease of installation. These factors have been evaluated in the bar charts prepared by W. Moskowitz of GE's Missile and Space Vehicle Dept, Philadelphia. Data are for five types using 10-32 internal threads. Part of this information is based on estimates of the operating personnel concerning the number of assembly operations, tolerances required during installation, and relative ease of installation.

Applications of Helical Wire Inserts

Paul E. Wolfe

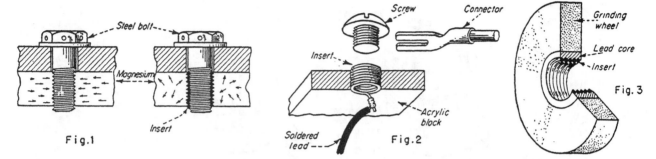

Fig. 1

Fig. 2

Fig. 3

Fig. 1—Galvanic action between steel bolt and magnesium part, (Left), attacks thread causing part failure. Stainless steel insert, (Right), reduces galvanic action to a negligible amount while strengthening threads.

Fig. 2—Insert used as an electrical connection as well as a thread reinforcement. Unit is threaded into plastic and tang is bent to form soldering lug.

Fig. 3—Direct connection of grinding wheel onto a threaded shaft by using a wire insert. Washers and nut are not required thus simplifying assembly.

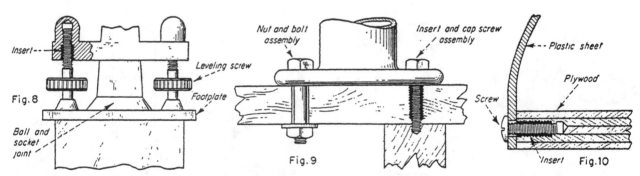

Fig. 8

Fig. 9

Fig. 10

Fig. 8—Wear and backlash can be reduced on adjusting threads. Clamping strength is not needed but intermittent thread travel makes reinforcement desirable.

Fig. 9—Combination of inserts and capscrews permits installation of machinery and other equipment on wood floors and walls. Access to opposite face of the wood is not a factor nor are joists and other obstructions.

Fig. 10—Plastic-to-wood connector. Repeated assembly and disassembly does not affect protected threads in wood or plastic parts.

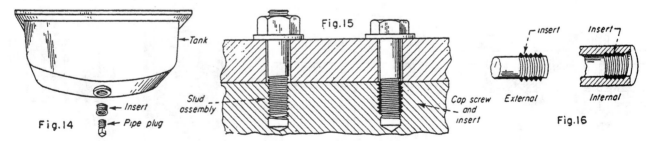

Fig. 14

Fig. 15

Fig. 16

Fig. 14—Seizure and corrosion of pipe threads on compression, fuel and lubricant tanks, pipe lines, fittings, pumps and boilers are prevented by using an insert.

Fig. 15—Stud (Left) transfers thread wear from the tapped hole and into the expendable threads on the stud and nut. Interference fit in the tapped hole is mandatory. Insert (Right) prevents wear and makes stud unnecessary. Lower cost cap screw can be used.

Fig. 16—Thread series can be changed from special to standard, from fine to coarse or vice versa, or corrected in case of a production error by redrilling and retapping. Inserts giving desired thread are then used.

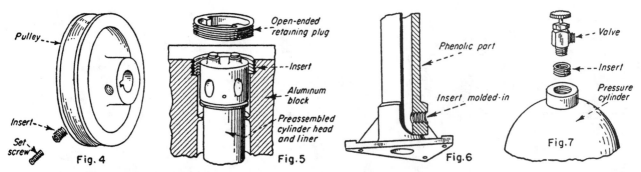

Fig. 4—Loosening of the set screw by vibration is reduced by using an insert. As pulley is a soft metal die casting set screw tightening often stripped threads.

Fig. 5—Insert withstands combustion thrust of diesel cylinder. It prevents heat seizure and scale on threaded plug making cylinder replacement and servicing easy.

Fig. 6—Phenolic part insert forms strong thread without tapping and drilling. Insert is resilient, does not crack phenolic or set up local stress concentrations.

Fig. 7—Enlargement of taper pipe threads in necks of pressure vessels, caused by frequent inspection and interchange of fittings, can be minimized.

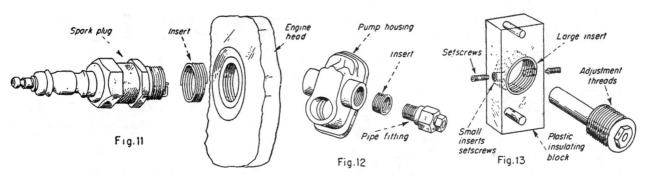

Fig. 11—Threads are protected from stripping in new aluminum engine heads by inserts. Also can be used to repair stripped spark plug holes in engine heads.

Fig. 12—Insert prevents pipe fittings from peeling chips out of tapped aluminum threads. Introduction of

chips into the lines could cause a malfunction.

Fig. 13—Center insert serves as brakeband to lock adjustable bushing. Small inserts keep set screws from stripping plastic when tightened. Also, adjustment threads can not be marred by end of set screw.

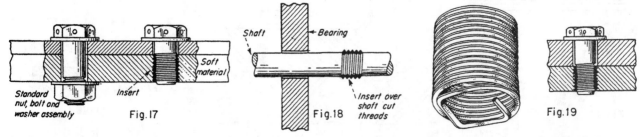

Fig. 17—Assembly weight may be reduced. Left view shows the standard method of attaching front and mid frame of a compressor. Lockwire is used after assembly is completed. Right view is new method resulting in weight and space economies.

Fig. 18—Assembly of a shaft through a bearing is sim-

plified by adding external insert over cut shaft threads. Machining the full shaft length is also unnecessary.

Fig. 19—Square tang insert—called screw lock—automatically locks the screw so that lock washers, nuts or wires are unnecessary. Insert locks itself into parent material without need for pins, rings, or staking.

SECTION 16

WASHERS

Ideas for Flat Washers

You can do more with washers than you may think.
Here are 10 ideas that may save the day next time you
need a simple, quick, inexpensive design.

Robert O. Parmley

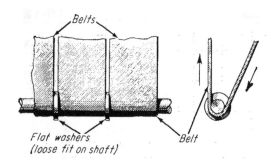

1 Are your belts overlapping? A flat washer, loosely fitted to the shaft, separates them.

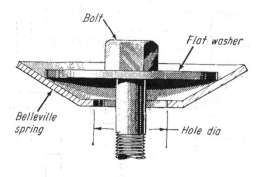

3 Need to hold odd-shaped parts? A flat washer and Belleville spring make a simple anchor.

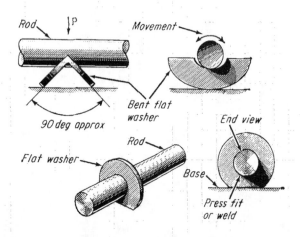

2 How about a rod support? A bent washer permits some rocking; if welded, the support is stable.

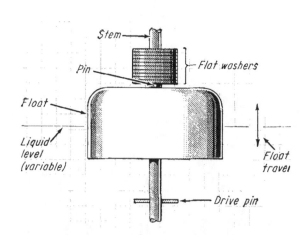

4 Got a weight problem? Adding or subtracting flat washers can easily control float action.

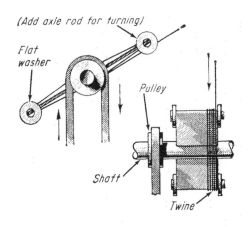

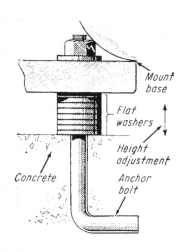

5 How about some simple flanges? Here the washers guide the twine and keep it under control.

8 Does your floor tilt? Stacked washers can level machines, or give a stable height adjustment.

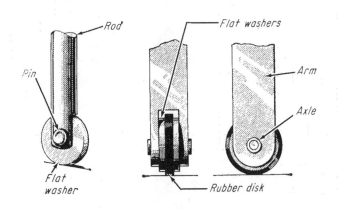

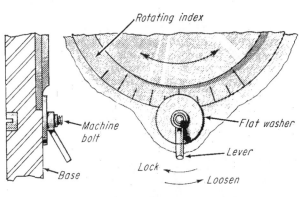

6 Need some wheels? Here flat washer is the wheel. A rubber disk quiets the assembly.

9 Here's a simple lock. A machine bolt, a washer, and a wing or lever nut make a strong clamp.

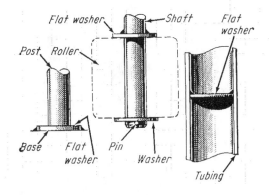

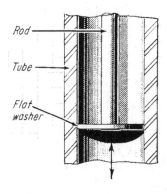

7 Want to avoid machining? Washers can make anchors, stop shoulders, even reduce tubing ID.

10 Need a piston in a hurry? For light service, tube, rod, and washer will be adequate.

Versatile Flat Washers:
can be used in a variety of unusual applications

Robert O. Parmley

Stiffen machine mount

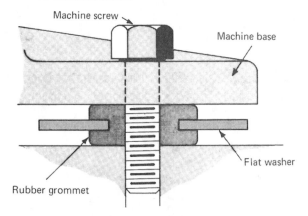

Machine screw

Machine base

Flat washer

Rubber grommet

Stabilize a point

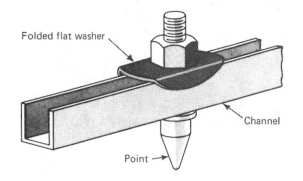

Folded flat washer

Channel

Point

Support a shaft

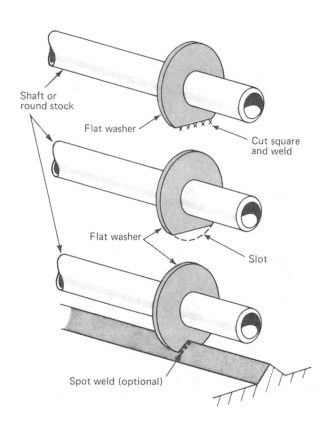

Shaft or round stock

Flat washer

Cut square and weld

Flat washer

Slot

Spot weld (optional)

Act as a valve seat

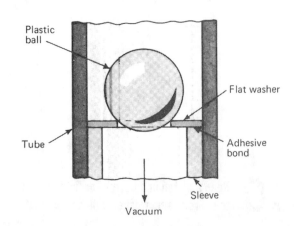

Plastic ball

Flat washer

Tube

Adhesive bond

Sleeve

Vacuum

Guide wire or tubing

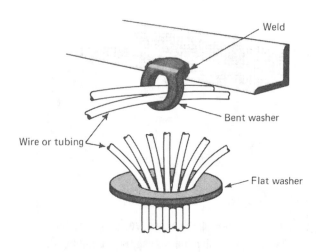

Weld

Bent washer

Wire or tubing

Flat washer

Washers are usually thought of as bearing surfaces placed under bolt heads. But they can be used in a variety of ways that could simplify a design or be an immediate fix until a designed part is available.

Reinforce a roller

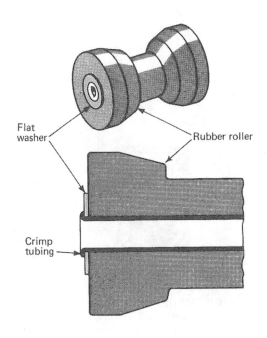

Flat washer

Rubber roller

Crimp tubing

Form a shock absorber

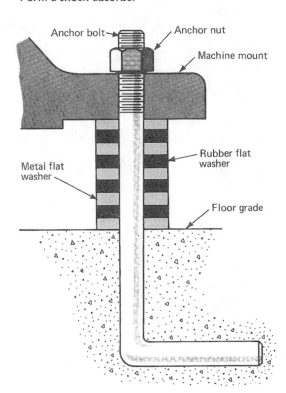

Anchor bolt

Anchor nut

Machine mount

Metal flat washer

Rubber flat washer

Floor grade

Act as a pulley flange

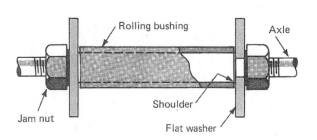

Rolling bushing

Axle

Jam nut

Shoulder

Flat washer

Form a pulley

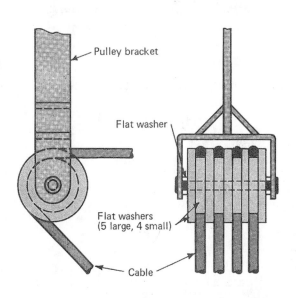

Pulley bracket

Flat washer

Flat washers (5 large, 4 small)

Cable

Act as a bearing surface

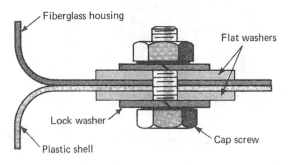

Fiberglass housing

Flat washers

Lock washer

Plastic shell

Cap screw

Jobs for Flat Rubber Washers

Rubber washers are more versatile than you think. Here are some odd jobs they can do that may make your next design job easier.

Robert O. Parmley

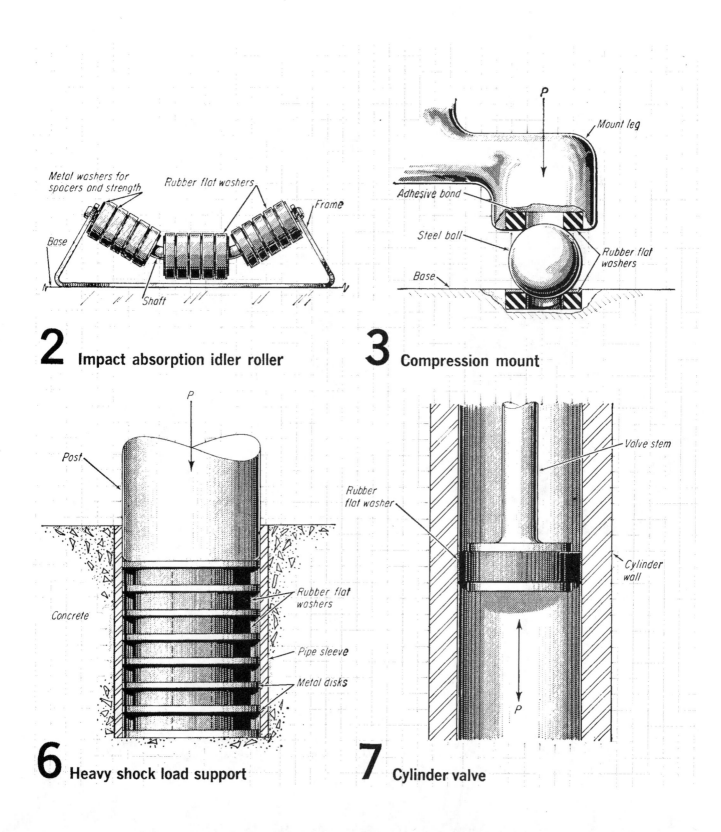

2 Impact absorption idler roller

3 Compression mount

6 Heavy shock load support

7 Cylinder valve

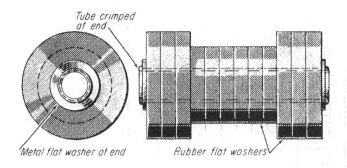

1 Step roller

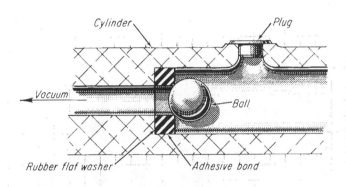

4 Compression ball seat

5 Hose bib retainer

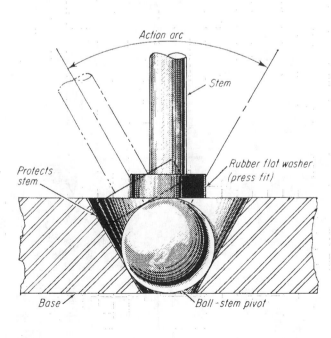

8 Protective bumper

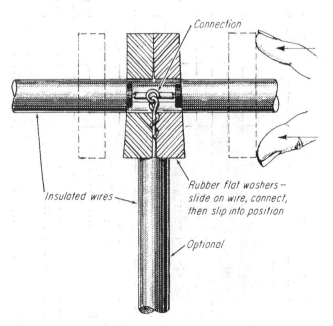

9 Expansion isolator

Take Another Look at Serrated Washers

They're a stock item and come in a variety of sizes.
With a little thought they can do a variety of jobs.
Here are just eight.

Robert O. Parmley

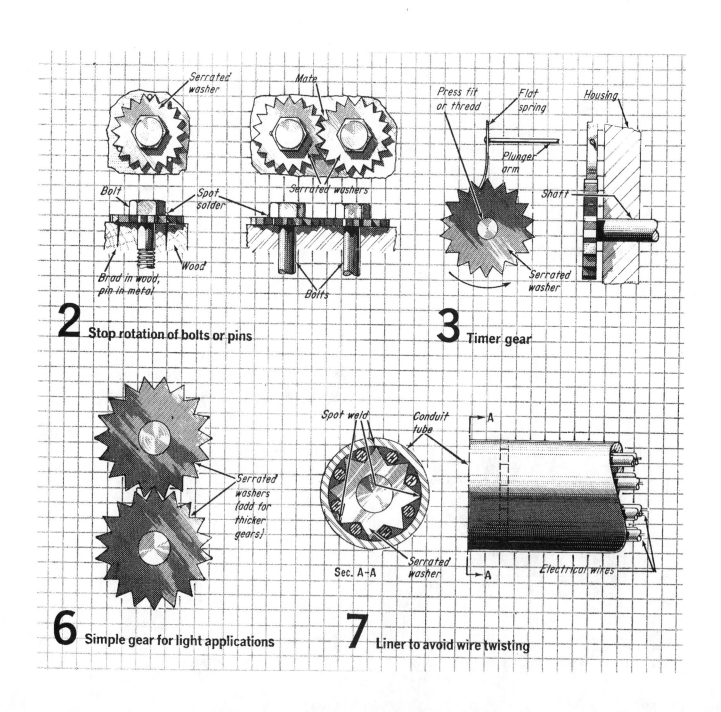

2 Stop rotation of bolts or pins

3 Timer gear

6 Simple gear for light applications

7 Liner to avoid wire twisting

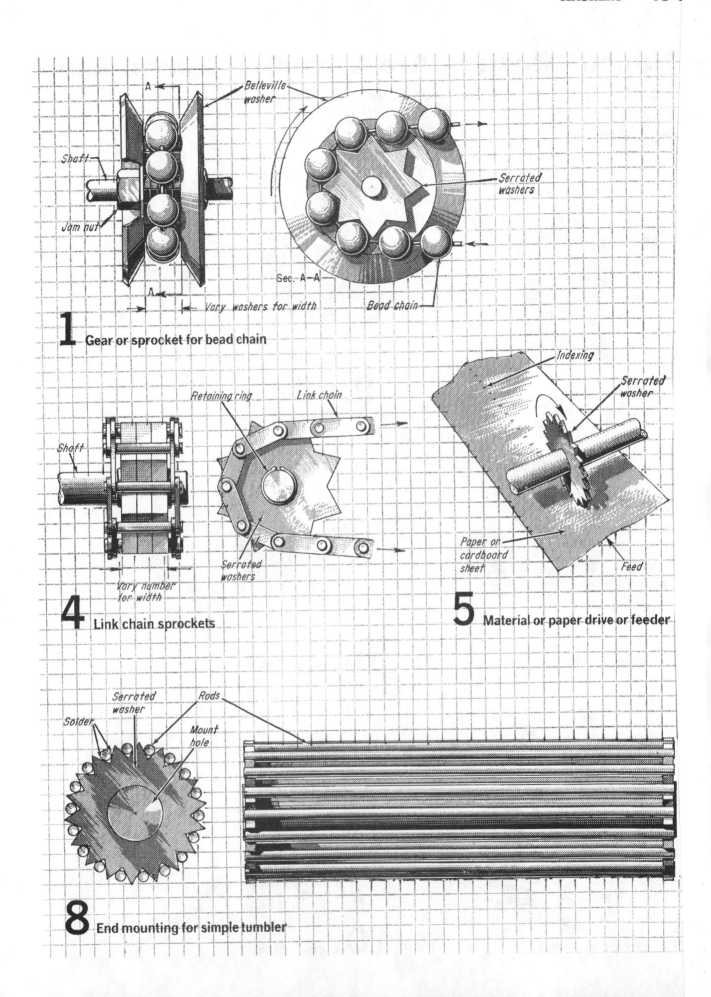

1 Gear or sprocket for bead chain

Belleville washer
Shaft
Jam nut
Serrated washers
Sec. A-A
Bead chain
Vary washers for width

4 Link chain sprockets

Retaining ring
Link chain
Shaft
Serrated washers
Vary number for width

5 Material or paper drive or feeder

Indexing
Serrated washer
Paper or cardboard sheet
Feed

8 End mounting for simple tumbler

Serrated washer
Rods
Solder
Mount hole

Dished Washers Are Versatile Components

Let these ideas spur your own design creativity. Sometimes commercial Belleville washers will suit; other wise you can easily dish your own.

Robert O. Parmley

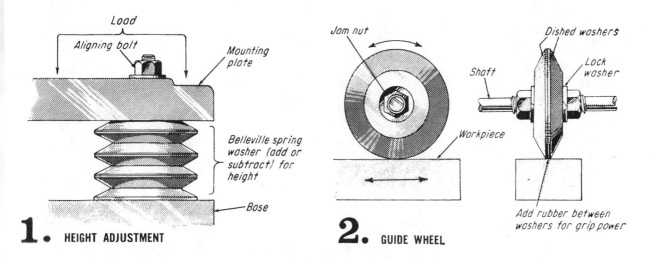

1. HEIGHT ADJUSTMENT

2. GUIDE WHEEL

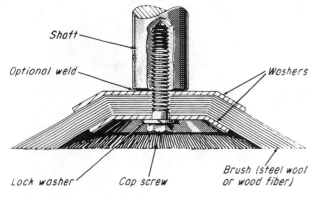

5. BRUSH RETAINER

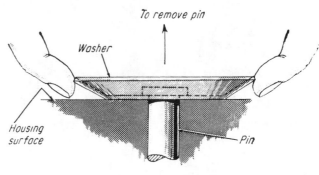

6. EASY PIN REMOVAL

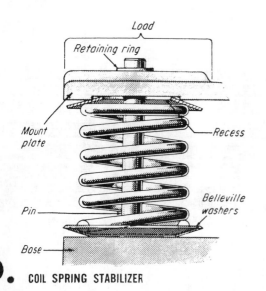

9. COIL SPRING STABILIZER

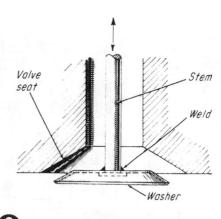

10. SIMPLE VALVE

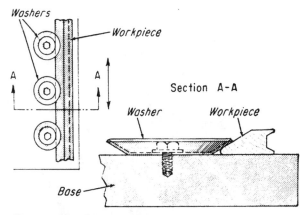

3. **ALIGNING BUTTONS**
will rotate if holding screw is shouldered

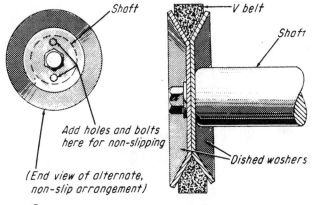

4. **V-BELT PULLEY**

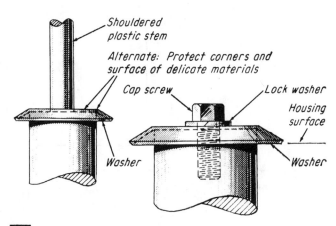

7. **END AND CORNER PROTECTION**

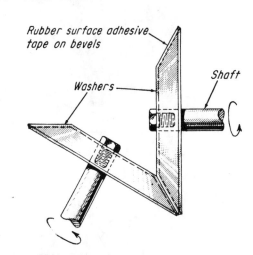

8. **SIMPLE BEVEL DRIVE**

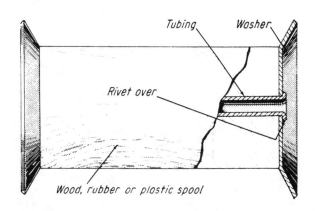

11. **FLARED SPOOL-FLANGES**

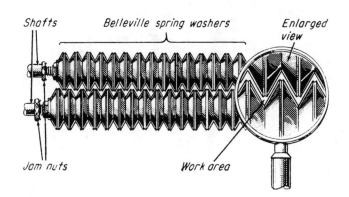

12. **CORRUGATING ROLLERS FOR PAPER OR CARDBOARD**

Design Problems Solved with Belleville Spring Washers

Robert O. Parmley

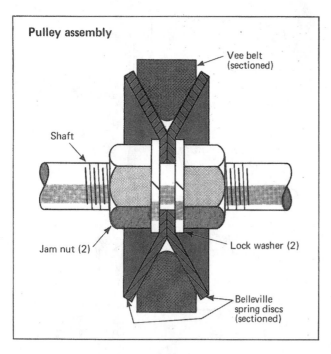

Pulley assembly

Vee belt (sectioned)

Shaft

Jam nut (2)

Lock washer (2)

Belleville spring discs (sectioned)

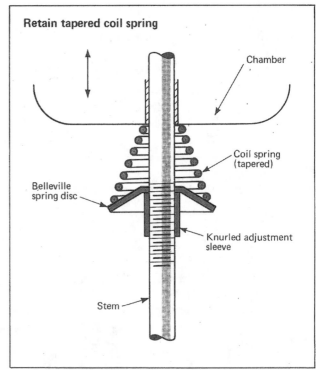

Retain tapered coil spring

Chamber

Coil spring (tapered)

Belleville spring disc

Knurled adjustment sleeve

Stem

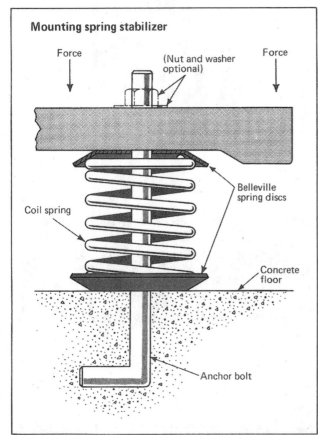

Mounting spring stabilizer

Force

(Nut and washer optional)

Force

Belleville spring discs

Coil spring

Concrete floor

Anchor bolt

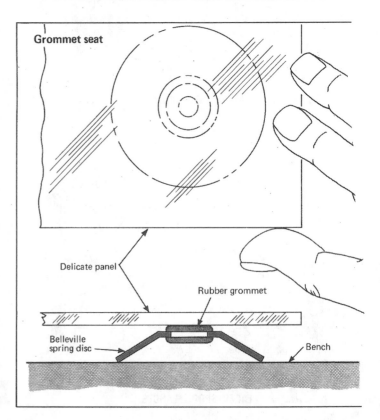

Grommet seat

Delicate panel

Rubber grommet

Belleville spring disc

Bench

Belleville springs are a versatile component that offer a wide range of applications. There are many places where these components can be used and their availability as a stock item should be considered when confronted with a design problem that requires a fast solution.

Secure anchor bolt

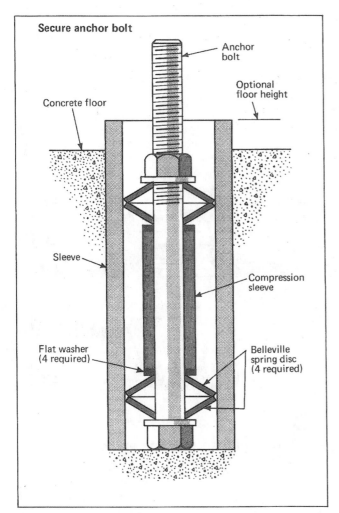

Anchor bolt

Optional floor height

Concrete floor

Sleeve

Compression sleeve

Flat washer (4 required)

Belleville spring disc (4 required)

Clamp fixture spring

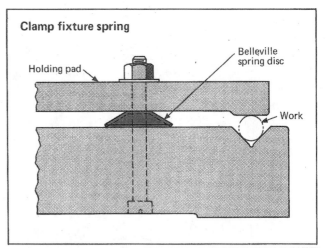

Belleville spring disc

Holding pad

Work

Lock retainer

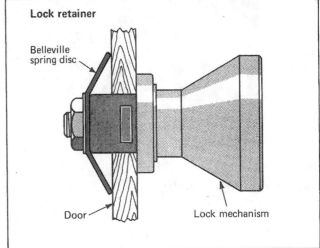

Belleville spring disc

Door

Lock mechanism

Machine leg spring mounts

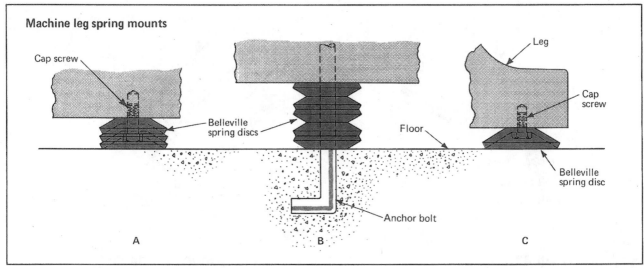

Cap screw

Belleville spring discs

Leg

Cap screw

Floor

Belleville spring disc

Anchor bolt

A   B   C

Creative Ideas for Cupped Washers

A standard off-the-shelf item with more uses than many ever considered.

Robert O. Parmley

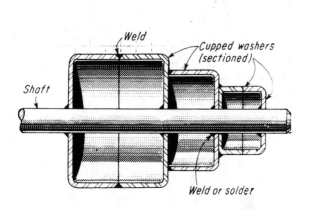

2 Simple step pulley

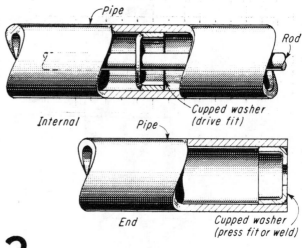

3 Rod aligner and pipe-end bearing

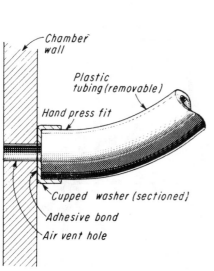

6 Tubing connector

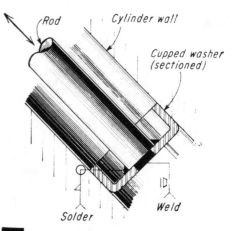

7 Simple piston for cylinder

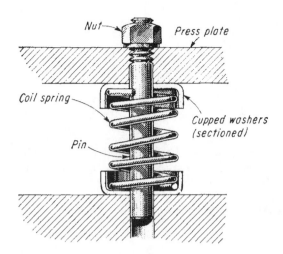

1 Coil spring stabilizer and compression brake

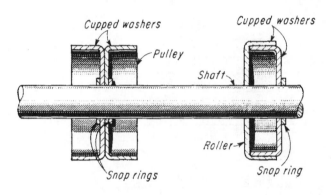

4 Simple pulley and roller

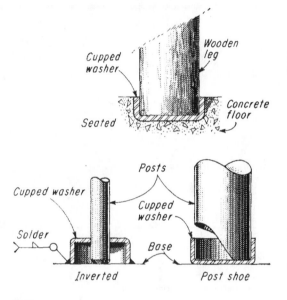

5 Post anchors and supports

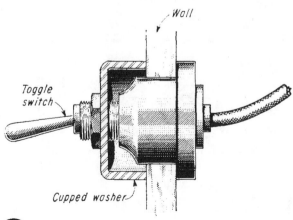

8 Toggle switch housing

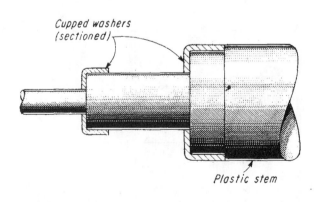

9 Protection for step shoulders

SEM Applications

N. Dale Long

When a split lockwasher is called for in a screw fastening, a flat washer is invariably necessary. The ways of assembling them illustrated below are strict requirements in military specifications—especially for electronic equipment. Commercial requirements usually vary—depending upon either the designer's decision or product-cost restrictions. For good quality and reliable service, however, the fastening methods shown here can be depended on to pay off.

ASSEMBLY OF FLAT WASHERS AND SPLIT LOCK WASHERS

Flat washers should be placed between . . .

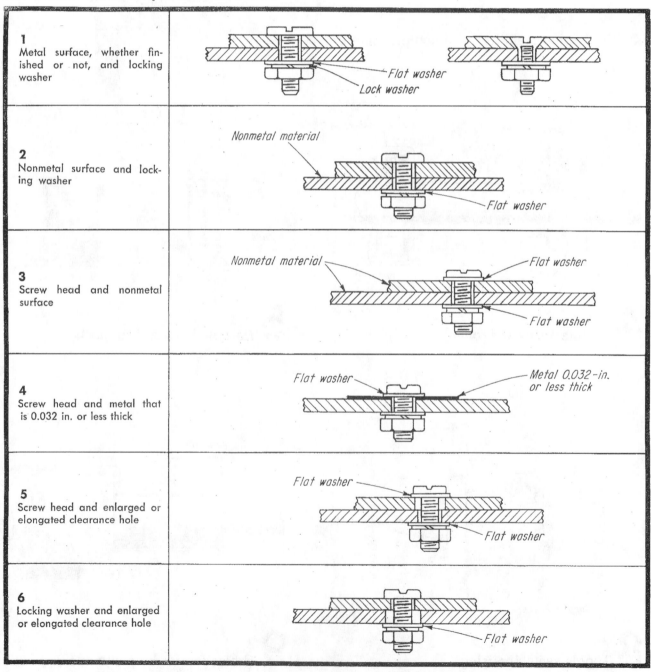

1
Metal surface, whether finished or not, and locking washer

2
Nonmetal surface and locking washer

3
Screw head and nonmetal surface

4
Screw head and metal that is 0.032 in. or less thick

5
Screw head and enlarged or elongated clearance hole

6
Locking washer and enlarged or elongated clearance hole

SEM Standard Tables

Pan Head Machine Screw

Fillister Head Type D Tapping Screw

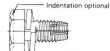

Indentation optional

Hex Head [Note (1)] Type F Tapping Screw

Hex Washer Head Type T Tapping Screw

Hex Cap Screw

Representative Examples of Helical Spring Lock Washer Sems

DIMENSIONS OF HELICAL SPRING LOCK WASHERS FOR SEMS WITH MACHINE AND TAPPING SCREWS HAVING MACHINE SCREW THREAD DIAMETER-PITCH COMBINATIONS

Nominal Size [Note (2)] or Basic Screw Diameter		Washer Inside Diameter		Pan Head Screw				Fillister Head Screw				Hex Head Screw [Note (1)]				Hex Washer Head Screw			
				Washer Section		Washer Outside Diameter		Washer Section		Washer Outside Diameter		Washer Section		Washer Outside Diameter		Washer Section		Washer Outside Diameter	
				Width	Thickness			Width	Thickness			Width	Thickness			Width	Thickness		
		Max.	Min.	Min.	Min.	Max.	Min.	Min.	Min.	Max.	Min.	Min.	Min.	Max.	Min.	Min.	Min.	Max.	Min.
2	0.0860	0.080	0.075	0.035	0.020	0.156	0.145	0.030	0.020	0.146	0.135	0.030	0.020	0.146	0.135	0.035	0.020	0.156	0.145
3	0.0990	0.091	0.086	0.040	0.025	0.178	0.166	0.035	0.020	0.168	0.156	0.035	0.020	0.168	0.156	0.040	0.025	0.178	0.166
4	0.1120	0.106	0.101	0.047	0.031	0.208	0.195	0.035	0.020	0.184	0.171	0.040	0.025	0.192	0.181	0.047	0.031	0.208	0.195
5	0.1250	0.118	0.113	0.047	0.031	0.220	0.207	0.035	0.020	0.196	0.183	0.047	0.031	0.220	0.207	0.047	0.031	0.220	0.207
6	0.1380	0.129	0.124	0.062	0.034	0.261	0.248	0.047	0.031	0.231	0.218	0.047	0.031	0.231	0.218	0.062	0.034	0.261	0.248
8	0.1640	0.155	0.149	0.078	0.031	0.319	0.305	0.047	0.031	0.257	0.243	0.055	0.040	0.271	0.259	0.078	0.031	0.319	0.305
10	0.1900	0.179	0.173	0.093	0.047	0.373	0.359	0.055	0.040	0.297	0.283	0.062	0.047	0.311	0.297	0.093	0.047	0.373	0.359
12	0.2160	0.203	0.196	0.109	0.062	0.429	0.414	0.062	0.047	0.335	0.320	0.070	0.056	0.351	0.336	0.109	0.062	0.429	0.414
1/4	0.2500	0.238	0.230	0.125	0.062	0.496	0.480	0.077	0.063	0.400	0.384	0.109	0.062	0.464	0.448	0.125	0.062	0.496	0.480
5/16	0.3125	0.298	0.290	0.156	0.078	0.618	0.602	0.109	0.062	0.524	0.508	0.125	0.078	0.556	0.540	0.156	0.078	0.618	0.602
3/8	0.3750	0.361	0.353	0.171	0.093	0.711	0.695	0.125	0.062	0.619	0.603	0.141	0.094	0.651	0.635	0.171	0.093	0.711	0.695
7/16	0.4375	0.420	0.411	...	...	...	...	...	...	...	...	0.156	0.109	0.740	0.723	...	...	...	...
1/2	0.5000	0.482	0.473	...	...	...	...	...	...	...	...	0.171	0.125	0.834	0.815	...	...	...	...

Pan Head Type AB Tapping Screw

Fillister Head Type BF Tapping Screw

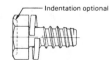

Indentation optional

Hex Head [Note (1)] Type B Tapping Screw

Hex Washer Head Type BT Tapping Screw

Representative Examples of Helical Spring Lock Washer Sems

DIMENSIONS OF HELICAL SPRING LOCK WASHERS FOR SEMS WITH SCREWS HAVING TYPE B TAPPING SCREW THREAD DIAMETER-PITCH COMBINATIONS

Nominal Size [Note (2)] or Basic Screw Diameter		Washer Inside Diameter		Pan Head Screw				Fillister Head Screw				Hex Head Screw [Note (1)]				Hex Washer Head Screw			
				Washer Section		Washer Outside Diameter		Washer Section		Washer Outside Diameter		Washer Section		Washer Outside Diameter		Washer Section		Washer Outside Diameter	
				Width	Thickness			Width	Thickness			Width	Thickness			Width	Thickness		
		Max.	Min.	Min.	Min.	Max.	Min.	Min.	Min.	Max.	Min.	Min.	Min.	Max.	Min.	Min.	Min.	Max.	Min.
4	0.1120	0.101	0.096	0.047	0.031	0.201	0.190	0.035	0.020	0.179	0.166	0.040	0.025	0.187	0.176	0.047	0.031	0.201	0.190
5	0.1250	0.112	0.107	0.050	0.034	0.218	0.207	0.035	0.020	0.190	0.177	0.047	0.031	0.214	0.201	0.050	0.034	0.218	0.207
6	0.1380	0.121	0.116	0.062	0.034	0.253	0.240	0.047	0.031	0.223	0.210	0.047	0.031	0.223	0.210	0.062	0.034	0.253	0.240
7	0.1510	0.135	0.130	0.062	0.034	0.267	0.254	0.047	0.031	0.237	0.224	0.047	0.031	0.237	0.224	0.062	0.034	0.267	0.254
8	0.1640	0.144	0.138	0.078	0.031	0.308	0.294	0.047	0.031	0.246	0.232	0.055	0.040	0.262	0.248	0.078	0.031	0.308	0.294
10	0.1900	0.162	0.156	0.081	0.056	0.332	0.318	0.055	0.040	0.280	0.266	0.062	0.047	0.294	0.280	0.081	0.056	0.332	0.318
12	0.2160	0.188	0.181	0.081	0.056	0.358	0.343	0.062	0.047	0.320	0.305	0.070	0.056	0.336	0.321	0.081	0.056	0.358	0.343
1/4	0.2500	0.217	0.209	0.120	0.062	0.465	0.449	0.077	0.063	0.379	0.363	0.109	0.062	0.443	0.427	0.120	0.062	0.465	0.449
5/16	0.3125	0.278	0.270	0.125	0.078	0.536	0.520	0.109	0.062	0.504	0.488	0.125	0.078	0.536	0.520	0.125	0.078	0.536	0.520
3/8	0.3750	0.338	0.330	0.141	0.094	0.628	0.612	0.125	0.062	0.596	0.580	0.141	0.094	0.628	0.612	0.141	0.094	0.628	0.612
7/16	0.4375	0.397	0.388	...	...	...	...	...	...	...	...	0.156	0.109	0.716	0.700	...	...	...	...
1/2	0.5000	0.460	0.451	...	...	...	...	...	...	...	...	0.171	0.125	0.812	0.793	...	...	...	...

GENERAL NOTES:
(a) For additional requirements, refer to Sections 2 and 3.
(b) Dimensions of helical spring lock washers applicable to "not recommended for new design" status round and truss head screw sems and Type A tapping screw sems are documented for reference purposes in Appendix B.

NOTES:
(1) The regular trimmed or upset hex head screws shall apply for the washers shown. Where upset large hex head screws in sizes No. 4, 5, 8, 12, and 1/4 in. are specified by the purchaser, the washers shown for the corresponding size hex washer head screw shall apply. Refer to appropriate tables for hex head tapping screws in ASME B18.6.4.
(2) Where specifying nominal size in decimals, zeros preceding the decimal and in the fourth place shall be omitted.

SEM Standard Tables (continued)

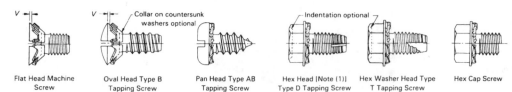

| Flat Head Machine Screw | Oval Head Type B Tapping Screw | Pan Head Type AB Tapping Screw | Hex Head [Note (1)] Type D Tapping Screw | Hex Washer Head Type T Tapping Screw | Hex Cap Screw |

Representative Examples of External Tooth Lock Washer Sems

DIMENSIONS OF EXTERNAL TOOTH LOCK WASHERS FOR SEMS

Nominal Size [Note (2)] or Basic Screw Diameter		Flat and Oval Head Screws			Pan Head Screw				Hex Head and Hex Cap Screws [Note (1)]				Hex Washer Head Screw			
		Washer Thickness		V Flush to Minus Tolerance of	Washer Thickness		Washer Outside Diameter		Washer Thickness		Washer Outside Diameter		Washer Thickness		Washer Outside Diameter	
		Max.	Min.		Max.	Min.	Max.	Min.	Max.	Min.	Max.	Min.	Max.	Min.	Max.	Min.
2	0.0860	...	...	...	0.016	0.010	0.180	0.170	0.016	0.010	0.180	0.170	0.016	0.010	0.180	0.170
3	0.0990	0.014	0.009	0.019	0.016	0.010	0.205	0.195	0.016	0.010	0.205	0.195	0.016	0.010	0.205	0.195
4	0.1120	0.014	0.009	0.022	0.018	0.012	0.230	0.220	0.018	0.012	0.230	0.220	0.018	0.012	0.230	0.220
5	0.1250	0.019	0.015	0.024	0.020	0.014	0.255	0.245	0.020	0.014	0.255	0.245	0.020	0.014	0.255	0.245
6	0.1380	0.020	0.016	0.026	0.022	0.016	0.285	0.270	0.022	0.016	0.285	0.270	0.022	0.016	0.317	0.306
7	0.1510	0.020	0.016	0.047	0.022	0.016	0.285	0.270	0.022	0.016	0.285	0.270	0.022	0.016	0.317	0.306
8	0.1640	0.020	0.016	0.030	0.023	0.018	0.320	0.305	0.023	0.018	0.320	0.305	0.023	0.018	0.317	0.306
10	0.1900	0.025	0.019	0.036	0.024	0.018	0.381	0.365	0.024	0.018	0.381	0.365	0.024	0.018	0.406	0.395
12	0.2160	0.025	0.019	0.041	0.027	0.020	0.410	0.395	0.027	0.020	0.410	0.395	0.027	0.020	0.406	0.395
1/4	0.2500	0.025	0.019	0.047	0.028	0.023	0.510	0.494	0.028	0.023	0.475	0.460	0.028	0.023	0.580	0.567
5/16	0.3125	0.028	0.023	0.060	0.034	0.028	0.610	0.588	0.034	0.028	0.580	0.567	0.034	0.028	0.654	0.640
3/8	0.3750	0.034	0.028	0.072	0.040	0.032	0.760	0.740	0.040	0.032	0.660	0.640	0.040	0.032	0.760	0.740

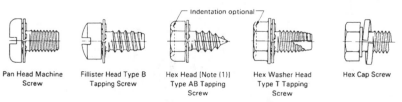

| Pan Head Machine Screw | Fillister Head Type B Tapping Screw | Hex Head [Note (1)] Type AB Tapping Screw | Hex Washer Head Type T Tapping Screw | Hex Cap Screw |

Representative Examples of Internal Tooth Lock Washer Sems

DIMENSIONS OF INTERNAL TOOTH LOCK WASHERS FOR SEMS

Nominal Size [Note (3)] or Basic Screw Diameter		Pan, Fillister, Hex [Note (1)], and Hex Washer [Note (2)] Head Screws			
		Washer Thickness		Washer Outside Diameter	
		Max.	Min.	Max.	Min.
2 [Note (2)]	0.0860	0.016	0.010	0.185	0.175
3 [Note (2)]	0.0990	0.016	0.010	0.225	0.215
4 [Note (2)]	0.1120	0.018	0.012	0.268	0.258
5 [Note (2)]	0.1250	0.018	0.012	0.268	0.258
6	0.1380	0.022	0.016	0.288	0.278
7	0.1510	0.022	0.016	0.288	0.278
8	0.1640	0.023	0.018	0.338	0.327
10	0.1900	0.024	0.018	0.383	0.372
12	0.2160	0.027	0.020	0.408	0.396
1/4	0.2500	0.028	0.023	0.478	0.466
5/16	0.3125	0.034	0.028	0.610	0.597
3/8	0.3750	0.040	0.032	0.692	0.678

GENERAL NOTES:
(a) For additional requirements, refer to Sections 2 and 4.
(b) Dimensions of internal tooth lock washers applicable to "not recommended for new design" status round and truss head screw sems and Types A and C tapping screw sems are documented for reference purposes in Appendix C.

NOTES:
(1) The regular trimmed or upset hex head screws shall apply except for sizes No. 4, 5, 8, 12, and 1/4 in., which shall have upset large hex head screws. Refer to appropriate tables for hex head machine and tapping screws in ASME B18.6.3 and ASME B18.6.4, respectively. Refer to appropriate tables for hex cap screws in ASME B18.2.1.
(2) Hex washer head sems are not available in sizes smaller than No. 6.
(3) Where specifying nominal size in decimals, zeros preceding the decimal and in the fourth place shall be omitted.

SEM Standard Tables (continued)

Pan Head Type AB
Tapping Screw and
Type L Regular Washer

Fillister Head Type B
Tapping Screw and
Type L Narrow Washer

Hex Head [Note (1)]
Machine Screw and
Type H Regular Washer

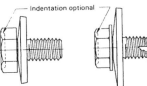
Hex Washer Head Type
T Tapping Screw and
Type H Wide Washer

Hex Cap Screw

Representative Examples of Conical Spring Washer Sems

DIMENSIONS OF CONICAL SPRING WASHERS FOR SEMS

Nominal Size [Note (2)] or Basic Screw Diameter		Washer Series	Washer Outside Diameter		Type L Washer					Type H Washer				
					Thickness			Crown Height		Thickness			Crown Height	
			Max.	Min.	Basic	Max.	Min.	Max.	Min.	Basic	Max.	Min.	Max.	Min.
6	0.1380	Narrow	0.320	0.307	0.025	0.029	0.023	0.025	0.015	0.035	0.040	0.033	0.025	0.015
		Regular	0.446	0.433	0.030	0.034	0.028	0.025	0.015	0.040	0.046	0.037	0.025	0.015
		Wide	0.570	0.557	0.030	0.034	0.028	0.031	0.021	0.040	0.046	0.037	0.029	0.019
8	0.1640	Narrow	0.383	0.370	0.035	0.040	0.033	0.025	0.015	0.040	0.046	0.037	0.025	0.015
		Regular	0.508	0.495	0.035	0.040	0.033	0.030	0.020	0.045	0.050	0.042	0.026	0.016
		Wide	0.640	0.620	0.035	0.040	0.033	0.037	0.027	0.045	0.050	0.042	0.040	0.030
10	0.1900	Narrow	0.446	0.433	0.035	0.040	0.033	0.025	0.015	0.050	0.056	0.047	0.025	0.015
		Regular	0.570	0.557	0.040	0.046	0.037	0.027	0.017	0.055	0.060	0.052	0.026	0.016
		Wide	0.765	0.743	0.040	0.046	0.037	0.036	0.026	0.055	0.060	0.052	0.034	0.024
12	0.2160	Narrow	0.446	0.433	0.040	0.046	0.037	0.025	0.015	0.055	0.060	0.052	0.025	0.015
		Regular	0.640	0.620	0.040	0.046	0.037	0.033	0.023	0.055	0.060	0.052	0.026	0.016
		Wide	0.890	0.868	0.045	0.050	0.042	0.044	0.034	0.064	0.071	0.059	0.033	0.023
1/4	0.2500	Narrow	0.515	0.495	0.045	0.050	0.042	0.025	0.015	0.064	0.071	0.059	0.025	0.015
		Regular	0.765	0.743	0.050	0.056	0.047	0.033	0.023	0.079	0.087	0.074	0.032	0.022
		Wide	1.015	0.993	0.055	0.060	0.052	0.040	0.030	0.079	0.087	0.074	0.039	0.029
5/16	0.3125	Narrow	0.640	0.620	0.055	0.060	0.052	0.026	0.016	0.079	0.087	0.074	0.026	0.016
		Regular	0.890	0.868	0.064	0.071	0.059	0.041	0.031	0.095	0.103	0.090	0.029	0.019
		Wide	1.140	1.118	0.064	0.071	0.059	0.044	0.034	0.095	0.103	0.090	0.040	0.030
3/8	0.3750	Narrow	0.765	0.743	0.071	0.079	0.066	0.025	0.015	0.095	0.103	0.090	0.025	0.015
		Regular	1.015	0.993	0.071	0.079	0.066	0.043	0.033	0.118	0.126	0.112	0.033	0.023
		Wide	1.265	1.243	0.079	0.087	0.074	0.047	0.037	0.118	0.126	0.112	0.045	0.035
7/16	0.4375	Narrow	0.890	0.868	0.079	0.087	0.074	0.028	0.018	0.128	0.136	0.122	0.026	0.016
		Regular	1.140	1.118	0.095	0.103	0.090	0.041	0.031	0.128	0.136	0.122	0.038	0.028
		Wide	1.530	1.493	0.095	0.103	0.090	0.059	0.049	0.132	0.140	0.126	0.049	0.039
1/2	0.5000	Narrow	1.015	0.993	0.100	0.108	0.094	0.031	0.021	0.142	0.150	0.136	0.030	0.020
		Regular	1.265	1.243	0.111	0.120	0.106	0.043	0.033	0.142	0.150	0.136	0.037	0.027
		Wide	1.780	1.743	0.111	0.120	0.106	0.062	0.052	0.152	0.160	0.146	0.052	0.042

Pan, Fillister, Hex [Note (1)], Hex Washer Head, and Hex Cap Screws

SECTION 17

O-RINGS

8 Unusual Applications for O-Rings

Playing many different roles, O-rings can perform as protective devices, hole liners, float stops, and other key design-components.

Robert O. Parmley

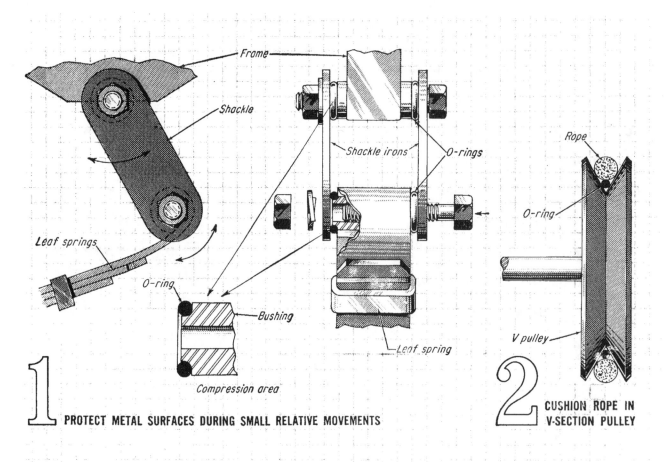

1 PROTECT METAL SURFACES DURING SMALL RELATIVE MOVEMENTS

Frame

Shackle

Shackle irons

O-rings

Leaf springs

O-ring

Bushing

Leaf spring

Compression area

2 CUSHION ROPE IN V-SECTION PULLEY

Rope

O-ring

V pulley

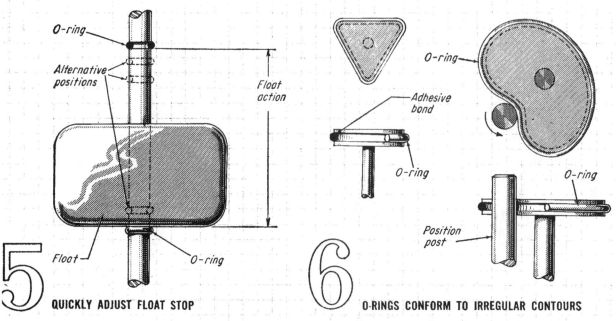

5 QUICKLY ADJUST FLOAT STOP

O-ring

Alternative positions

Float action

Float

O-ring

6 O-RINGS CONFORM TO IRREGULAR CONTOURS

O-ring

Adhesive bond

O-ring

Position post

O-ring

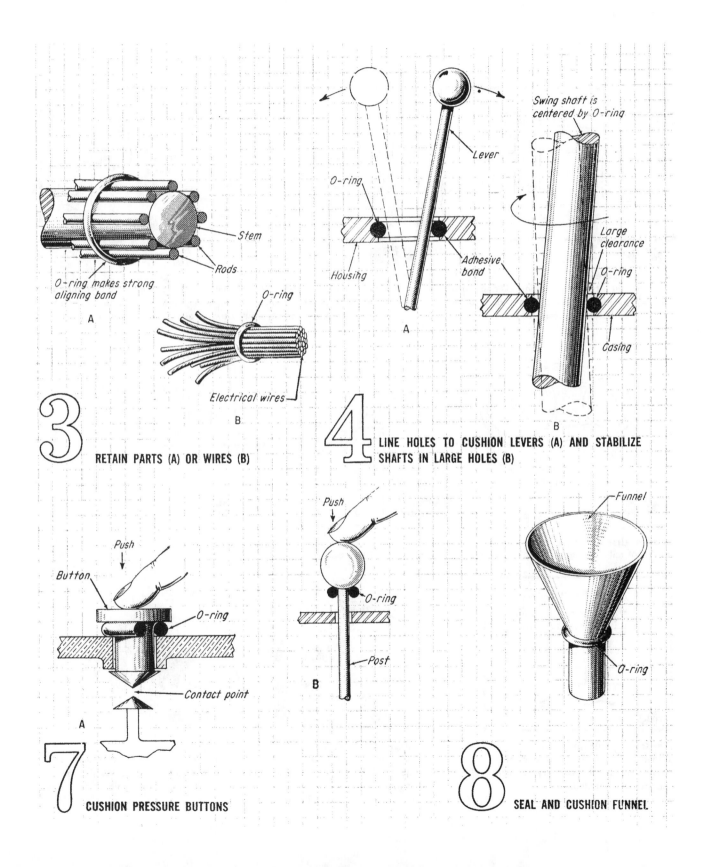

3 RETAIN PARTS (A) OR WIRES (B)

O-ring makes strong aligning band

A

Stem

Rods

O-ring

Electrical wires

B

4 LINE HOLES TO CUSHION LEVERS (A) AND STABILIZE SHAFTS IN LARGE HOLES (B)

Lever

O-ring

Housing

Adhesive bond

A

Swing shaft is centered by O-ring

Large clearance

O-ring

Casing

B

7 CUSHION PRESSURE BUTTONS

Push

Button

O-ring

Contact point

A

Push

O-ring

Post

B

8 SEAL AND CUSHION FUNNEL

Funnel

O-ring

16 Unusual Applications for the O-Ring

This handy little component finds a place in pumps, drives, glands, shock-mounts, pivots, knobs, valves and seals.

James F. Machen

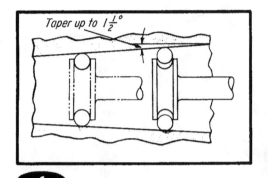

1

Tapered bore . . .
in diecasting, plus loose-fitting O-ring, gives low-cost pump for low-pressure applications. Example: carburetor accelerator-pump.

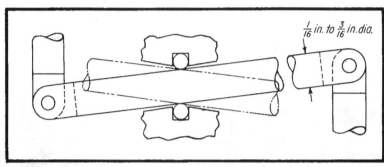

2

Sealed pivot . . .
allows transmittal of multidirectional, mechanical movement to hydraulically or pneumatically isolated system. For high-temperature seals, silicone rubber can often solve the problem—but always guard against excessive "set.'

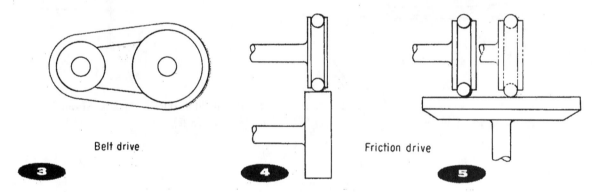

Belt drive

3

Friction drive

4

5

Simple drives . . .
utilize not only O-ring but its physical properties also—high friction and elasticity.

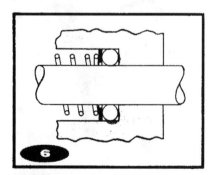

6

Single-ring gland . . .
is ideal for low pressures and high-viscosity fluids. If necessary, another ring may be installed.

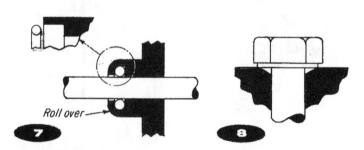

Roll over

7

8

Shaft seal . . .
may be held by rolling the thin body-wall over the O-ring. Bolt seal (8) is squeezed into countersink when bolt is tightened. Cross-sectional area of countersink must not be less than that of O-ring since molded rubber is practically incompressible when confined.

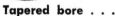

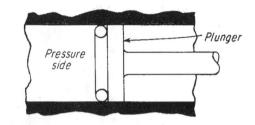

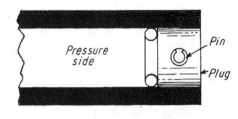

9

10

One-way pressure . . .
applications require O-ring seals to be supported on pressure side only. Seal may be movable (9) as in grease gun, or static (10) as in pipe plug. Anchor ring to plunger and plug for greater convenience and reliability.

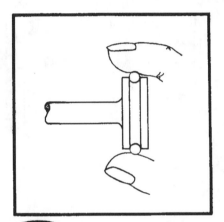

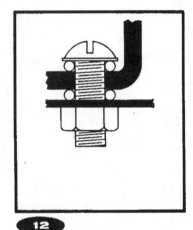

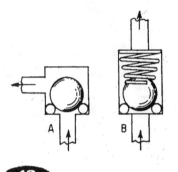

11

12

13

Friction grip . . .
on knob not only allows better grip but insulates fingers from heat or electricity. It also improves appearance on both mock-ups and working models.

Miniature shock-mount . . .
will isolate equipment from vibrations in accordance with behavior of visco-elastic materials

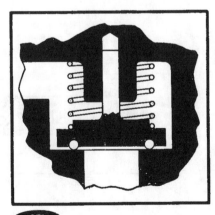

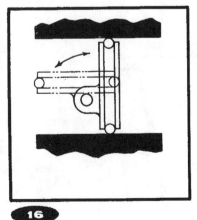

14

Checkvalves . . .
may have ball free (A); or spring-loaded (B). Back pressure will always force ball onto seat provided that gravity first helps locate ball on seat. Heavier-duty checkvalve (14) can be opened to allow back pressure to escape if necessary for shutdown etc.

15

16

High-pressure checkvalve . . .
shown here cannot allow release of back-pressure but could be easily modified to do this by letting valve stem protrude.

Butterfly valve . . .
can become a checkvalve if it is unbalanced; otherwise, it will act as normal two-way valve.

Look at O-Rings Differently

Sure they're seals, but they can also do a variety of other jobs
as well as more sophisticated pieces of hardware

Robert O. Parmley

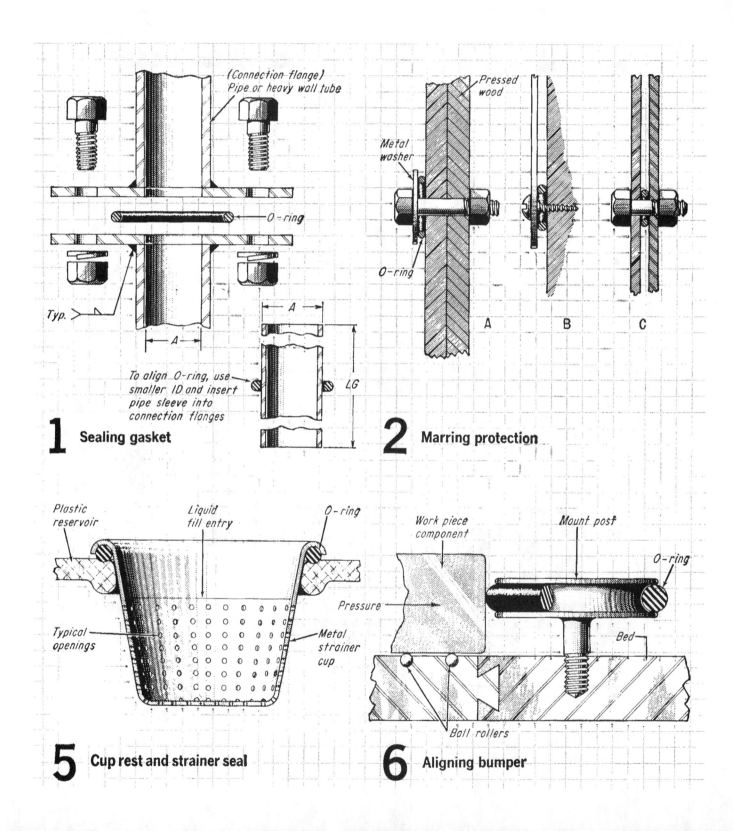

1 Sealing gasket

2 Marring protection

5 Cup rest and strainer seal

6 Aligning bumper

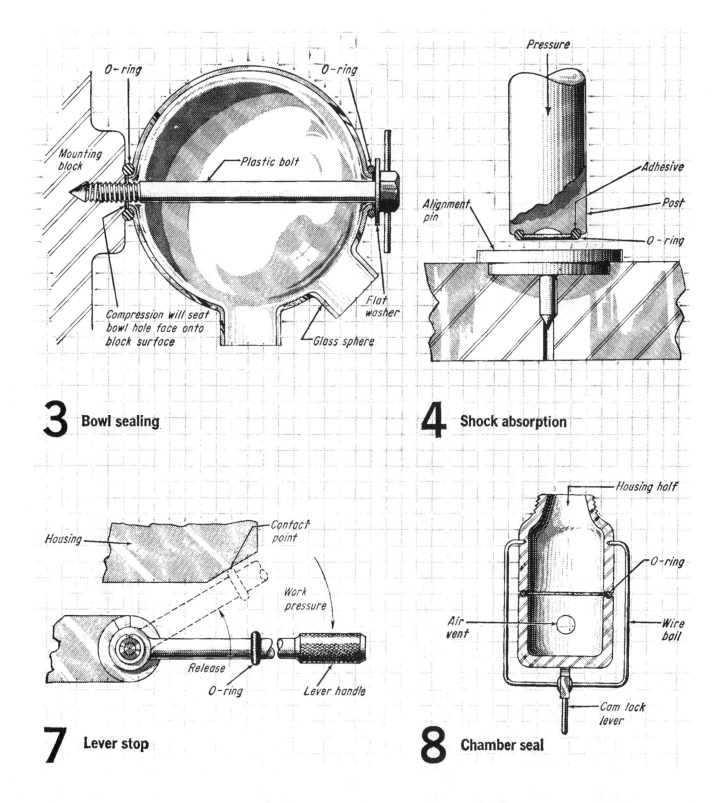

3 Bowl sealing

O-ring

O-ring

Mounting block

Plastic bolt

Compression will seat bowl hole face onto block surface

Flat washer

Glass sphere

4 Shock absorption

Pressure

Adhesive

Post

O-ring

Alignment pin

7 Lever stop

Housing

Contact point

Work pressure

Release

O-ring

Lever handle

8 Chamber seal

Housing half

O-ring

Air vent

Wire bail

Cam lock lever

O-Rings Solve Design Problems I

Rubber rings provide for thermal expansion, protect surfaces, seal pipe ends
and connections, and prevent slipping.

Robert O. Parmley

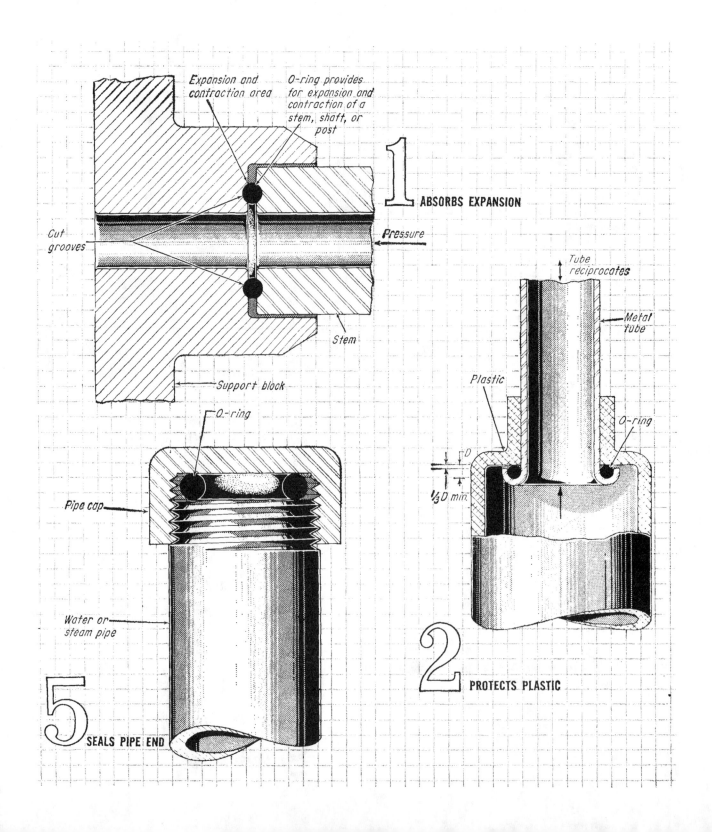

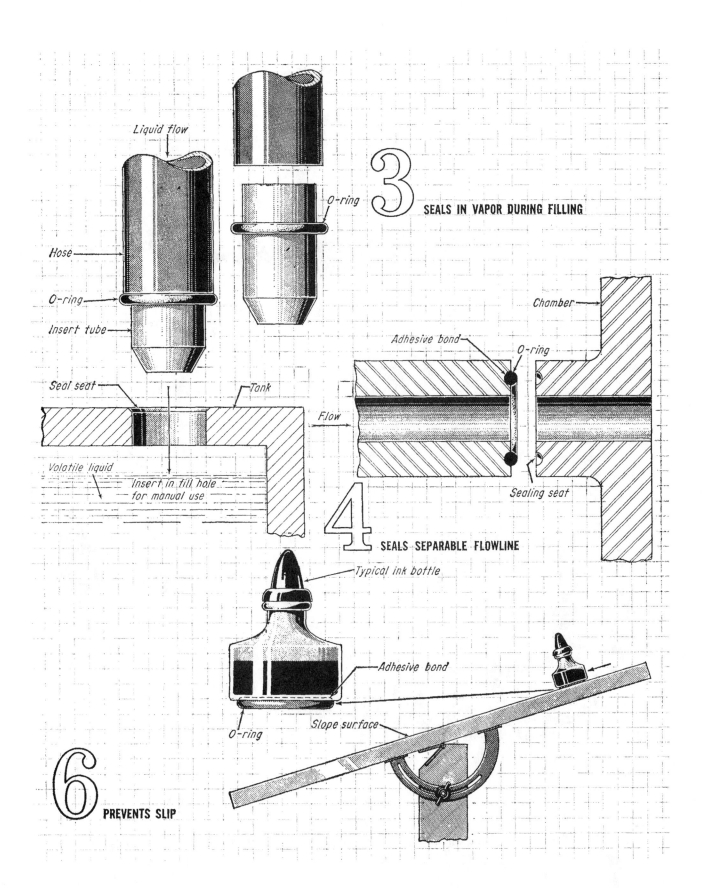

Liquid flow

O-ring

3 SEALS IN VAPOR DURING FILLING

Hose

O-ring

Insert tube

Chamber

Adhesive bond

O-ring

Seal seat

Tank

Flow

Volatile liquid

Insert in fill hole
for manual use

Sealing seat

4 SEALS SEPARABLE FLOWLINE

Typical ink bottle

Adhesive bond

Slope surface

O-ring

6 PREVENTS SLIP

O-Rings Solve Design Problems II

More examples of how rubber rings provide seals for shafts, lids, nozzles, and elbows, and also protect corners, cushion metal surfaces.

Robert O. Parmley

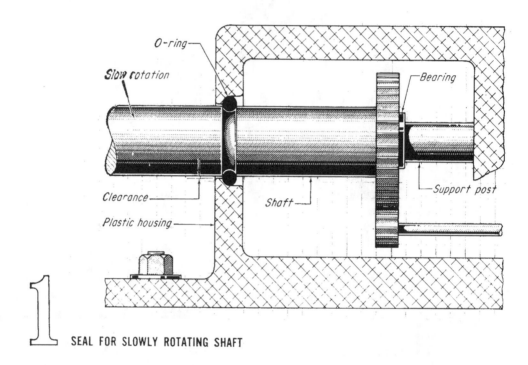

1 SEAL FOR SLOWLY ROTATING SHAFT

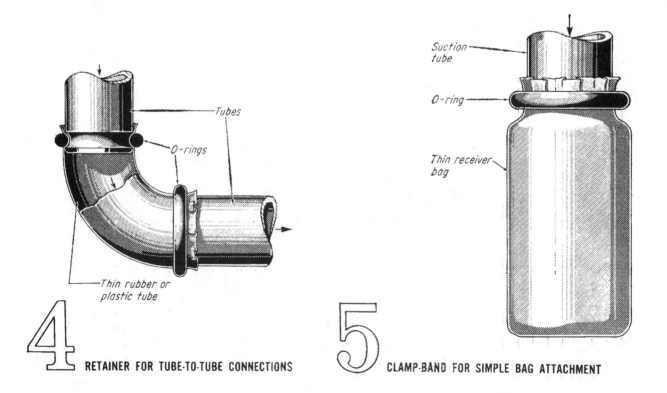

4 RETAINER FOR TUBE-TO-TUBE CONNECTIONS

5 CLAMP-BAND FOR SIMPLE BAG ATTACHMENT

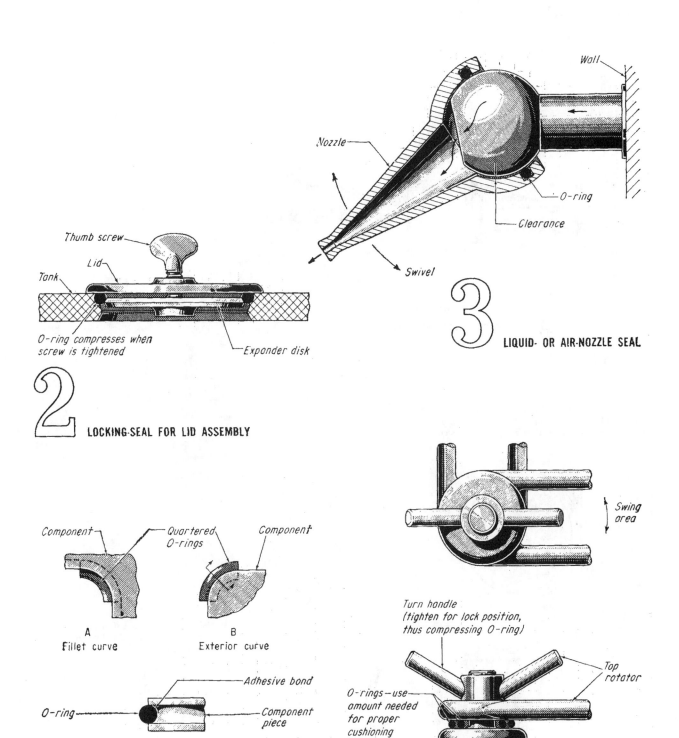

3 LIQUID- OR AIR-NOZZLE SEAL

Nozzle

Wall

O-ring

Clearance

Swivel

Thumb screw

Lid

Tank

O-ring compresses when screw is tightened

Exponder disk

2 LOCKING-SEAL FOR LID ASSEMBLY

Component

Quartered O-rings

Component

A
Fillet curve

B
Exterior curve

Adhesive bond

O-ring

Component piece

Typical section

6 PROTECTIVE MOLDING MADE FROM O-RING SEGMENTS

Swing area

Turn handle (tighten for lock position, thus compressing O-ring)

Top rotator

O-rings—use amount needed for proper cushioning (no metal wear)

Bottom rotator

7 CUSHION-RING FOR SWIVEL OR LIGHTWEIGHT ROTATING COMPONENTS

7 More Applications for O-Rings

For an encore to the roundup in the previous issue, O-rings are shown here performing in valves, on guide wheels, and as cushioning, etc.

Robert O. Parmley

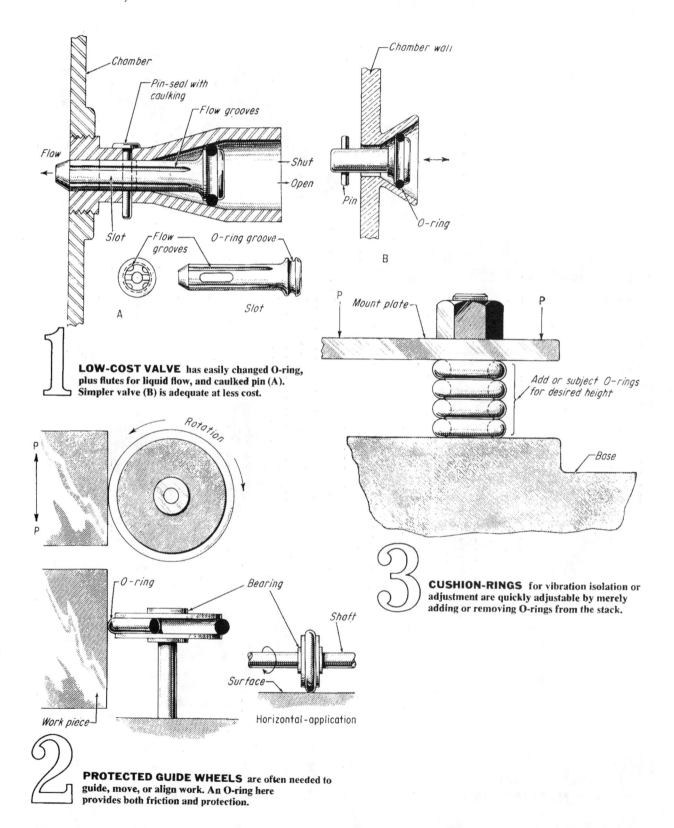

1 LOW-COST VALVE has easily changed O-ring, plus flutes for liquid flow, and caulked pin (A). Simpler valve (B) is adequate at less cost.

2 PROTECTED GUIDE WHEELS are often needed to guide, move, or align work. An O-ring here provides both friction and protection.

3 CUSHION-RINGS for vibration isolation or adjustment are quickly adjustable by merely adding or removing O-rings from the stack.

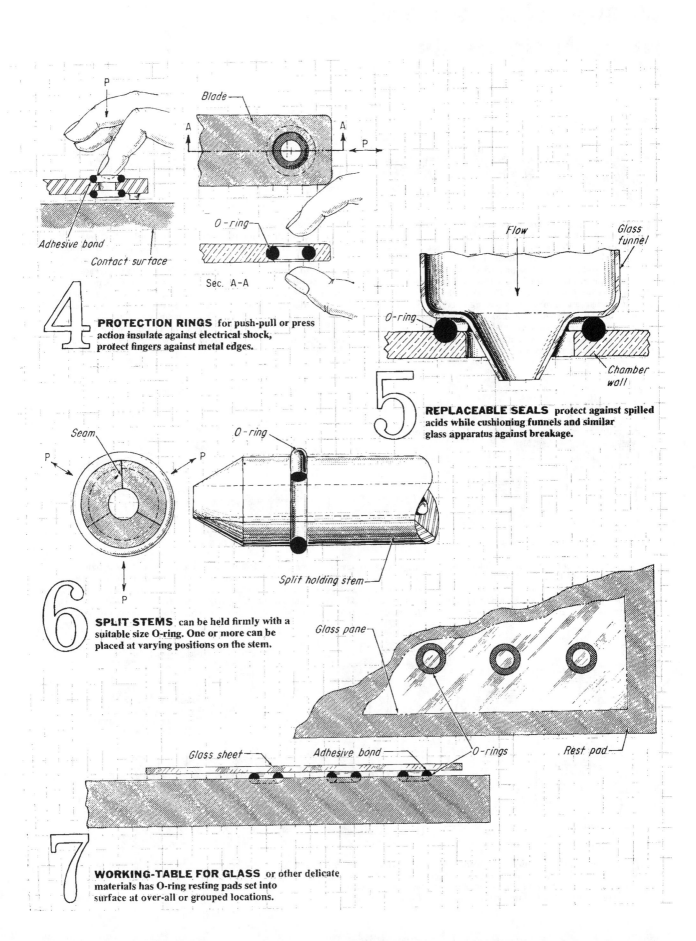

4 PROTECTION RINGS for push-pull or press action insulate against electrical shock, protect fingers against metal edges.

P

Blade

A — — A

P

O-ring

Sec. A-A

Adhesive bond

Contact surface

5 REPLACEABLE SEALS protect against spilled acids while cushioning funnels and similar glass apparatus against breakage.

Flow

Glass funnel

O-ring

Chamber wall

6 SPLIT STEMS can be held firmly with a suitable size O-ring. One or more can be placed at varying positions on the stem.

Seam

O-ring

P   P

P

Split holding stem

7 WORKING-TABLE FOR GLASS or other delicate materials has O-ring resting pads set into surface at over-all or grouped locations.

Glass pane

O-rings

Rest pad

Glass sheet

Adhesive bond

O-rings

Design Recommendations for O-Ring Seals

J. H. Swartz

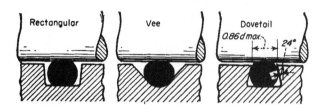

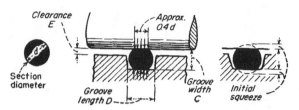

1 Rectangular grooves are recommended for most applications, whether static or dynamic. Slightly sloping sides (up to 5 deg) facilitate machining with form tools. Where practical, all groove surfaces should have the same degree of finish as the rod or cylinder against which the O-ring operates. The Vee type groove is used for static seals and is especially effective against low pressures. The dovetail groove reduces operating friction and minimizes starting friction. The effectiveness of the seal with this groove is critical depending upon: pressure, ring squeeze and angle of undercut. In general, the groove volume should exceed the maximum ring volume by at least 15 percent.

2 To insure a positive seal, a definite initial squeeze or interference of the ring is required. As a rule, this squeeze is approximately 10 percent of the O-ring cross sectional diameter d. This results in a ring contact distance of approximately 40 percent under zero pressure and can increase as much as 80 percent of the cross section diameter depending on pressure and composition of the ring. Starting friction can be reduced somewhat by decreasing the amount of squeeze but such a seal would be only moderately effective at pressures above 500 psi. Table I lists the recommended dimensions and tolerances for O-ring grooves for both static and dynamic applications.

EXTERNAL GROOVES

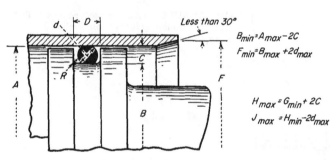

$$B_{min} = A_{max} - 2C$$
$$F_{min} = B_{max} + 2d_{max}$$

$$H_{max} = G_{min} + 2C$$
$$J_{max} = H_{min} - 2d_{max}$$

INTERNAL GROOVES

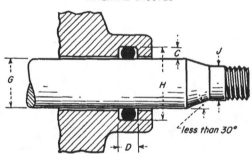

3 On small diameters, to facilitate machining, O-ring grooves should be located on the ram or rod rather than on an inside surface. For larger diameters, grooves can be machined either way. One important factor is that the rubbing surfaces must be extremely smooth. The recommended dimensional data in Table I and listed under dynamic seals should be used for these applications. All cylinders and rods should have a gradual taper to prevent damage to the O-ring during assembly. Equations are listed for calculating limiting dimensions for both external and internal grooves.

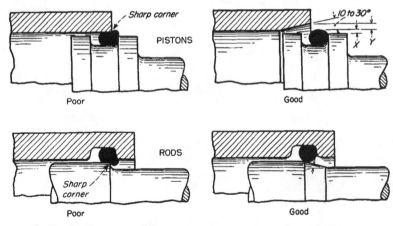

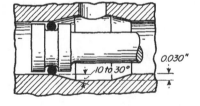

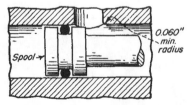

6 To facilitate assembly, all members which slide over O-rings should be chamfered or tapered at an angle less than 30 degrees. An alternative method is to use a generous radius. Such details prevent any possibility of pinching or cutting the O-ring during assembly.

7 Undercut all sharp edges, or cross-drilled ports over which O-rings must pass. While under pressure, rings should not pass over ports or grooves.

Specification AN 6227 or J. I. C. O-Ring Dash Number	Nominal Ring Section Diameter	d Actual Section Diameter	For Static Seals		For Dynamic Seals		D Groove Length**	R Minimum Radius	$2E$ Diametral Clearance (maximum)	Eccentricity (maximum)
			Diametral Squeeze* (minimum)	C Groove Width +0.000 −0.005	Diametral Squeeze* (minimum)	C Groove Width +0.000 −0.001				
1 to 7	1/16	0.070±0.003	0.015	0.052	0.010	0.057	3/32	1/64	0.005	0.002
8 to 14	3/32	0.103±0.003	0.017	0.083	0.010	0.090	9/64	1/64	0.005	0.002
15 to 27	1/8	0.139±0.004	0.022	0.113	0.012	0.123	3/16	1/32	0.006	0.003
28 to 52	3/16	0.210±0.005	0.032	0.173	0.017	0.188	9/32	3/64	0.007	0.004
53 to 88	1/4	0.275±0.006	0.049	0.220	0.029	0.240	3/8	1/16	0.008	0.005
AN 6230 or J. I. C. gaskets 1 to 52	1/8	0.139±0.004	0.022	0.113	———	———	3/16	1/32	0.006	0.003

Table I—Dimensional Data for Standard AN or J.I.C. O-Rings and Gaskets

Note: All dimensions are in inches.
* Diametral squeeze is the minimum interference between O-Ring cross section diameter d and gland width C.
** If space is limited, the groove length D can be reduced to a distance equal to the maximum O-Ring diameter d plus the static seal squeeze.

FACE SEAL GROOVES

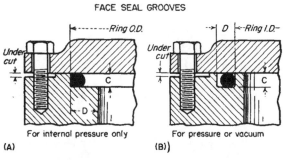

(A) For internal pressure only
(B) For pressure or vacuum

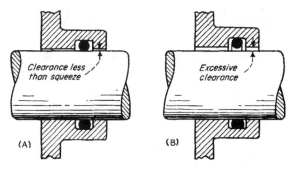

(A) Clearance less than squeeze
(B) Excessive clearance

4 For static face seals, two types of grooves are shown. Type (A) is more commonly used because of simpler machining. Groove depths listed in Table I under static seals apply to this application. In high pressure applications where steel flanges are used, slight undercutting of one face (not exceeding 0.010 in.) minimizes possible O-ring extrusion.

5 Radial clearances should never exceed one-half of the recommended O-ring squeeze even where the pressure does not require the use of a close fit between sliding parts. Under these conditions, if the shaft is eccentric (A), the ring will still maintain its sealing contact. (B) Excessive clearance results in the loss of sealing contact of the O-ring.

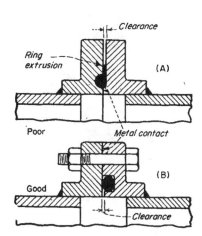

Poor (A)
Good (B)

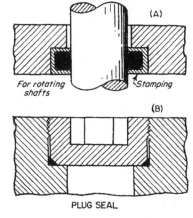

For rotating shafts (A)
Stamping (B)
PLUG SEAL

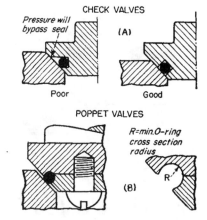

CHECK VALVES
Pressure will bypass seal (A)
Poor
Good
POPPET VALVES
R=min.O-ring cross section radius
(B)

8 Metal-to-metal contact of the inner mating surfaces (A) should be avoided. Clearances should be permitted only on inner surfaces (B).

9 Simple stamping (A) pressed in housing is for low speeds and pressures. (B) Chamfered corners of plug makes a recess for an O-ring.

10 Rectangular grooves (A) should be normal to the sealing surface. Special grooves (B) avoid the washout of O-rings during pressure surges.

O-Ring Seals for Pump Valves

Robert O. Parmley

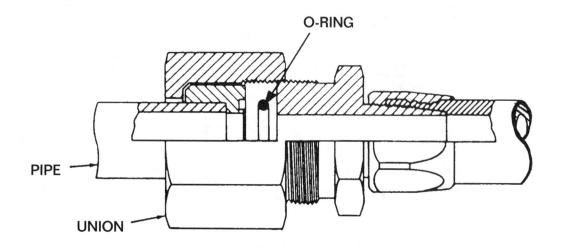

A-Combination Pump Valve

The CPV O-ring seal fitting (a Navy standard) uses an O-ring which is inserted in the packing-gland recess on the face or the union which has been silver-brazed to the end of a pipe. The union and pipe are sometimes called a "tailpiece."

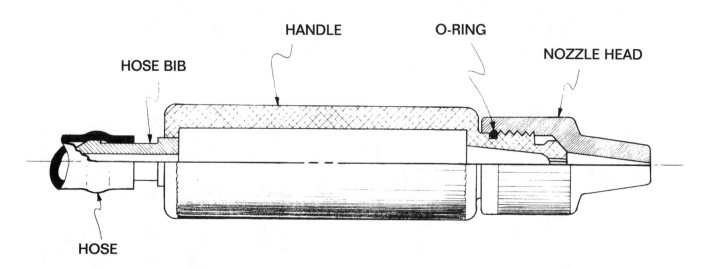

B-Hand-Adjusted Pump Nozzle

The discharge end (nozzle head) of this portable pump unit has the spray adjusted by manually turning the nozzle head. The O-ring maintains a positive, water-tight seal for any adjusted position.

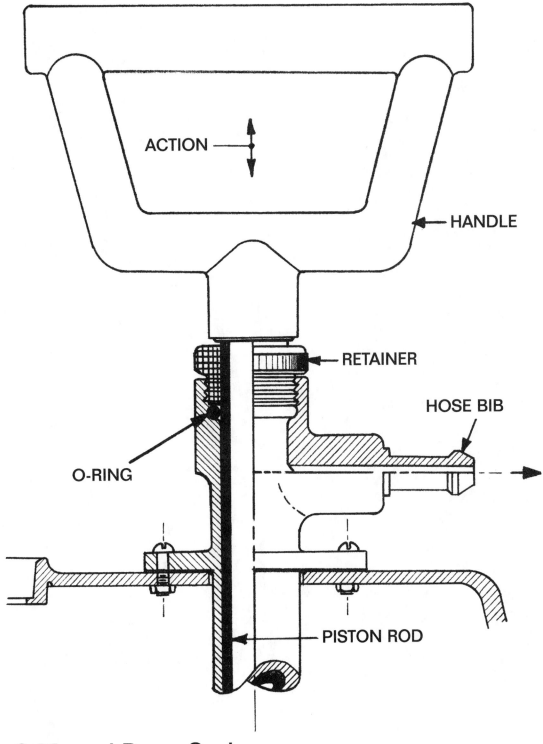

C-Manual Pump Seal

The O-ring, which is seated by the threaded retainer, provides a water-tight seal for the up & down action of the piston rod.

SECTION 18

RETAINING RINGS

Comparisons of Retaining Rings Versus Typical Fasteners

A variety of basic applications show how these rings simplify design and cut costs.

Howard Roberts

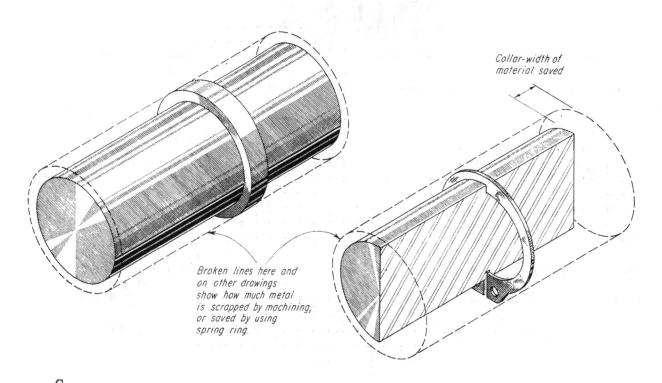

Collar-width of material saved

Broken lines here and on other drawings show how much metal is scrapped by machining, or saved by using spring ring.

1 **MACHINED SHOULDERS** are replaced with savings in material, tools and time. Grooving for ring can be done during a cut-off, or other machining operation.

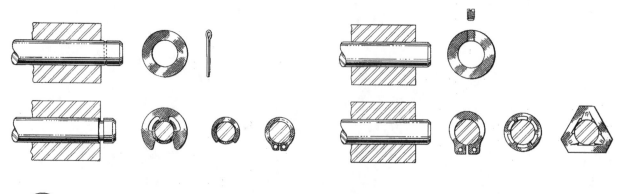

2 **RINGS THAT CAN REPLACE** cotter pin and washer are economical since only one part is required and pin-spreading operation is not needed thus cutting time and costs.

3 **WHEN COLLAR AND SETSCREW** are substituted by ring, risk of screw vibrating loose is avoided. Also, no damage to shaft by screw point occurs — a frequent cause of trouble.

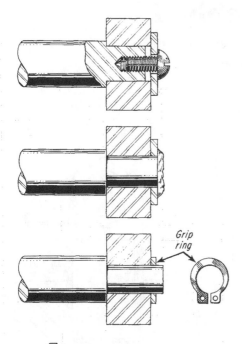

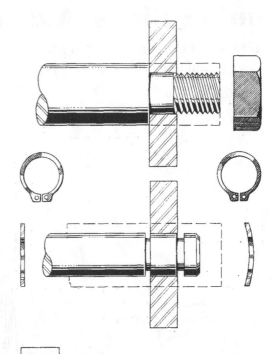

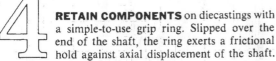

4 **RETAIN COMPONENTS** on diecastings with a simple-to-use grip ring. Slipped over the end of the shaft, the ring exerts a frictional hold against axial displacement of the shaft.

5 **SHOULDER AND NUT** are replaced by two retaining rings. A flat ring replaces the shoulder, while a bowed ring holds the component on shaft for resilient end-play take-up.

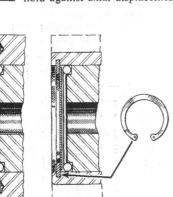

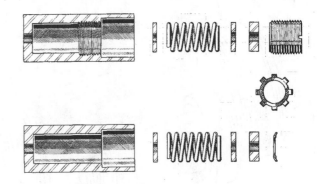

6 **COVER-PLATE ASSEMBLY** has been re-designed (lower drawing) to avoid use of screws and machined cover-plate. Much thinner wall can be used—no drilling or tapping.

7 **THREADED INTERNAL FASTENERS** are costly because of expensive internal threading operation. Simplify by substituting a self-locking retaining ring—see lower drawing.

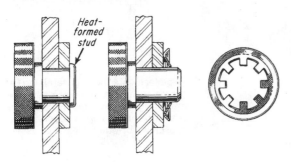

8 **HEAT-FORMED STUD** provides a shoulder against retained parts but must be scrapped if the parts must be disassembled for service. Self-locking ring can be easily removed.

Retaining Rings Aid Assembly, I

By functioning as both a shoulder and as a locking device, these versatile fasteners reduce machining and the number and complexity of parts in an assembly.

Robert O. Parmley

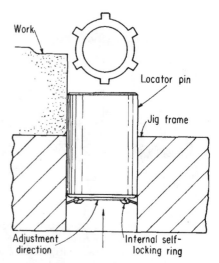

Internal self-locking ring supports a locator. Elevation of the pin may be altered in the entry direction only; the pin won't push down into the frame

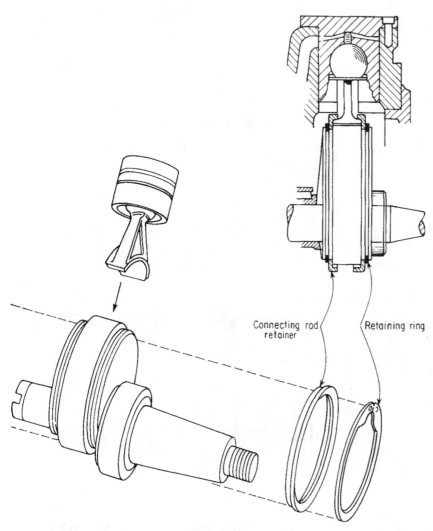

Slow-moving piston of hydraulic motor is assembled to the crank throw by two retainers. These are held in place by two retaining rings that fit into grooves in the crankthrow

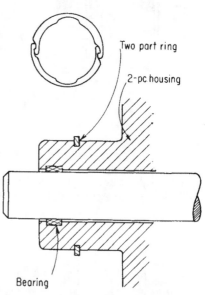

Two-piece interlocking retaining ring serves to hold a two-piece assembly on a rotating shaft, and is more simple than a threaded cap, a couple of capscrews or other means of assembly

Courtesy: *American Machinist*: Published by Penton Media, Inc.

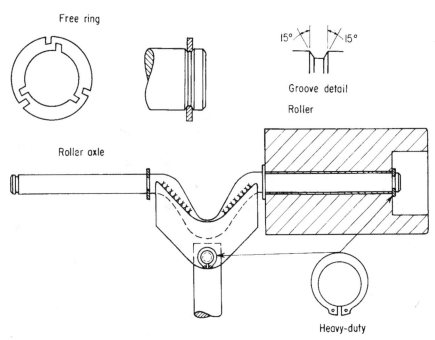

Two types of rings may be used on one assembly. Here permanent-shoulder rings provide a uniform axle step for each roller, without spotwelding or the like. Heavy-duty rings keep the rollers in place

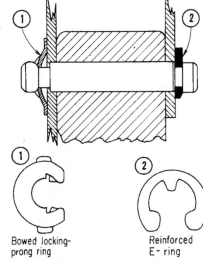

Snug assembly of side members to a casting with cored hole is secured with two rings: 1—spring-like ring has high thrust capacity, eliminates springs, bow washers, etc; 2—reinforced E-ring acts as a retaining shoulder or head. Each ring can be dismantled with a screwdriver

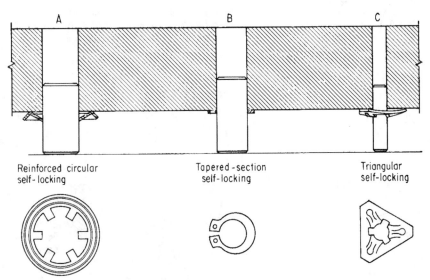

These three examples show self-locking retaining rings used as adjustable stops on support members (pins made to commercial tolerances): A—external ring provides positive grip, and arched rim adds strength; B—ring is adjustable in both directions, but frictional resistance is considerable, and C—triangular ring with dished body and three prongs will resist extreme thrust. Both A and C have one-direction adjustment only

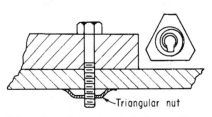

Triangular retaining nut eliminates the need for tapping mounting holes and using a large nut and washer. Secure mounting of small motors and devices can be obtained in this manner

Retaining Rings Aid Assembly, II

Here are eight thought-provoking uses for retaining rings.

Robert O. Parmley

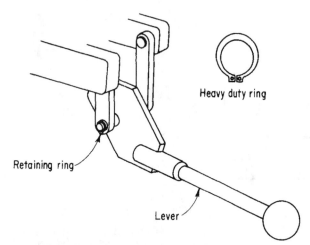

Axial pin for lever is secured with a heavy-duty ring, making a neat, strong assembly

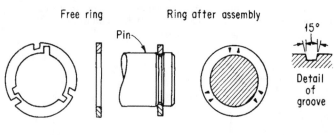

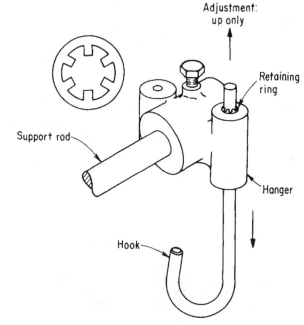

Hanger hook is held at desired height by a self-locking external ring. A multitude of adjustments can be made without trouble

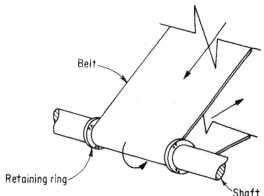

Belt alignment is assured by using permanent shoulder retaining rings. The rings are crimped into the shaft grooves for a permanent, clean, and inexpensive flange. A retainer ring of this type has a high capacity for thrust loads

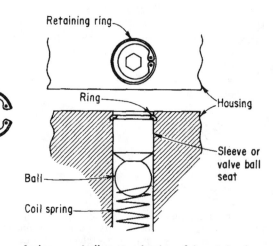

A sleeve or ball seat valve is safely retained by a ring that acts like a wedge in the outer groove. Rigid end-play take-up is provided

Courtesy: *American Machinist*: Published by Penton Media, Inc.

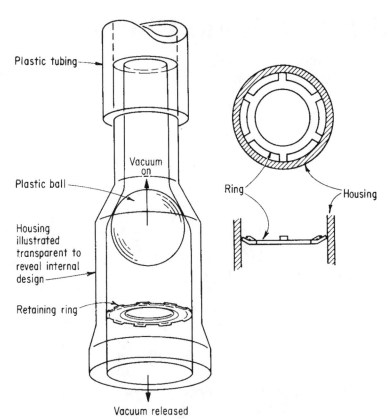

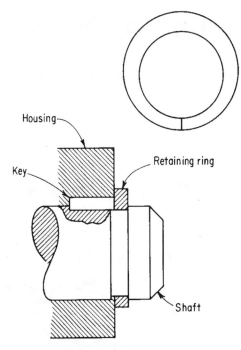

Internal self-locking ring supports the plastic ball valve when the vacuum is released, thus providing a support during the "off" cycle. Air or liquid is released when ball is at rest and exits through the areas between the grip points of the ring, which is adjustable at entry position

Tamper-proof lock for a shaft in a housing provides location of the shaft and at the same time retains the key. Heavy axial loading and permanent retention of the key are double values in this application

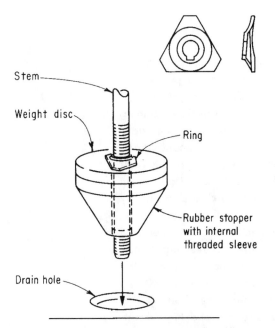

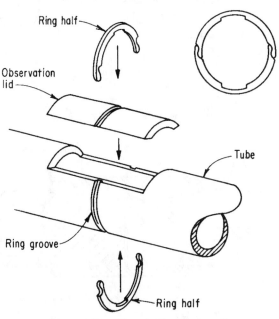

Triangular retainer nut positions and unifies components of the tank drain assembly. The triangular nut eliminates the need for a large standard nut and lockwasher or spring-type component and simplifies the design

Observation lid on tubing makes it possible to inspect wiring at will. The two-part balanced retainer ring has identical semicircular halves, which are held together by the interlocking prongs at the free ends

Coupling Shafts with Retaining Rings

These simple fasteners can provide an original way around certain design snags.
For example, here are eight ways they're used to solve shaft-coupling problems.

Robert O. Parmley

Pin, Sleeve, and Ring

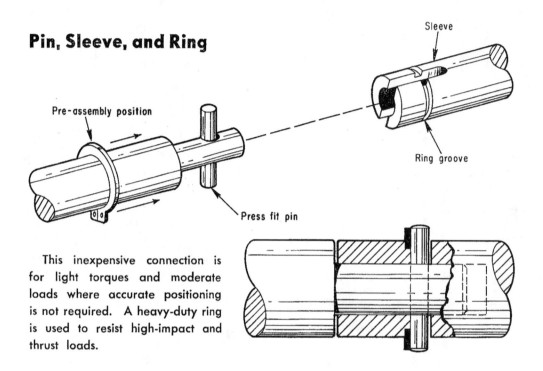

This inexpensive connection is for light torques and moderate loads where accurate positioning is not required. A heavy-duty ring is used to resist high-impact and thrust loads.

Sleeve, Key, and Ring

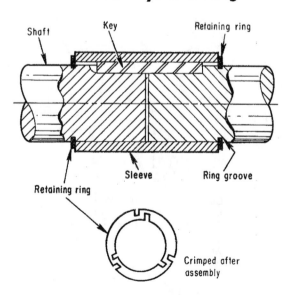

Crimping the retaining ring into the groove produces a permanent, simple, and clean connection. This method is used to avoid machining shoulders in expensive materials, and to permit use of smaller-diameter shafts. When the ring is compressed into a V-shaped groove on the shaft, the notches permanently deform into small triangles, causing a reduction of the inner and outer diameters of the ring. Thus, the fastener tightly grips the groove, and provides a 360-deg shoulder around the shaft. Good torsional strength and high thrust-load capacity is provided by this connection.

Courtesy: *Machine Design*: Published by Penton Media, Inc.

Two-Shaft Splice

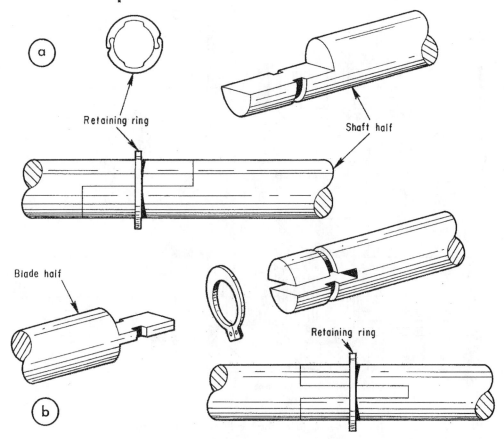

A balanced two-part ring provides an attractive appearance in addition to withstanding high rotational speeds and heavy thrust loads, *a*. The one-piece ring, *b*, secures the shafts in a high-torque capacity design.

End-Flange Connection

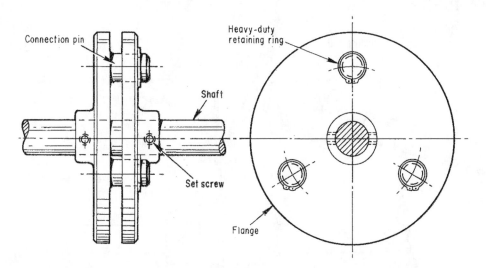

This assembly for heavy-duty service requires minimum machining. Ring thickness should be substantial, and extra ring-section height is desirable.

Collar, Rings, and Threaded Shaft

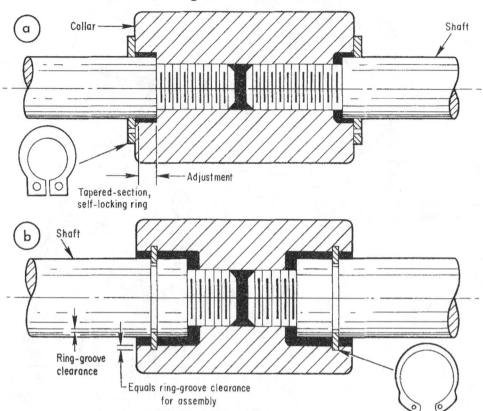

For a connection that requires axial shaft adjustment, the self-locking ring requires no groove, *a*. An alternate solution, *b*, uses an inverted-lug ring seated in an internal groove. Extra ring-section height provides a good shoulder. The ring is uniformly concentric with housing and shaft.

Coupler and Ring

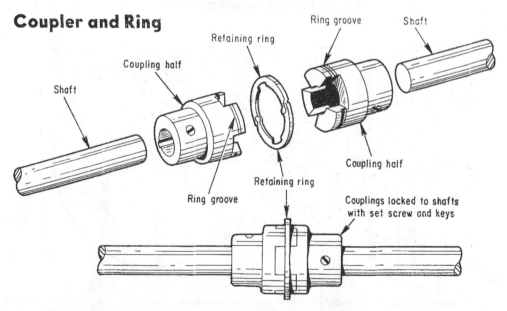

Where attractive appearance is desired in a dependable locking device, this connector and ring can be used.

Slotted Sleeve with Tapered Threads

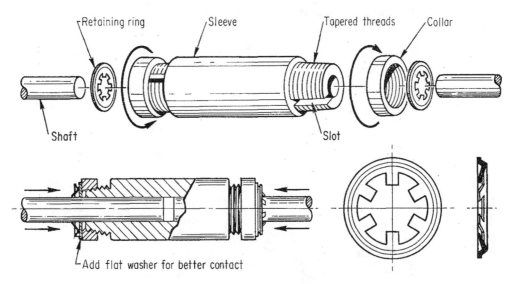

A slotted sleeve with tapered threads connects shafts which cannot be machined. Prongs on the retaining ring provide positive shaft gripping to stop collar movement. The arched rim adds extra strength.

Bossed Coupling and Rings

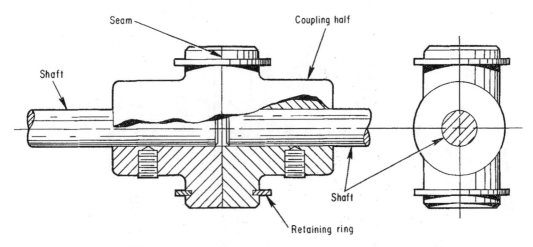

An alternate solution for coupling unmachined shafts uses bossed coupling halves with locking retaining rings.

The Versatile Retaining Ring

A design roundup of some unusual applications of retaining rings.

Robert O. Parmley

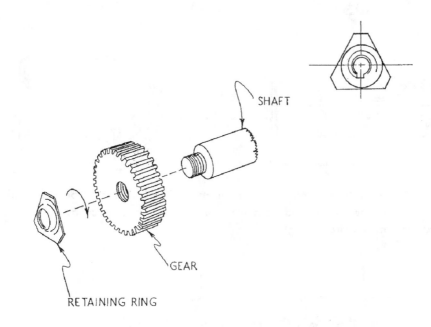

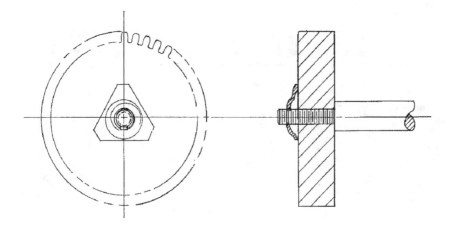

Fig 1 The assembly of a hubless gear and threaded shaft may be accomplished by using a triangular nut retaining component which eliminates the need for a large standard nut and lock washer or other spring type part. The dished body of the triangular nut flattens under torque to lock the gear to the shaft.

Excerpted from *Assembly Engineering*, February 1968 © Business News Media, Troy, MI, USA

Every engineer is familiar with the use of retaining rings in product assembly. Applications for this type of fastening device range from miniature electronic assemblies to heavy duty equipment. In spite of this widespread use, many opportunities for taking advantage of these versatile fastening components often are overlooked. However, when a value engineering approach is taken and the basic function of retaining rings as easily assembled locating and locking devices is kept clearly in mind it will be found that these simple fasteners can provide a unique solution to difficult assembly problems.

This roundup of 8 unusual applications illustrates how different types of retaining rings have been used to simplify assembly and reduce manufacturing costs. The captions under the drawings give the details involved in each case.

The author wishes to acknowledge with appreciation the cooperation he received from the Truarc Retaining Rings Division of Waldes Kohinoor, Inc. in developing these assembly designs.

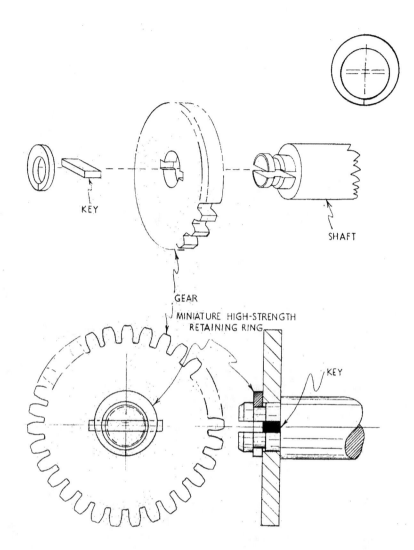

Fig 2 This heavy duty hubless gear and shaft is designed for high torque and end thrusts. The retaining ring seated in a square groove and the key in slot provide a tamper-proof lock. This design is recommended for permanent assemblies in which the ring may be subjected to heavy loads from either or both axial directions. An angled groove can be provided which has one wall cut at a 40° angle to the shaft axis. This will permit the ring to be removed without damage.

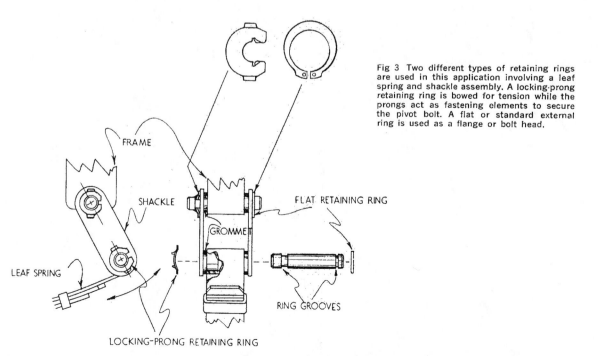

Fig 3 Two different types of retaining rings are used in this application involving a leaf spring and shackle assembly. A locking-prong retaining ring is bowed for tension while the prongs act as fastening elements to secure the pivot bolt. A flat or standard external ring is used as a flange or bolt head.

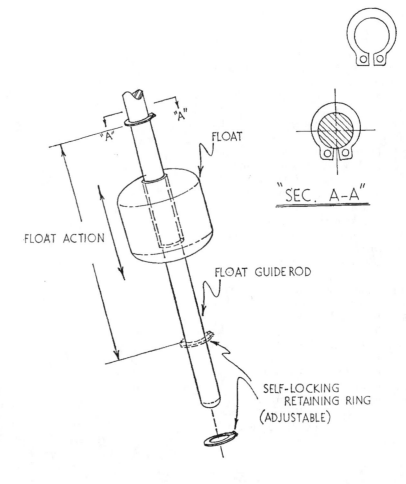

"SEC. A-A"

FLOAT

FLOAT ACTION

FLOAT GUIDE ROD

SELF-LOCKING
RETAINING RING
(ADJUSTABLE)

Fig 4 The self-locking retaining rings used
in this application provide stops for a float.
The rings are adjustable on the guide rod and
yet the friction force produced by the heavy
spring pressure makes axial displacement
from the light weight hollow float impossible.

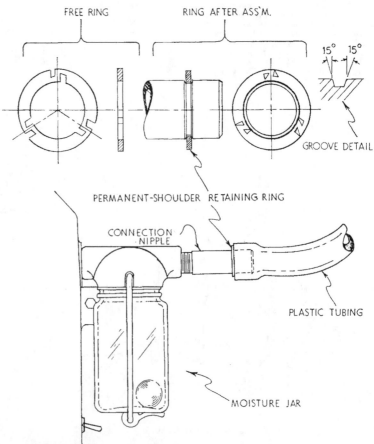

FREE RING

RING AFTER ASS'M.

15° 15°

GROOVE DETAIL

PERMANENT-SHOULDER RETAINING RING

CONNECTION
NIPPLE

PLASTIC TUBING

MOISTURE JAR

Fig 5 Retaining rings provide a uniform circular shoulder for small diameter parts such
as the pipe nipple shown here. In this case
the retaining ring shoulder is used as a
stop for the plastic tube. The wall thickness
of the nipple should be at least three times
as thick as the depth of the groove. When
assembling the ring in the groove, the nipple
should be supported by inserting a mandrel
or rod.

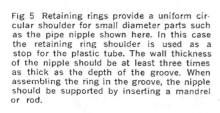

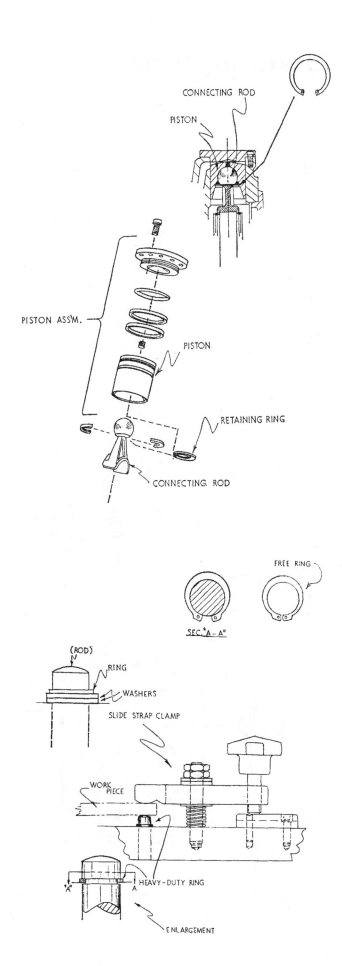

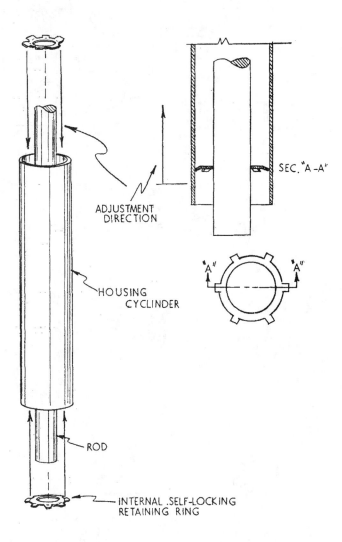

Fig 6 This internal retaining ring is a key part in the assembly of a connecting rod and piston for a hydraulic motor. The ring's lug holes make rapid assembly and disassembly possible when the proper pliers are used. The piston assembly in this case is slow moving and is not subject to heavy cycle loading.

Fig 7 Internal self-locking rings can act as a support carrier when the I.D. of a sleeve or housing cylinder is too large to center and stabilize small rods or conduit. The rings are adjustable in the entry direction only, however, and a sufficient number should be used to secure the rod.

Fig 8 The heavy duty external retaining ring shown here controls the elevation or position of a support post in a holding clamp. This type of ring is ideal for heavy duty applications where extreme loading conditions are encountered. By adding washers under the ring the elevation of the support post can be adjusted as required.

The Multiple-Purpose Retaining Ring

A roundup of ten unusual ways for putting retaining rings to work in assembly jobs.

Robert O. Parmley

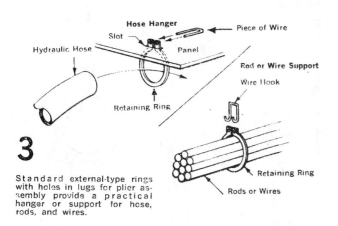

1 Special lugless ring of external type is used to offset center-line of shaft which is station-ary or rotating slowly in hub or journal.

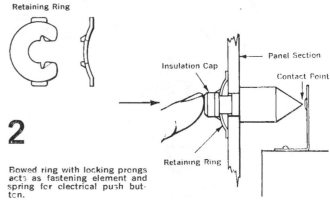

2 Bowed ring with locking prongs acts as fastening element and spring for electrical push but-ton.

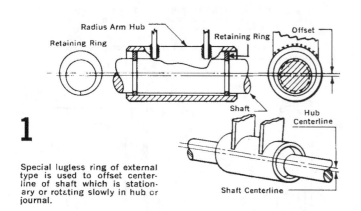

3 Standard external-type rings with holes in lugs for plier as-sembly provide a practical hanger or support for hose, rods, and wires.

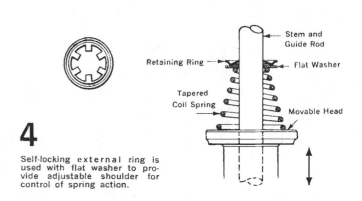

4 Self-locking external ring is used with flat washer to pro-vide adjustable shoulder for control of spring action.

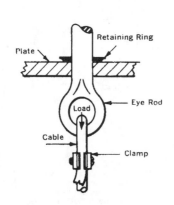

5 Self-locking external ring con-trols position of eye rod in an adjustable cable hanger ar-rangement.

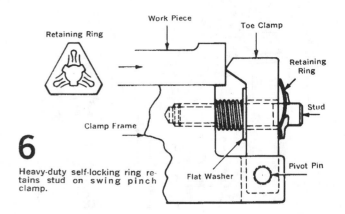

6 Heavy-duty self-locking ring re-tains stud on swing pinch clamp.

Excerpted from *Assembly Engineering*, July 1966 © Business News Media, Troy, MI, USA

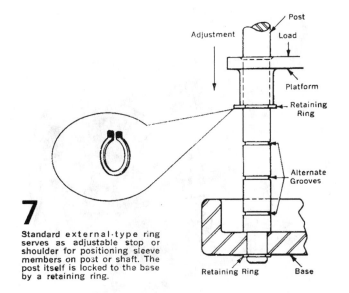

7

Standard external-type ring serves as adjustable stop or shoulder for positioning sleeve members on post or shaft. The post itself is locked to the base by a retaining ring.

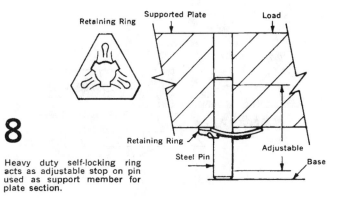

8

Heavy duty self-locking ring acts as adjustable stop on pin used as support member for plate section.

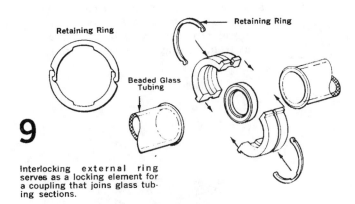

9

Interlocking external ring serves as a locking element for a coupling that joins glass tubing sections.

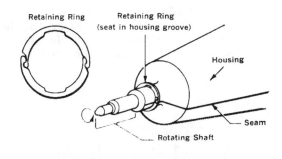

10

Interlocking external ring locks two-piece housing that fits around a rotating shaft.

More Work for Round Retaining Rings

Try this low-cost fastener for locking shafts and other parts. It will also work as a shaft step for bearings and an actuating ring for switches.

Dominic J. Lalera

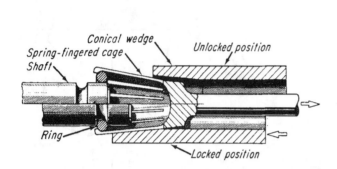

1 **LOCK A SHAFT** by forcing a retaining ring over the groove in the shaft. In locking position, the spring-fingered cage is actuated by the conical wedge.

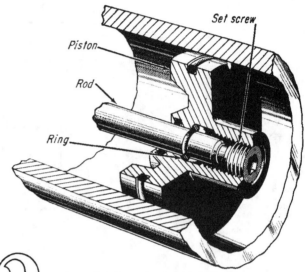

2 **PISTON IS LOCKED** in place on the rod when drawn into place by means of a setscrew. To remove, slide the piston away from the ring, then remove the ring.

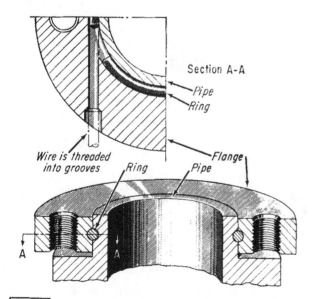

5 **FLANGE ASSEMBLY** is permanently fastened by threading the wire into the mating grooves through the flange. Flange can rotate if wire doesn't protrude.

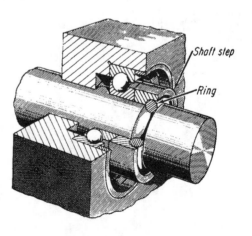

6 **THIS SHAFT STEP** for a rotating bearing is quickly and simply made by grooving the shaft to accept a spring ring. Counterbore the shaft step to mate.

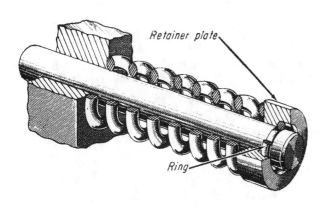

Retainer plate

Ring

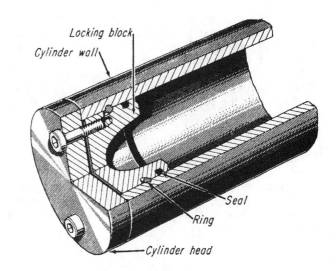

Locking block

Cylinder wall

Seal

Ring

Cylinder head

 THIS SPRING-HELD shaft lock is a basic application for retaining rings. The best groove dimensions for round spring rings are readily available from suppliers.

ASSEMBLE CYLINDER HEADS and similar parts to thin walls by means of a retaining ring and a locking block. Tightening the screws expands the ring.

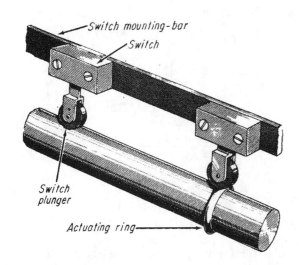

Switch mounting-bar

Switch

Switch plunger

Actuating ring

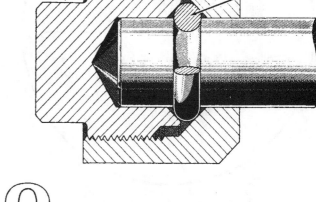

Male threaded wedge

Internal wedge

Ring

SWITCH ACTUATORS of round retaining rings offer a simple solution when permanent shaft steps would present assembly problems. Close the ring gap.

 THREE-PIECE WEDGE lets the shaft move freely until the wedge is tightened by screwing it in. The round retaining ring is then forced into the groove.

Deflections of Perpendicularly Loaded Split Circular Rings

M. M. Lemcoe

FORMULAS FOR THE DEFLECTION of a split uniform circular ring perpendicular to the plane of the ring are given for various positions of the load. Methods of developing those formulas are demonstrated.

In Fig. 1 is shown a split uniform circular ring of radius R, loaded with a force P applied perpendicular to the plane of the ring at the point B. At the point Q, the bending moment M and the twisting moment T due to the load P are respectively:

$$M = P R \sin (\beta - \theta) \tag{1}$$
$$T = P R [1 - \cos (\beta - \theta)] \tag{2}$$

Also, if there were a unit load at the point A, there would be at point Q a bending moment m and a twisting moment t due to that unit load. These are given by the following formula

$$m = R \sin (\alpha - \theta) \tag{3}$$
$$t = R [1 - \cos (\alpha - \theta)] \tag{4}$$

From strain energy considerations, the deflection Δ of the point A can be formulated. If E is the modulus of elasticity and G is the shear modulus and if I is the moment of inertia about the neutral axis of a cross-section, and

J is the polar moment of inertia of a cross-section, the formula is

$$\Delta = \int_o^\phi \frac{M m R d\theta}{E I} + \int_o^\phi \frac{T t R d\theta}{G J} \tag{5}$$

The angle ϕ equals the smaller of angles α or β. Substituting Eqs (1), (2), (3) and (4) into Eq (5) gives:

$$\Delta = P R^3 \int_o^\phi \left\{ \frac{\sin (\alpha - \theta) \sin (\beta - \theta)}{E I} + \frac{[1 - \cos (\alpha - \theta)] [1 - \cos (\beta - \theta)]}{G J} d\theta \right\}$$

From the trigonometric identities

$$\sin (\alpha - \theta) \sin (\beta - \theta) = \tfrac{1}{2} [\cos (\alpha - \beta) - \cos (\alpha + \beta - 2\theta)]$$
$$\cos (\alpha - \theta) \cos (\beta - \theta) = \tfrac{1}{2} [\cos (\alpha - \beta) + \cos (\alpha + \beta - 2\theta)]$$

The formula for Δ becomes:

$$\Delta = \frac{P R^3}{2 E I} \left[\int_o^\phi \cos (\alpha - \beta) d\theta - \int_o^\phi \cos (\alpha + \beta - 2\theta) d\theta \right]$$
$$+ \frac{P R^3}{2 G J} \left[\int_o^\phi \cos (\alpha - \theta) d\theta - \int_o^\phi \cos (\alpha + \beta - 2\theta) d\theta + \int_o^\phi [1 - \cos (\alpha - \theta)] d\theta - \int_o^\phi \cos (\beta - \theta) d\theta \right] \tag{6}$$

Integrating Eq (6) gives

$$\Delta = \frac{P R^3}{E I} \left[\frac{\phi \cos (\alpha - \beta)}{2} + \frac{\sin (\alpha + \beta - 2\phi) - \sin (\alpha - \beta)}{4} \right]$$
$$+ \frac{P R^3}{G J} \left[\frac{\phi \cos (\alpha - \beta)}{2} - \frac{\sin (\alpha + \beta - 2\phi) - \sin (\alpha + \beta)}{4} + \phi + \sin (\alpha - \phi) - \sin \alpha + \sin (\beta - \theta) - \sin \beta \right] \tag{7}$$

From Eq (7) formulas can be developed for Δ for various positions of points A and B.

Formulas for various positions of points A and B are given below.

Fig. 1—Split circular ring loaded by a force perpendicular to the ring at point B. The deflection Δ at point A is to be calculated.

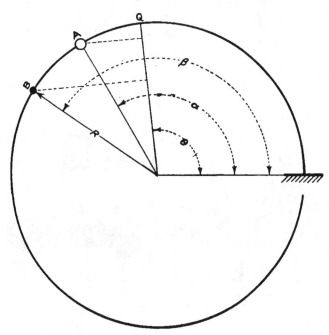

Δ = Deflection at Point A

P = Load at Point B

R = Radius of Ring

E = Modulus of Elasticity

G = Shear Modulus

J = Cross-Section Polar Moment of Inertia

I = Cross-Section Polar Moment of Inertia About Neutral Axis

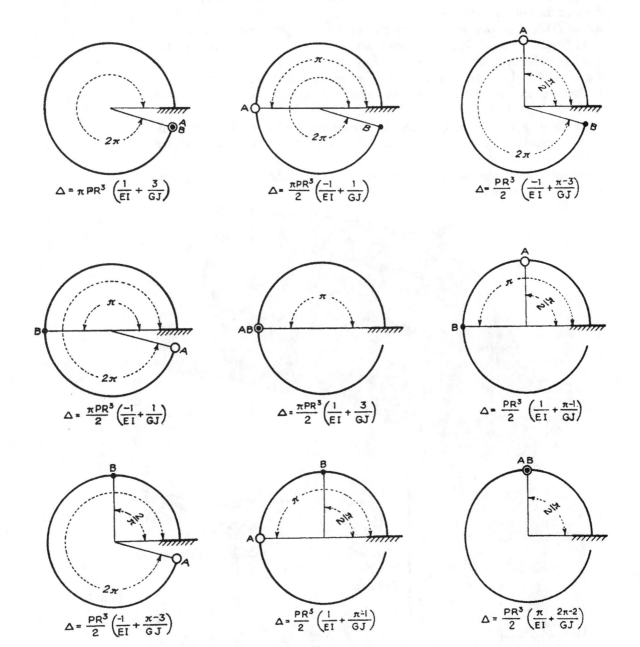

$$\Delta = \pi P R^3 \left(\frac{1}{EI} + \frac{3}{GJ} \right)$$

$$\Delta = \frac{\pi P R^3}{2} \left(\frac{-1}{EI} + \frac{1}{GJ} \right)$$

$$\Delta = \frac{P R^3}{2} \left(\frac{-1}{EI} + \frac{\pi-3}{GJ} \right)$$

$$\Delta = \frac{\pi P R^3}{2} \left(\frac{-1}{EI} + \frac{1}{GJ} \right)$$

$$\Delta = \frac{\pi P R^3}{2} \left(\frac{1}{EI} + \frac{3}{GJ} \right)$$

$$\Delta = \frac{P R^3}{2} \left(\frac{1}{EI} + \frac{\pi-1}{GJ} \right)$$

$$\Delta = \frac{P R^3}{2} \left(\frac{-1}{EI} + \frac{\pi-3}{GJ} \right)$$

$$\Delta = \frac{P R^3}{2} \left(\frac{1}{EI} + \frac{\pi-1}{GJ} \right)$$

$$\Delta = \frac{P R^3}{2} \left(\frac{\pi}{EI} + \frac{2\pi-2}{GJ} \right)$$

Improve Design with Retaining Rings

Waldes Kohinoor supplied the rings for the assembly illustrated here. Look at the old design, look at the new design. What did retaining rings do? Well, by changing the design to retaining rings these advantages are achieved:

1) Six beveled rings—one at each end of the three shaft bores—replace 24 hex-head bolts and eliminate drilling and tapping 24 holes in the cast housing.

2) Special gaskets needed to provide a proper seal between the end cap and housing have been replaced by less expensive standard O-rings. Six facing operations on the outside of the casting, required for the gasket seals, have been eliminated. (O-ring grooves are an integral part of the redesigned cover plates.)

3) Twelve external rings—six bowed, six flat—secure the inner races of the bearings. The rings are assembled in grooves machined simultaneously with the shaft cut-off and chamfering operations. They replace six threaded ring nuts and six lock washers and eliminate 12 ground diameters, six threading operations and six keyways on the shaft.

4) Six basic internal rings, installed in grooves machined in the housing, eliminate six machined shoulders and the need for holding close axial tolerances on the bearing bore and end cap.

Reusable following disassembly, the rings are assembled with special pliers, and can be removed for field service.

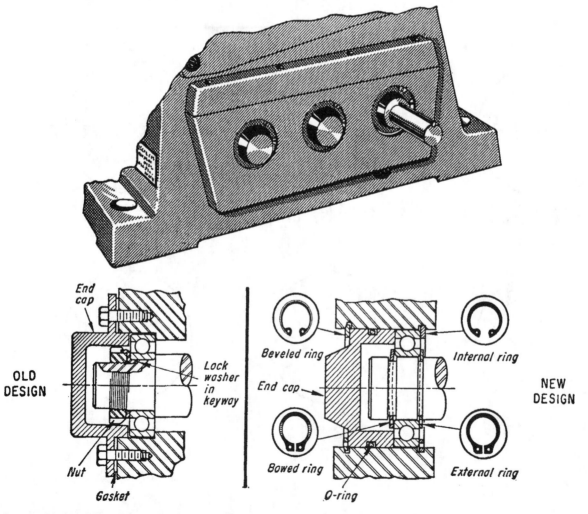

Courtesy: *Machine Design:* © Penton Media, Inc.

SECTION 19

BALLS

12 Ways to Put Balls to Work

Bearings, detents, valves, axial movements, clamps and other devices can all have a ball as their key element.

Louis Dodge

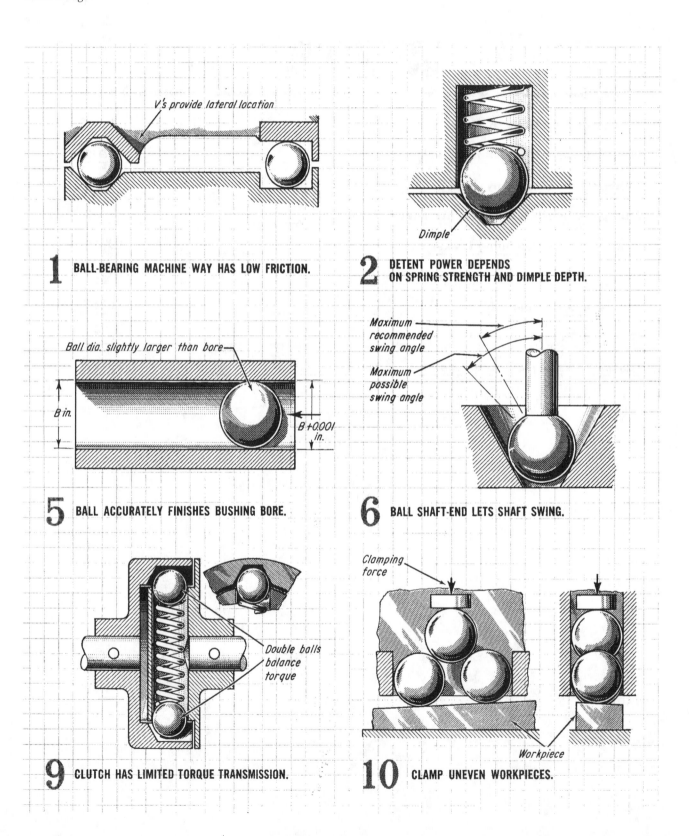

1 BALL-BEARING MACHINE WAY HAS LOW FRICTION.

V's provide lateral location

2 DETENT POWER DEPENDS ON SPRING STRENGTH AND DIMPLE DEPTH.

Dimple

5 BALL ACCURATELY FINISHES BUSHING BORE.

Ball dia. slightly larger than bore

B in.

B +0.001 in.

6 BALL SHAFT-END LETS SHAFT SWING.

Maximum recommended swing angle

Maximum possible swing angle

9 CLUTCH HAS LIMITED TORQUE TRANSMISSION.

Double balls balance torque

10 CLAMP UNEVEN WORKPIECES.

Clamping force

Workpiece

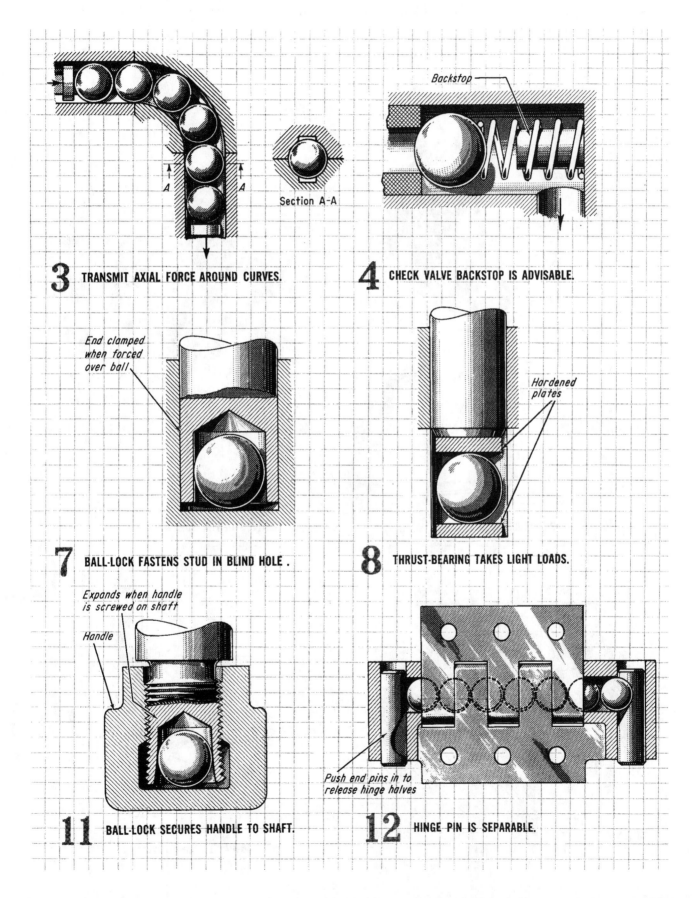

3 TRANSMIT AXIAL FORCE AROUND CURVES.

Section A-A

4 CHECK VALVE BACKSTOP IS ADVISABLE.

Backstop

7 BALL-LOCK FASTENS STUD IN BLIND HOLE.

End clamped when forced over ball

8 THRUST-BEARING TAKES LIGHT LOADS.

Hardened plates

11 BALL-LOCK SECURES HANDLE TO SHAFT.

Expands when handle is screwed on shaft

Handle

12 HINGE PIN IS SEPARABLE.

Push end pins in to release hinge halves

How Soft Balls Can Simplify Design

Balls of flexible material can perform as latches, stops for index disks, inexpensive valves and buffers for compression springs.

Robert O. Parmley

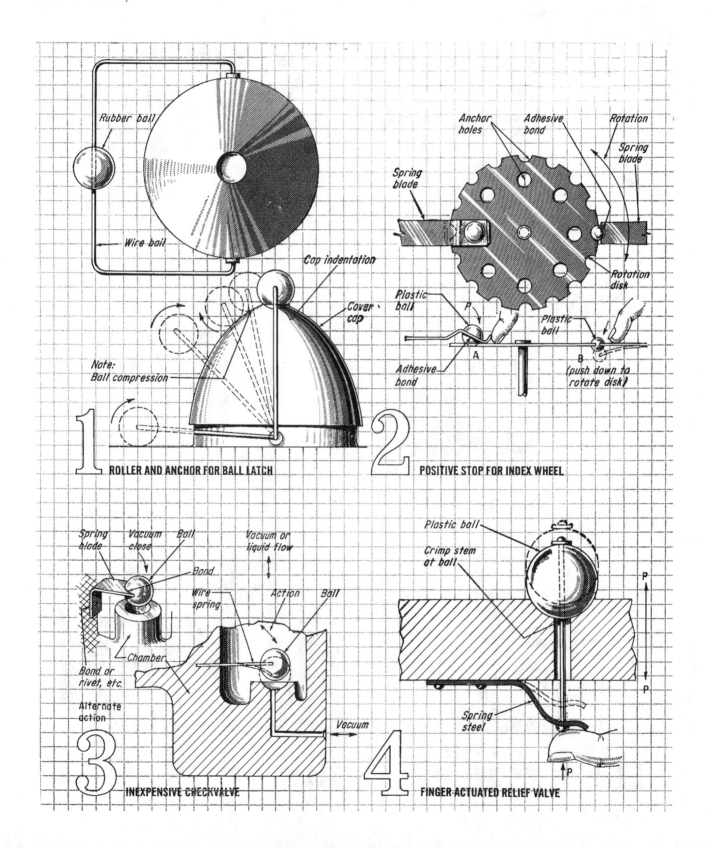

1 ROLLER AND ANCHOR FOR BALL LATCH

2 POSITIVE STOP FOR INDEX WHEEL

3 INEXPENSIVE CHECKVALVE

4 FINGER-ACTUATED RELIEF VALVE

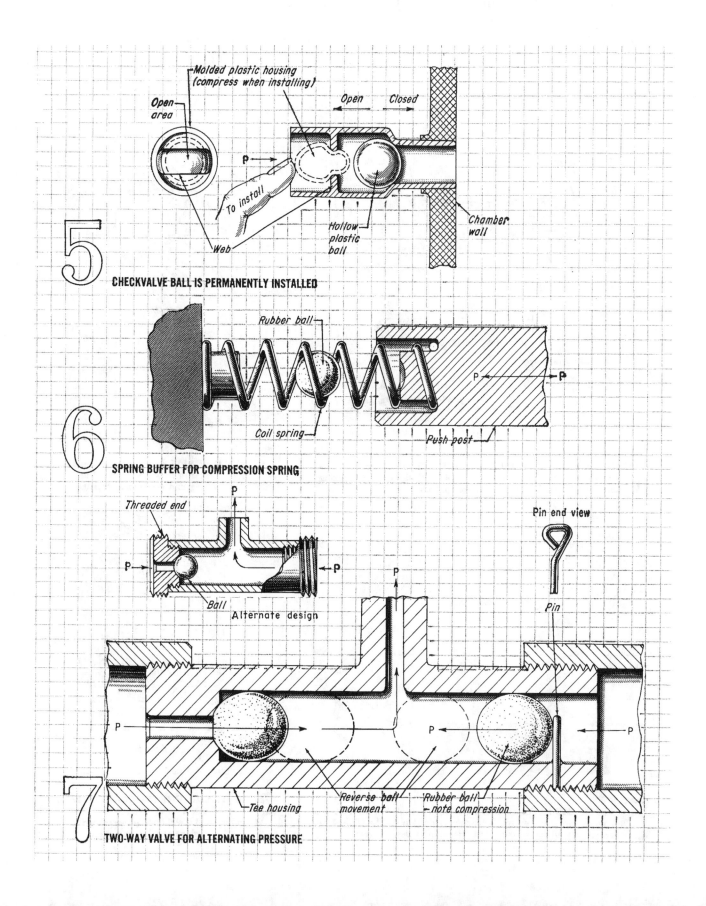

5

CHECKVALVE BALL IS PERMANENTLY INSTALLED

6

SPRING BUFFER FOR COMPRESSION SPRING

7

TWO-WAY VALVE FOR ALTERNATING PRESSURE

Rubber Balls Find Many Jobs

Plastic and rubber balls, whether solid or hollow, can find a variety of
important applications in many designs.

Robert O. Parmley

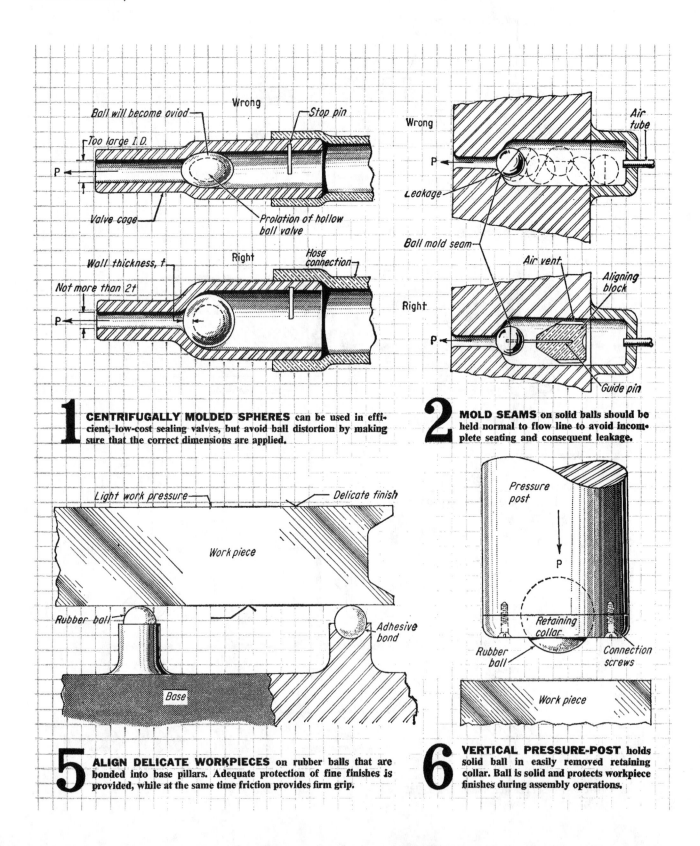

1 **CENTRIFUGALLY MOLDED SPHERES** can be used in efficient, low-cost sealing valves, but avoid ball distortion by making sure that the correct dimensions are applied.

2 **MOLD SEAMS** on solid balls should be held normal to flow line to avoid incomplete seating and consequent leakage.

5 **ALIGN DELICATE WORKPIECES** on rubber balls that are bonded into base pillars. Adequate protection of fine finishes is provided, while at the same time friction provides firm grip.

6 **VERTICAL PRESSURE-POST** holds solid ball in easily removed retaining collar. Ball is solid and protects workpiece finishes during assembly operations.

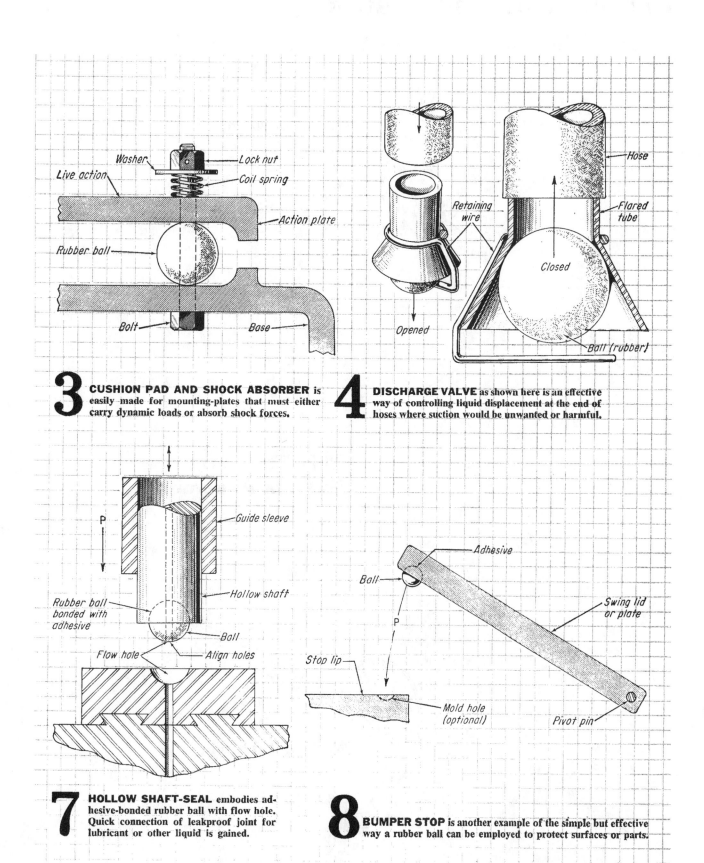

3 **CUSHION PAD AND SHOCK ABSORBER** is easily made for mounting-plates that must either carry dynamic loads or absorb shock forces.

4 **DISCHARGE VALVE** as shown here is an effective way of controlling liquid displacement at the end of hoses where suction would be unwanted or harmful.

7 **HOLLOW SHAFT-SEAL** embodies adhesive-bonded rubber ball with flow hole. Quick connection of leakproof joint for lubricant or other liquid is gained.

8 **BUMPER STOP** is another example of the simple but effective way a rubber ball can be employed to protect surfaces or parts.

Multiple Use of Balls in Milk Transfer System

Source: Bender Machine Works, Inc.
Illustrated by: Robert O. Parmley

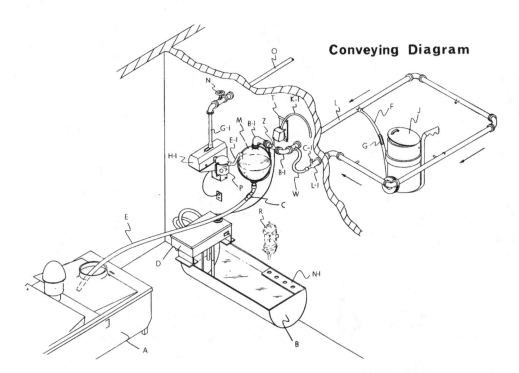

Conveying Diagram

Milk Transfer System Assembly

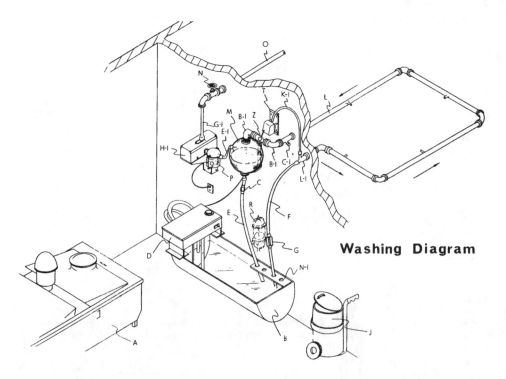

Washing Diagram

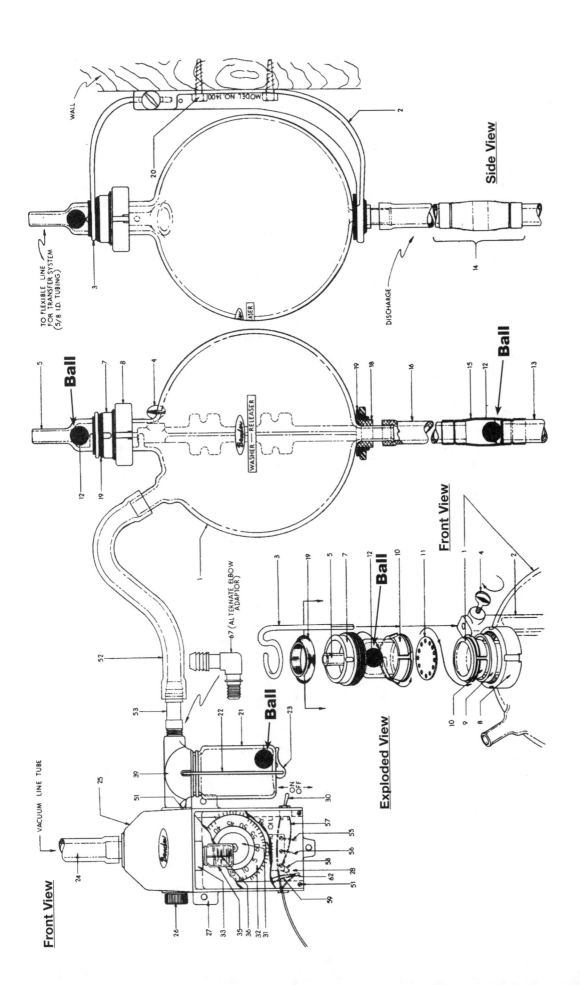

Four plastic balls, located at key positions within the system, act as positive check valves as they respond to the vacuum pulsations.

Use of Balls in Reloading Press

Inventor: E. E. Lawrence
Draftsman: R. O. Parmley

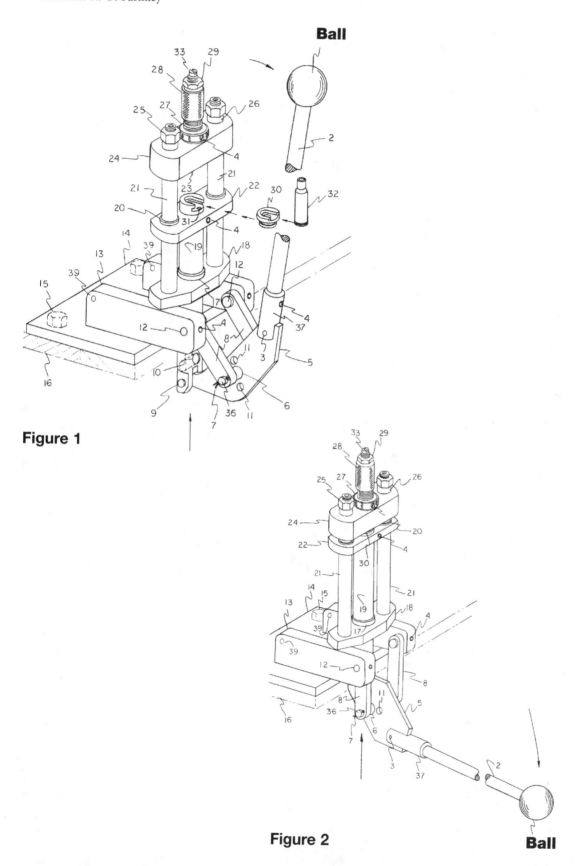

Ball

Figure 1

Figure 2

Ball

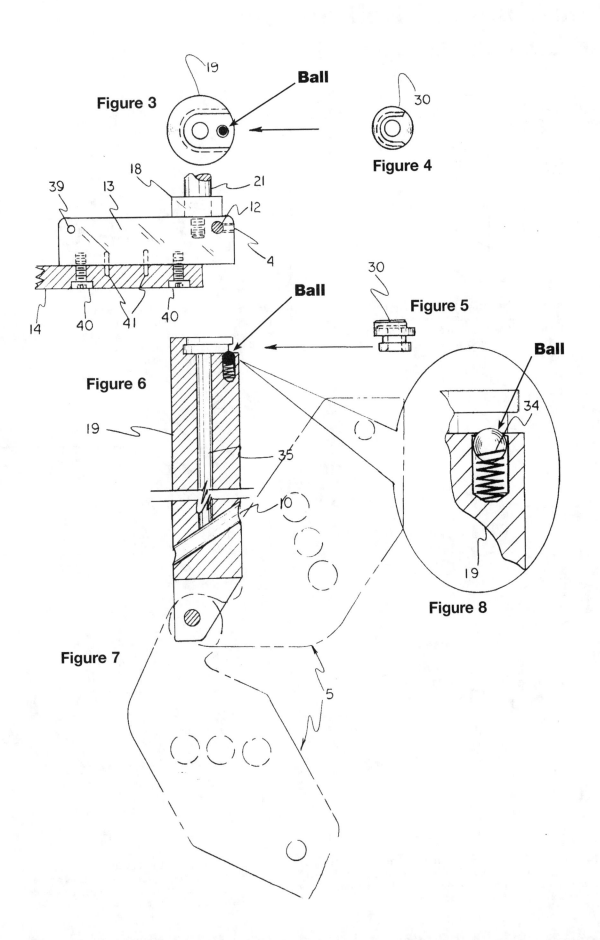

Figure 3

Ball

Figure 4

Figure 5

Ball

Figure 6

Ball

Figure 8

Figure 7

Nine Types of Ball Slides for Linear Motion

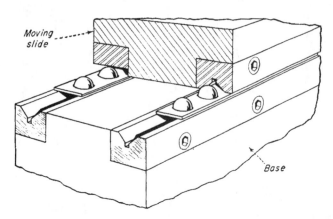

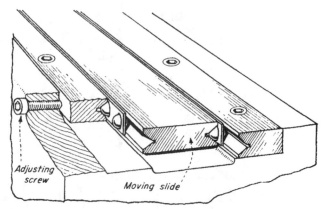

1 V grooves and flat surface make simple horizontal ball slide for reciprocating motion where no side forces are present and a heavy slide is required to keep balls in continuous contact. Ball cage insures proper spacing of balls; contacting surfaces are hardened and lapped.

2 Double V grooves are necessary where slide is in vertical position or when transverse loads are present. Screw adjustment or spring force is required to minimize looseness in the slide. Metal-to-metal contact between the balls and grooves insure accurate motion.

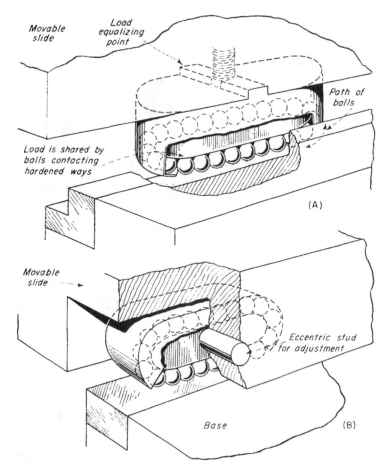

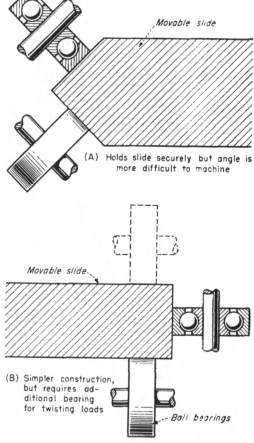

3 Ball cartridge has advantage of unlimited travel since balls are free to recirculate. Cartridges are best suited for vertical loads. (A) Where lateral restraint is also required, this type is used with a side preload. (B) For flat surfaces cartridge is easily adjusted.

4 Commercial ball bearings can be used to make a reciprocating slide. Adjustments are necessary to prevent looseness of the slide. (A) Slide with beveled ends, (B) Rectangular-shaped slide.

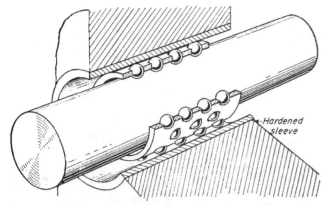

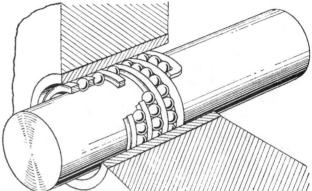

5 Sleeve bearing consisting of a hardened sleeve, balls and retainer, can be used for reciprocating as well as oscillating motion. Travel is limited similar to that of Fig. 6. This type can withstand transverse loads in any direction.

6 Ball reciprocating bearing is designed for rotating, reciprocating or oscillating motion. Formed-wire retainer holds balls in a helical path. Stroke is about equal to twice the difference between outer sleeve and retainer length.

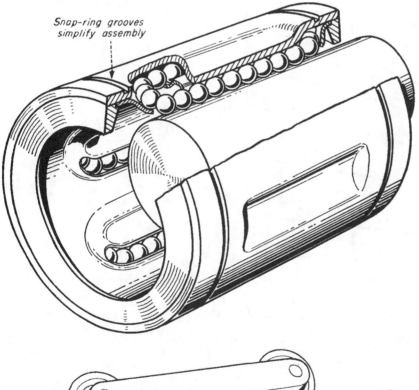

Snap-ring grooves simplify assembly

7 Ball bushing with several recirculating systems of balls permit unlimited linear travel. Very compact, this bushing simply requires a bored hole for installation. For maximum load capacity a hardened shaft should be used.

8 Cylindrical shafts can be held by commercial ball bearings which are assembled to make a guide. These bearings must be held tightly against shaft to prevent looseness.

9 Curvilinear motion in a plane is possible with this device when the radius of curvature is large. However, uniform spacing between grooves is important. Circular - sectioned grooves decrease contact stresses.

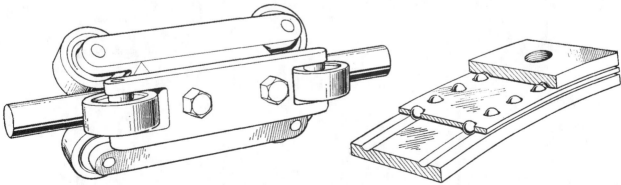

Stress on a Bearing Ball

These curves indicate permissible loads when seat is spherical or flat, steel or aluminum.

Jerome E. Ruzicka

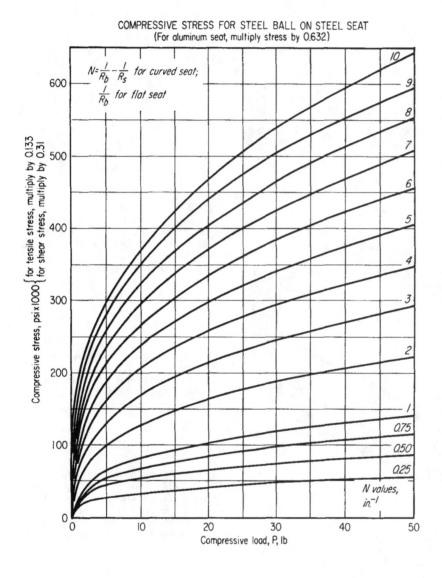

COMPRESSIVE STRESS FOR STEEL BALL ON STEEL SEAT
(For aluminum seat, multiply stress by 0.632)

$N = \dfrac{1}{R_b} - \dfrac{1}{R_s}$ for curved seat;

$\dfrac{1}{R_b}$ for flat seat

Compressive stress, psi x 1000 {for tensile stress, multiply by 0.133} {for shear stress, multiply by 0.31}

N values, in.$^{-1}$

Compressive load, P, lb

When a design uses steel bearing balls to support a load, it is important to know what stresses result. They are charted on this page. On the continuing page is a diagram that identifies symbols—for applications where the seat is spherical or flat—and also a chart that will help calculate maximum permissible loads.

Symbols used with curves

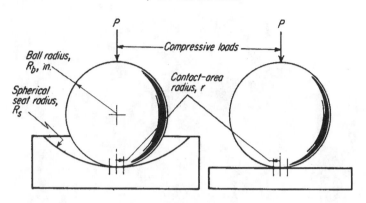

CONTACT RADIUS FOR STEEL BALL ON STEEL SEAT
(For aluminum seat, multiply radius by 1.25)

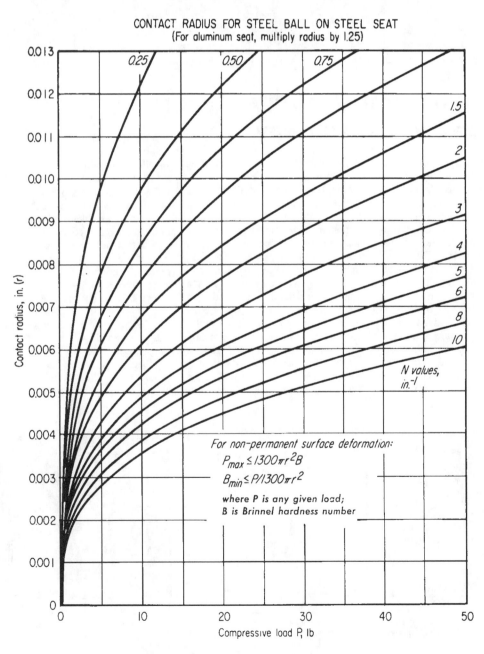

For non-permanent surface deformation:

$$P_{max} \leq 1300\pi r^2 B$$

$$B_{min} \leq P/1300\pi r^2$$

where P is any given load;
B is Brinnel hardness number

N values, in.$^{-1}$

Contact radius, in. (r)

Compressive load P, lb

Compact Ball Transfer Units

Masses of ball transfer units in airplane floor make it easy to shove cargo loads in any direction.

Compact ball transfer units roll loads every which way

An improved design of an oft-neglected device for moving loads—ball transfers—is opening up new applications in air cargo planes (photo above) and other materials handling jobs. It can serve in production lines to transfer sheets, tubes, bars, and parts.

Uses of established ball transfer units have been limited largely to furniture (in place of casters) and other prosaic duties. With new design that takes fuller advantage of their multiple-axis translation and instantaneous change of direction, ball transfer units can be realistically considered as another basic type of anti-friction bearing. The improved units are made by General Bearing Co., West Nyack, N.Y.

How they work. Essentially, ball transfers (photo below) are devices that translate omnidirectional linear motion into rolling motion to provide an unlimited number of axes of movement in any given plane. In such a unit, a large main ball rotates on its own center within a housing. This ball is supported by a circular group of smaller balls (drawing below) that roll under load and, in so doing, recirculate within the housing in endless chains.

These units are designed either as "ball up" or as "ball down." In the "ball down" units, design must provide a positive means of recirculating the support balls so they won't fall away under their own weight.

Variations. Many different configurations are available to suit the specific requirements of customers. Balls of carbon steel are most often used, but stainless steel balls are available for uses where corrosion may be a problem. Ball transfer units can be sealed to exclude dirt.

Where loads require that a number of ball transfers must simultaneously contact the load surface, a spring technique has been developed. Each ball transfer (drawing below) is spring-loaded. It starts to deflect when its own rated load is exceeded, allowing other ball transfers to pick up their share of the load. This concept also provides protection against major overloads in any ball transfer unit. □

Main ball shown, 1 in. dia., is supported by 70 smaller balls, hidden.

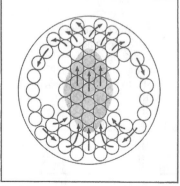

Shaded area seen from above shows load; arrows show ball circulation.

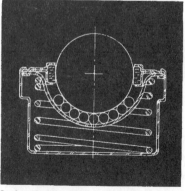

Spring loading assures even distribution of the load on the small balls.

Classic Uses of Balls in Valves

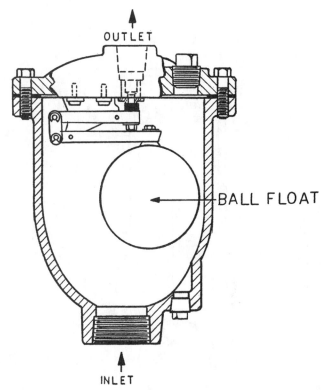

AIR RELEASE VALVE

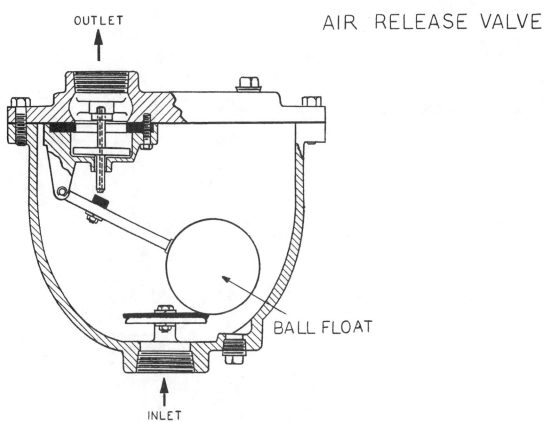

COMBINATION AIR VALVE

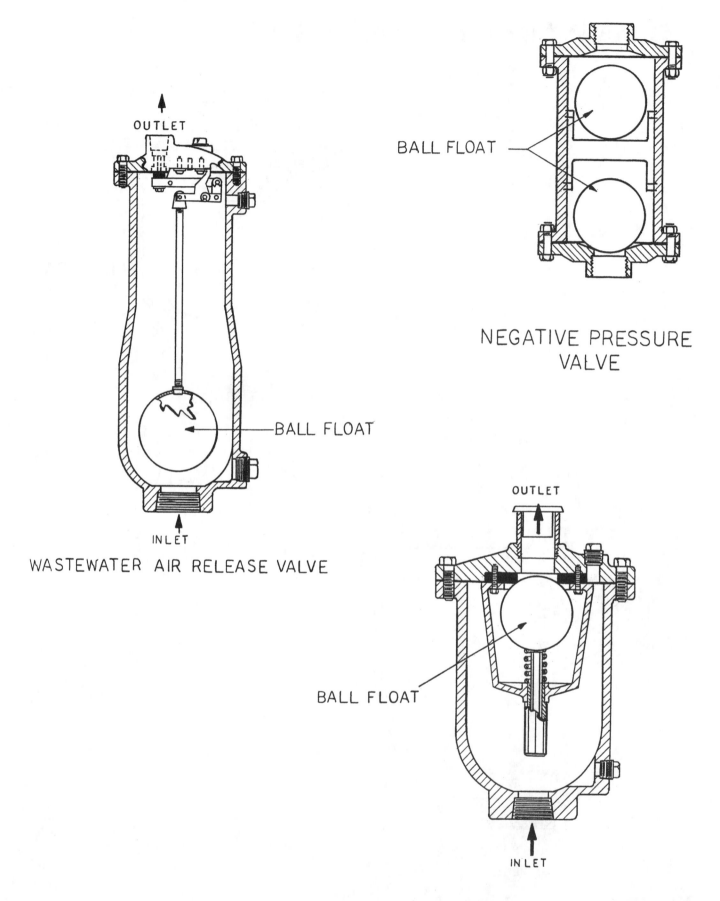

OUTLET

BALL FLOAT

INLET

WASTEWATER AIR RELEASE VALVE

BALL FLOAT

NEGATIVE PRESSURE
VALVE

OUTLET

BALL FLOAT

INLET

VACUUM BREAKER AIR VALVE

SECTION 20

BUSHINGS & BEARINGS

Going Creative with Flanged Bushings

These sintered bushings find a variety of jobs and are available in 88 sizes, from $1/8$ inch to $1^{5}/8$ inch internal diameter.

Robert O. Parmley

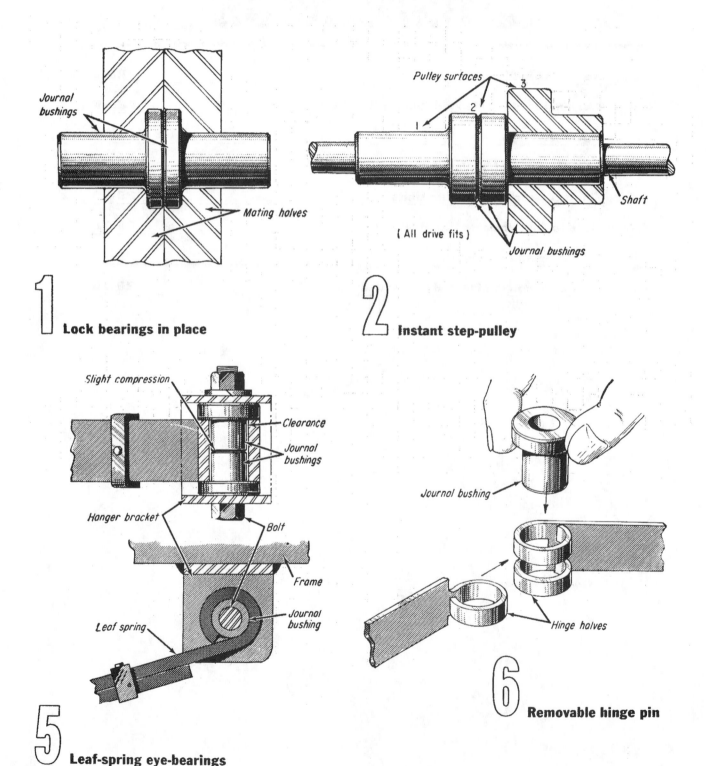

1 Lock bearings in place

Journal bushings

Mating halves

2 Instant step-pulley

Pulley surfaces

1 2 3

(All drive fits)

Journal bushings

Shaft

5 Leaf-spring eye-bearings

Slight compression

Clearance

Journal bushings

Hanger bracket

Bolt

Frame

Leaf spring

Journal bushing

6 Removable hinge pin

Journal bushing

Hinge halves

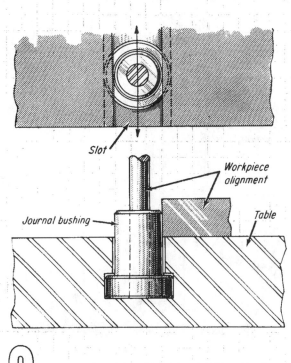

Slot

Workpiece alignment

Journal bushing

Table

3 Post or location-pin holder

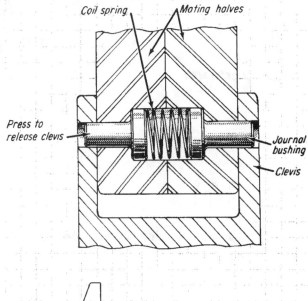

Coil spring

Mating halves

Press to release clevis

Journal bushing

Clevis

4 Spring-loaded pins

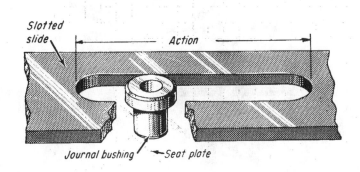

Slotted slide

Action

Journal bushing

Seat plate

7 Slider pin is self-lubricating

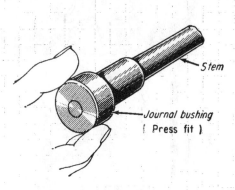

Stem

Journal bushing (Press fit)

8 Handle or knob

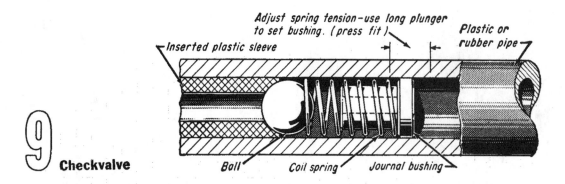

Adjust spring tension-use long plunger to set bushing. (press fit)

Inserted plastic sleeve

Plastic or rubber pipe

9 Checkvalve

Ball   Coil spring   Journal bushing

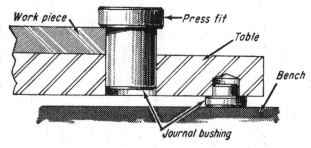

Work piece   Press fit   Table   Bench

10 Holding fixture, complete with feet

Journal bushing

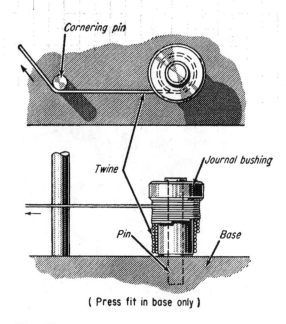

Cornering pin

Twine   Journal bushing

Pin   Base

(Press fit in base only)

11 Bobbin

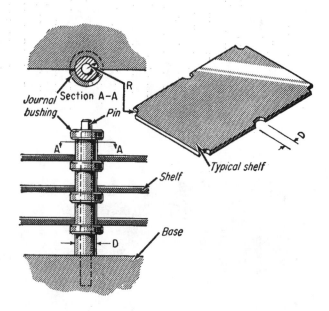

Journal bushing   Section A–A   R

Pin

A   A

Shelf

D   Base

Typical shelf   D

12 Shelf post

Seven Creative Ideas for Flanged Rubber Bushings

They're simple. Inexpensive, and often overlooked. Check your design for places where rubber bushings may be a solution to a design problem.

Robert O. Parmley

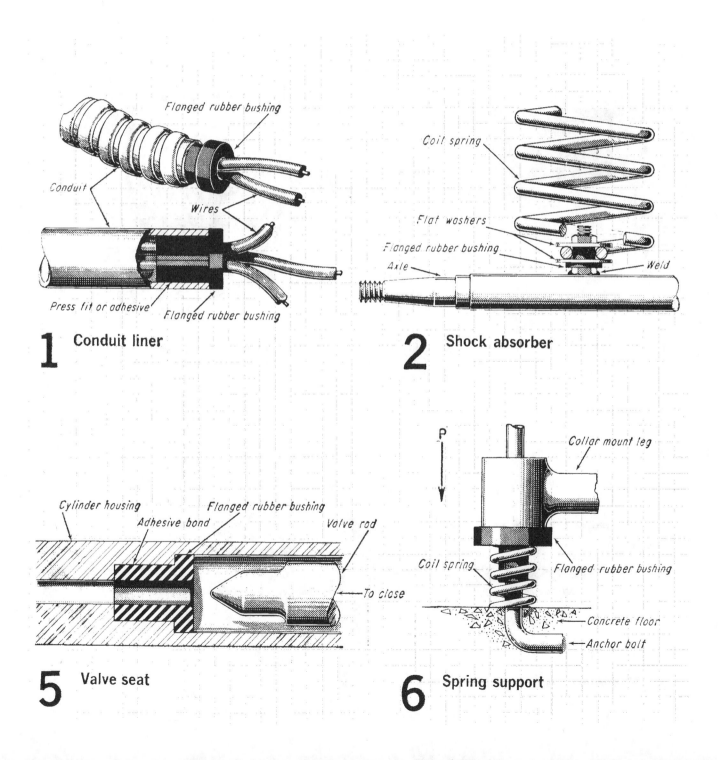

1 Conduit liner

2 Shock absorber

5 Valve seat

6 Spring support

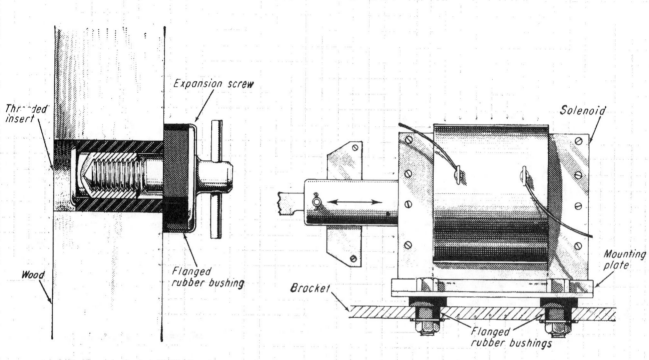

3 Seal expander

4 Cushion and noise absorber

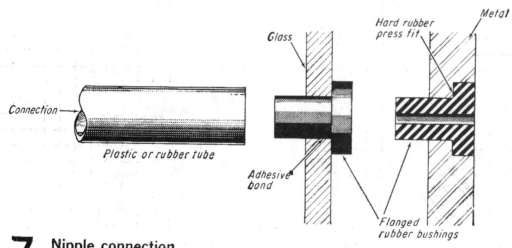

7 Nipple connection

Rotary-Linear Bearings

Rotary-linear bearings can take a beating from heat and shock. New designs find first use in steel converter, where heat cause expansion and shaft and jarring blows are applied.

A novel two-directional bearing designed especially for steelmaking equipment may find other uses where the loads are hot and heavy. In this design, a linear inclined bearing causes the bearing housing to roll over the roller bearings (sketch below, right) instead of sliding as in conventional designs.

Because of its demonstrated reliability, this assembly of pillow block and bearing is already being used in steel mills throughout the country. It was designed by Carl L. Dellinger, an engineer with Norma-Hoffmann Bearings Corp., Stamford, Conn., to overcome problems of friction along with thermal expansion.

Hot, heavy load. The prime use of the new design is in the trunnion that supports a steel converter. When 250 tons of molten metal at 3200 F are poured into a converter during the making of steel, the trunnion bearings are subjected to severe punishment by the very hot, very heavy load. The shaft that supports the converter heats up and expands, causing the bearings to slide inside their housings.

The wear caused by this action is bad enough. It means additional power is needed to overcome friction when the shaft is rotated during oxygen lancing operations. Worse yet, however, is the fact that when the sliding surfaces start to gall, the bearing will seize.

One-sided. Only one of the pair of horizontal shafts supporting the converter is equipped with the new bearing assembly. This shaft and bearing take up the expansion. The other shaft is supported by a conventional trunnion bearing.

Both bearings are packed with a lithium-based grease that contains molydisulfide and special extreme-pressure additives to operate in the 200 F temperature inside the bearing. The grease is held in the housing by a Buna N seal that keeps out contaminants.

Heat and shock. As the shafts expand, the spherical roller bearing at the end of one shaft rolls over the enclosed and inclined linear-motion bearings. These linear bearings are tilted 20 deg. from the axis of the shaft center line. Dellinger found that if the lines normal to the roller axis are allowed to intersect above the shaft center line, the housing can withstand greater torsional loads.

In the converter, severe torsional loads are produced during de-sculling operations (knocking out the cooled metal that adheres to the inside of the vessel after steel has been poured out). To remove this encrustation, the converter is rotated until it hits the base plate with the jarring impact of a truck hitting a brick wall. This shock load dislodges the steel, but it punishes the support bearings.

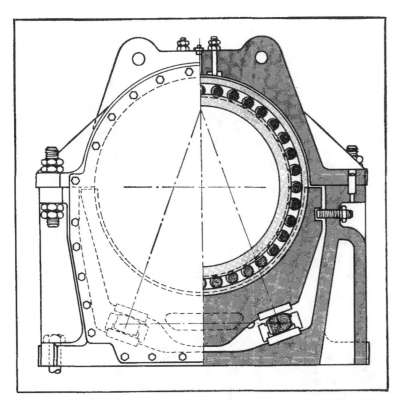

Spherical and linear motion bearings are contained in a sealed housing. The linear bearings are inclined for stability under torsional loading.

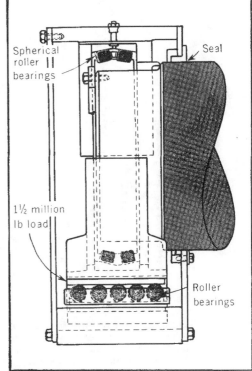

Linear roller bearings take up thermal expansion of shaft and housing, to eliminate sliding.

Unusual Applications of Miniature Bearings

R. H. Carter

Fig. 1—BALL-BEARING SLIDES. Six miniature bearings accurately support a potentiometer shaft to give low-friction straight line motion. In each end housing, three bearings are located 120 deg radially apart to assure alignment and freedom from binding of the potentiometer shaft.

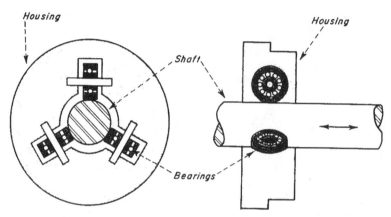

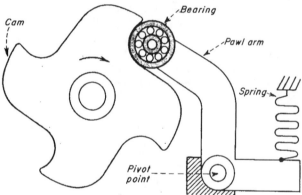

Fig. 2—CAM-FOLLOWER ROLLER. Index pawl on a frequency selector switch uses bearing for a roller. Bearing is spring loaded against cam and extends life of unit by reducing cam wear. This also retains original accuracy in stroke of swing of the pawl arm.

Fig. 3—SEAT FOR PIVOTS. Pivot-type bearings reduce friction in linkages especially when manually operated such as in pantographing mechanisms. Minimum backlash and maximum accuracy are obtained by adjusting the threaded pivot cones. Mechanism is used to support diamond stylus that scribes sight lines on the lenses of gunnery telescopes.

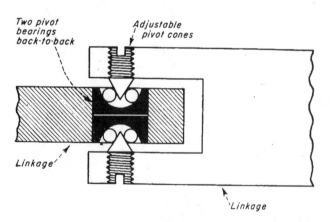

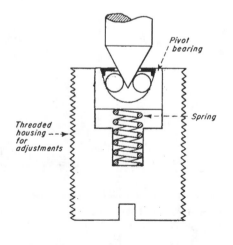

Fig. 4—SHOCK-ABSORBING PIVOT POINT. Bearing with spherical seat resting on spring acts as a pivot point and also absorbs mild shock loads. Used on a recording potentiometer that is temperature controlled. Spring applies uniform load over short distances and gives uniform sensitivity to the heat-sensing element. Close fit of bearing in housing is required.

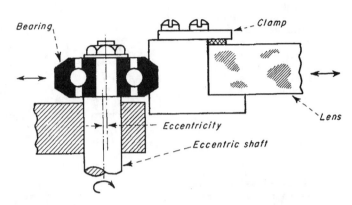

Fig. 5—PRECISE RADIAL ADJUSTMENTS obtained by rotating the eccentric shaft thus shifting location of bearing. Bearing has special-contoured outer race with standard inner race. Application is to adjust a lens with grids for an aerial survey camera.

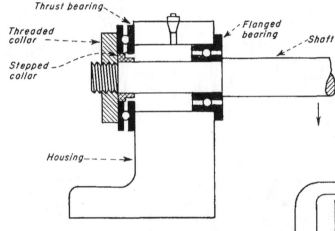

Fig. 6—SUPPORT FOR CANTILEVERED SHAFT obtained with combination of thrust and flanged bearings. Stepped collar provides seat for thrust bearing on the shaft but does not interfere with stationary race of thrust bearing when shaft is rotating.

Fig. 7—GEAR-REDUCTION UNIT. Space requirements reduced by having both input and output shafts at same end of unit. Output shaft is a cylinder with ring gears at each end. Cylinder rides in miniature ring bearings that have relative large inside diameters in comparison to the outside diameter.

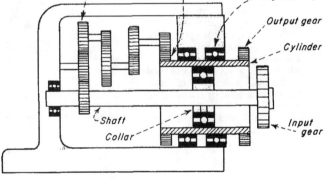

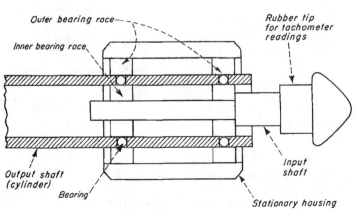

Fig. 8—BEARINGS USED AS GEARS. Manually operated tachometer must take readings up to 6000 rpm. A 10-to-1 speed reduction was obtained by having two bearings function both as bearings and as a planetary gear system. Input shaft rotates the inner race of the inner bearings, causing the output shaft to rotate at the peripheral speed of the balls. Bearings are preloaded to prevent slippage between races and balls. Outer housing is held stationary. Pitch diameters and ball sizes must be carefully calculated to get correct speed reduction.

Rolling Contact Bearing Mounting Units

FIG. 1—Pillow blocks are for supporting shafts running parallel to the surface on which they are mounted. Provision for lubrication and sealing are incorporated in the pillow block unit. Assembly and disassembly are easily accomplished. For extremely precise installations, mounting units are inadvisable.

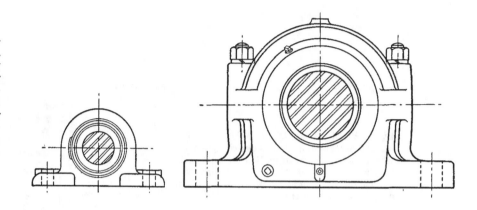

FIG. 2—Pillow blocks can be designed to prevent the transmission of noise to the support. One design (A) consists of a bearing mounted in rubber. The rubber in turn is firmly supported by a steel casing. Another design (B) is made of synthetic rubber. Where extra rigidity is required the synthetic rubber mount can be reinforced by a steel strap bolted around it.

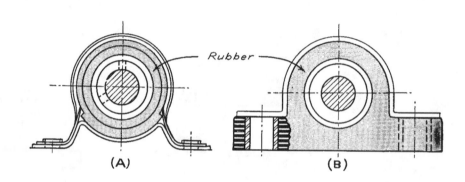

(A) (B)

FIG. 3—Changes in the temperature are accompanied by changes in the length of a shaft. To compensate for this change in length, the pillow block (B) supporting one end of the shaft is designed to allow the bearing to shift its position. The pillow block (A) at the other end should not allow for longitudinal motion.

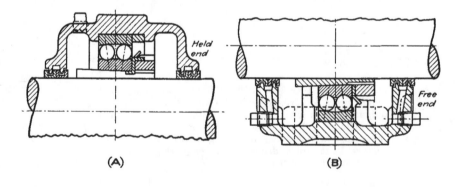

(A) (B)

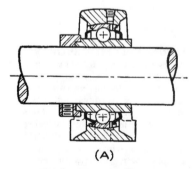

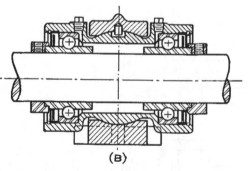

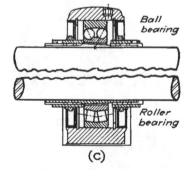

(A) (B) (C)

Fig. 4—Compensation for misalignment can be incorporated into pillow blocks in various ways. One design

(A) uses a spherical outer surface of the outer race. Design (B) uses a two-part housing. The spherical joint

compensates for misalignment. Another design (C) uses a spherical inner surface of the outer race.

Fig. 5—The cylindrical cartridge is readily adaptable to various types of machinery. It is fitted as a unit into a straight bored housing with a push fit. A shoulder in the housing is desirable but not essential. The advantages of a predesigned and preassembled unit found in pillow blocks also apply here.

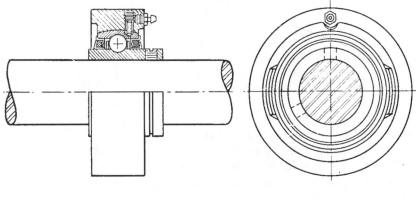

FIG. 6—The flange mounting unit is normally used when the machine frame is perpendicular to the shaft. The flange mounting unit can be assembled without performing the special boring operations required in the case of the cartridge. The unit is simply bolted into the housing when it is being installed.

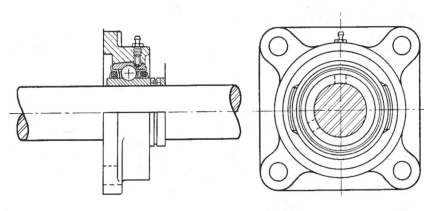

FIG. 7—The flange cartridge unit projects into the housing and is bolted in place through the flange. The projection into the housing absorbs a large part of the bearing loads. A further use of the cylindrical surface is the location of the mounting unit relative to the housing.

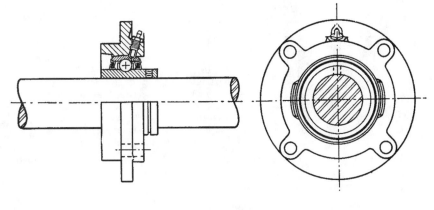

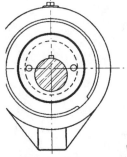

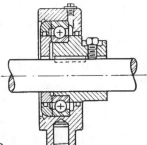

(A)

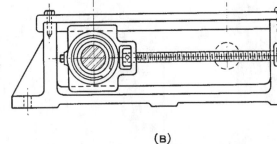

(B)

FIG. 8—Among specialized types of mounting units are (A) Eccentrics used particularly for cottonseed oil machinery and mechanical shakers and (B) Take-up units which make possible an adjustment in the position of the shaft for conveyor units. Many other types of special rolling contact bearing mounting units are made.

Eleven Ways to Oil Lubricate Ball Bearings

D. L. Williams

The method by which oil should be applied to a ball bearing depends largely on the surface speed of the balls. Where ball speeds are low, the quantity of oil present is of little importance, provided it is sufficient. Over-lubrication at low speeds is not likely to cause any serious temperature rise. However, as speeds increase, fluid friction due to churning must be avoided. This is done by reducing the amount of oil supplied and by having good drainage from the housing. At very high speeds, with light loads, the oil supply can be limited to a very fine mist.

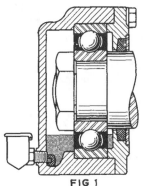

FIG 1

Fig. 1—*Oil Level System.* For moderate speeds, the bearing housing should be filled with oil to the lowest point of the bearing inner race. An oil cup is located to maintain this supply level. Wick acts as a filter when fresh oil is added. This system requires periodic attention.

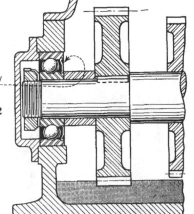

Shielded bearing

FIG 2

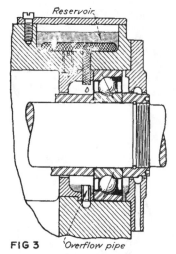

Reservoir

FIG 3 Overflow pipe

Fig. 3—*Drop Feed.* Oil may be fed in drops using either sight-feed oilers or an overhead reservoir and wick. Drains must be provided to remove excess oil. A short overflow stand-pipe, serves to maintain a proper oil level. It also retains a small amount of oil even though the reservoir should be empty.

Fig. 2—*Splash Feed* is used where rotating parts require oil for their own lubrication. Splash lubrication is not recommended for high speeds because of possible churning. Bearings should be protected from chips or other foreign material by using a shaft mounted slinger or shielded bearings.

Fig. 4—*Spray Feed.* With higher speeds, definite control of oil fed to bearings is important. This problem is more difficult for vertical bearings because of oil leakage. One method uses a tapered slinger to spray oil into the bearings. Oil flow is altered by the hole diam., the taper and oil viscosity.

Fig. 5—*Circulating Feed.* Most circulating systems are somewhat complicated and expensive but this is justified by their permanence and reliability. Oil reservoir is attached to the shaft and when rotated, the oil is forced upward where it strikes a scoop, flows through and onto the bearing.

FIG 4

Oil metering holes

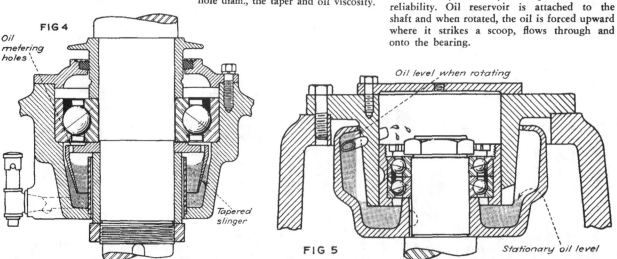

Tapered slinger

Oil level when rotating

Stationary oil level

FIG 5

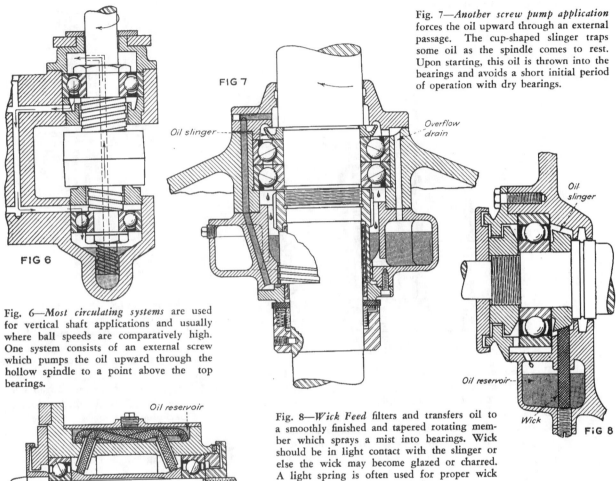

FIG 7

FIG 6

Oil slinger

Overflow drain

Oil slinger

Oil reservoir

Wick

FiG 8

Fig. 7—*Another screw pump application* forces the oil upward through an external passage. The cup-shaped slinger traps some oil as the spindle comes to rest. Upon starting, this oil is thrown into the bearings and avoids a short initial period of operation with dry bearings.

Fig. 6—*Most circulating systems* are used for vertical shaft applications and usually where ball speeds are comparatively high. One system consists of an external screw which pumps the oil upward through the hollow spindle to a point above the top bearings.

Fig. 8—*Wick Feed* filters and transfers oil to a smoothly finished and tapered rotating member which sprays a mist into bearings. Wick should be in light contact with the slinger or else the wick may become glazed or charred. A light spring is often used for proper wick pressure.

Oil reservoir

Oil slingers

FIG 9

Fig. 9—*Wick feeds* are used in applications of extremely high speeds with light loads and where a very small quantity of oil is required in the form of a fine mist. Slingers clamped on the outside tend to draw the mist through the bearings.

FIG 10

Air-oil mist

Air

Oil return

Fig. 10—*Air-Oil Mist.* Where the speeds are quite high and the bearing loads relatively light, the air-oil mist system has proven successful in many applications. Very little oil is consumed and the air flow serves to cool bearings.

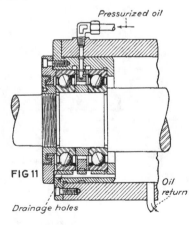

Pressurized oil

FIG 11

Oil return

Drainage holes

Fig. 11—*Pressure Jet.* For high speeds and heavy loads, the oil must often function as a coolant. This method utilizes a solid jet of cool oil which is directed into the bearings. Here adequate drainage is especially important. The oil jets may be formed integrally with the outer preload spacer.

Lubrication of Small Bearings

Examples of good modern practice in the lubrication of small bearings. These designs have the feature that no attention to lubrication is required over long periods. Several of them show applications of porous bronze bushings for long-life lubrication.

Herbert Chase

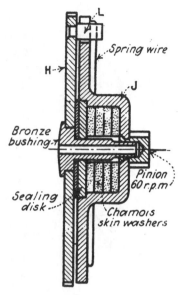

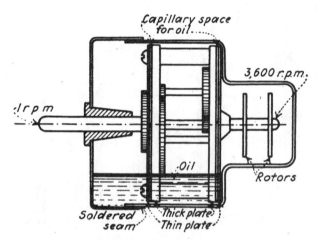

This electric clock mechanism is inclosed in an oil-tight case with only a single opening for the 1-r.p.m. shaft. The bushing for this shaft projects sufficiently far into the case so that the oil level is below its inner end regardless of how the case is tilted. Oil feeds by capillary action between the plates as indicated and works out along the shaft.

In this design, the main plate H of the rotor tends, because of magnetic attraction, to stay in the same place. The central bronze bushing is a press fit in the main rotor plate and turns with it. On the outside of the bushing is a floating flywheel assembly made of the cupped stamping J, the pinion, and the sealing disk, the latter inclosing a series of chamois-skin washers that have been soaked in oil. Oil seeps slowly through holes in the bushing, and thus to the bearing surfaces.

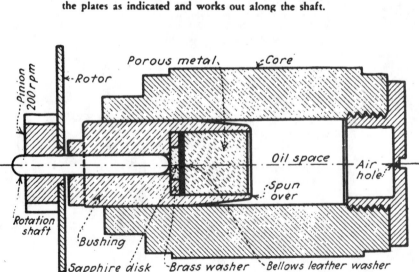

The rotor shaft turns in a phosphor bronze bushing pressed into the stainless steel core of the motor. The oil seeps in minute quantities through the cylindrical porous plug, through the bellows leather washer, around the floating thrust jewel or sapphire disk, and finally into the bore of the bushing in which the rotor shaft turns. The bearing clearance is held within the limits of 0.0008 to 0.0003 in.

Below, to the left is an oil resrevoir cast integrally; to the right, a drawn metal shell is screwed into the hub for an oil reservoir. In both designs, lubricant is fed through the porous metal bushing. Increased bearing temperature brings additional oil to the bearing surface because of expansion and increased oil fluidity.

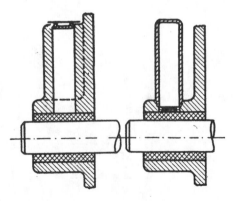

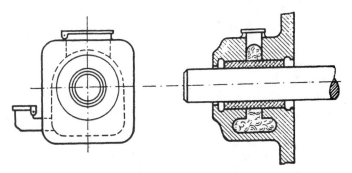

Below are shown three different designs of bearings for extremely light duty, as in clocks and meters. To the left is a self-aligning bearing having a porous bushing seated in a two-piece cadmium-plated steel shell which also holds a felt washer saturated with sufficient oil to last a year or more in ordinary service. In the center is shown a design wherein a pressed steel frame forms a spherical seat for the porous bearing. A light stamping incloses an oil-soaked felt washer that contains sufficient lubrication for the life of the motor. To the right is shown the bearing for an electric clock, the light cupped stamping that contains the oil-soaked felt being pressed over the bearing flange. In both of these designs, the lubricant in the felt is sufficient for the life of the motor.

In this design, an annular groove for oil or light grease is cored in the housing of the bearing. As indicated, oil-soaked wool waste can be packed in the cored recess. The concentric grooves at each end of the porous bushing are for the purpose of catching any end leakage. A refinement would be the addition of drilled holes to lead the oil back to the cellar. The additional oiler at the bottom is optional.

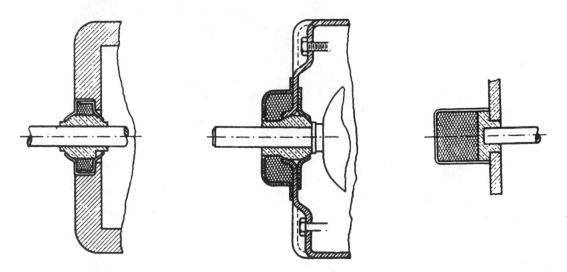

A cup-shaped stamping (below) holds the oil-soaked felt in the housing, and also acts as a dust shield. Lubrication of the bearing is through the porous bushing which is die-pressed from powdered metal. The lubricant is always fed to the outer wall of the bushing, which acts as a wick. To the right is shown an optional construction wherein an annular groove in the bearing housing forms the oil reservoir.

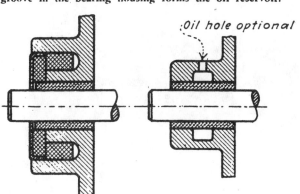

Oil hole optional

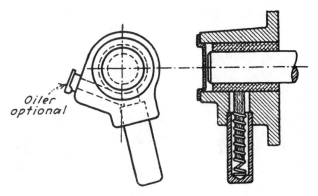

Oiler optional

In this wick-feed arrangement for oil or grease, capillary action feeds the lubricant to the surface of the porous bushing. For light service, sufficient lubricant is contained in the cup to last over a long period of time. The addition of the oiler shown in the end view is for the purpose of convenience so that the machine user will be more apt to give some attention to lubrication. Note the metal dust shield.

Cage Keeps Bearings in Line and Lubricated

Ball-to-ball friction in bearings generates heat, limits the speed, and shortens the life of linear-motion guides. To solve these problems, engineers at THK designed a caged-ball system that places recirculating bearings in a retainer, or cage, that separates and aligns the bearings. The cage creates a bearing chain with spaces, or grease pockets, to hold lubricant and keep the bearings from touching each other.

Separating the bearings provides several benefits. Because the bearings are not in contact with one another, there is much less metallic noise during operation. The lack of bearing-to-bearing contact also decreases wear and allows room for lubricant, two factors which promote long-term, maintenance-free operation. Meanwhile, the lack of bearing noise or vibration means rolling resistance is more uniform, keeping performance of the linear guide consistent. And reduced heat levels and low bearing stresses lets the linear guide move at higher speeds.

The caged-ball technology is especially well suited for medical devices and semiconductor-inspection applications, according to THK, and is available in a wide variety of linear-guide models.

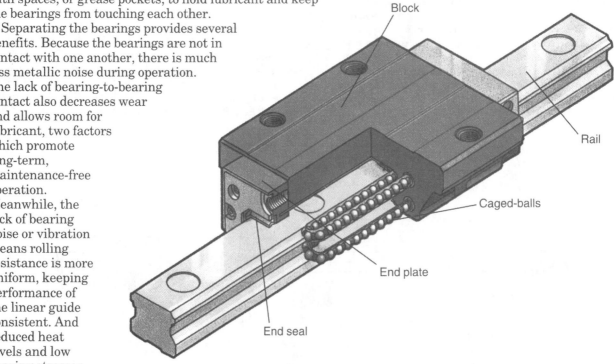

Block

Rail

Caged-balls

End plate

End seal

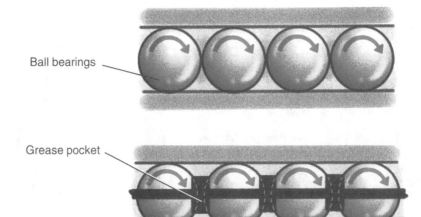

Ball bearings

Grease pocket

INDEX